Educative Accountability

Titles of Related Interest

Evers & Lakomski/Knowing Educational Administration

Evers & Lakomski/Exploring Educational Administration: Coherentist Applications and Critical Debates

Hodgkinson/Administrative Philosophy: Values and Motivations in Administrative Life

Educative Accountability

Theory, Practice, Policy and Research in Educational Administration

by

R. J. S. 'Mac' Macpherson
University of Tasmania

PERGAMON
An Imprint of Elsevier Science

UK	Elsevier Science Ltd, The Boulevard, Langford Lane, Kidlington, Oxford OX5 1GB, U.K.
USA	Elsevier Science Inc., 660 White Plains Road, Tarrytown, New York 10591-5153, U.S.A.
JAPAN	Elsevier Science Japan, Higashi Azabu 1-chome Building 4F, 1-9-15, Higashi Azabu, Minato-ku, Tokyo 106, Japan

Copyright © 1996 Elsevier Science Ltd

All Rights Reserved. No part of this publication may be reproduced, stored in a retrieval system or transmitted in any form or by any means: electronic, electrostatic, magnetic tape, mechanical, photocopying, recording or otherwise, without permission in writing from the copyright holders.

First edition 1996

Library of Congress Cataloging-in-Publication Data
Educative accountability: theory, practice, policy, and research in educational administration/edited by R. J. S. 'Mac' Macpherson.
p. cm.
Includes bibliographical references and index.
1. Educational accountability—Australia—Tasmania. 2. Educational accountability–England. 3. Educational accountability—Wales. 4. Educational accountability—United States. I. Macpherson, R. J. S.
LB2806.22.E38   1996
379.1'54—dc20   96-41875
                                                                                CIP

British Library Cataloguing in Publication Data
Educative accountability: theory, practice, policy, and research in educational administration
1. Educational accountability
I. Macpherson, R. J. S.
379.1'54

ISBN 0 08 042768 5

Printed and bound in Great Britain by Redwood Books Ltd, Trowbridge, Wilts

Contents

PREFACE	vii
ACKNOWLEDGEMENTS	ix

Chapter 1
The International Challenge of Accountability — 1

Chapter 2
Rehabilitating a Politically Incorrect Policy Issue — 13

Chapter 3
From Consensus to Conflict in England and Wales — 25

Chapter 4
The New Right Intervention in England and Wales — 39

Chapter 5
The Contested Policy Settlement in England and Wales — 59

Chapter 6
Accountability Policy Evolution in US Public Education — 83

Chapter 7
Testing, Testing, 1, 2, 3 ... Confusion in the Early 1990s — 105

Chapter 8
The Adequacy of US Policies and Policy Making — 129

Chapter 9
Accountability Policy Research in Tasmania — 151

Chapter 10
Epistemological Reflections on Policy Research — 175

Chapter 11
Accountability Policy Preferences in Tasmania — 193

Chapter 12
Tasmanian Parents' Policy Preferences and Implications — 215

Chapter 13
Tasmanian Teachers' Policy Preferences Concerning Accountability: Individualistic or Communitarian Professionalism? — 237

Chapter 14
Tasmanian Principal's Policy Preferences and Implications — 259

Chapter 15
System Administrators' Policy Preferences: Mediating Purposes and Politics while Supporting Community School Development — 283

Chapter 16
Reflections on Educative Accountability Policy: Theory, Practice and Research — 307

BIBLIOGRAPHY — 333

AUTHOR INDEX — 355

SUBJECT INDEX — 361

Preface

The purpose of this book is to provide practical ideas and educative policies for practitioners and theorists troubled by the persistent problem of accountability in public education. Educative accountability processes and criteria are defined as those believed to enhance the quality of learning, teaching, leadership and governance. This is an important issue in an era when governments seek mechanisms to measure the relative productivity of schools while school communities search for ways of discharging obligations in ways that both 'prove' and 'improve' education. We need fresh policy settlements in the area.

The text begins and ends in Tasmania, the Australian 'home' of the 'self-managing' school. In 1992 I was asked to help the public school system rethink its accountability policies and practices. A policy research project was mounted to supplement the ongoing systemic policy review processes. Policy, theory and practices have been advanced.

The project drew from and speaks to the challenges being faced in the United States and in England and Wales. It was aware of and reflected on how the field of educational administration is learning post-positivist and post-paradigmatic policy research methods, following the seminal contributions of Colin Evers, Gabriele Lakomski and others. The project developed policy options, measured support, and encouraged stakeholders to identify and interpret the area of overlap between three seemingly incompatible sets of interests; systemic or technical, professional and client priorities. The overlap between theories was found to be far greater than differences and comprise a communitarian theory of mutual accountability with immediate implications for democratic and limited school community government.

This book has five parts. In two relatively short chapters, Chapters 1 and 2, I explore the general issue of accountability in education in an international context and establish that it is a source of widespread concern to stakeholders. It is shown that any review of accountability policy implies the clarification of purposes, responsibilities, obligations, evaluation, reporting and 'high stakes' action and consequences. It is argued that these matters are too important for the health of society and education to be defined solely using political processes or in economically rational terms.

The second part, Chapters 3–5, is a case study of accountability policy in England and Wales since the collapse of the post-war consensus. There are many lessons drawn from the New Right intervention and the current uneasy current policy settlement. Given its hobgoblin status in many parts of the world, the current inspection system managed by the Office for Standards in Education is evaluated. Strengths and severe limits are found.

The third part of the book, Chapters 6–8, is a second case study. It traces the path of accountability policy development in the United States. Two matters are given special attention; the role of tests and the adequacy of policy research. Again, successes and profound problems are highlighted.

In the fourth part I locate the research project in its Tasmanian setting (Chapter 9) and reflect on the epistemology of the research design (Chapter 10). This is to clarify the extent to which the findings may be trusted and to advance the discussion on policy research methods. I try to make a case for epistemically critical research.

The fifth section of the text, Chapters 11–15, is a report of the findings, as interpreted and used by stakeholders in Tasmania. I then provide a provisional summary of the recommended policies in Chapter 16 as well as implications for theory, practice and further research.

I take full responsibility for the arguments that follow. Any quality they possess derives in considerable part from the people it has been my good fortune to work with over the years. And while I have been at pains to list the joint works that have influenced my thinking, and the papers that were developed for local use and later published, I feel they do poor justice to the value of the discussions they drew on and contributed to. In sum, this text and its published antecedents are poor fare compared to the 'real thing': co-operative policy research.

RJSM

Acknowledgements

As noted above, interaction with many colleagues over many years have helped shape the ideas and research discussed in this book. I am indebted to those who contributed to work-in-progress sessions at meetings of the Australian Council for Educational Administration, the British Educational Management and Administration Society, the Commonwealth Council for Educational Administration, the Australian Association of Researchers in Education, the British Educational Researchers in Education, the American Educational Research Association (especially the Politics of Education Association) and the European Council of Educational Research.

It is particularly appropriate to thank my colleagues at the University of Tasmania, officials in the Tasmanian Department of Education and the Arts, and the public school communities of Tasmania and the Australian Research Council for their support. I especially wish to single out Bruce Davis, Graham Harrington, Jan Baker, the district superintendents of Tasmania and the leaders of all stakeholder groups for their wisdom and commitment to public education.

There are some whose thinking and research has been particularly helpful in recent years, although I expect that many will want to contest some of the arguments presented in this book. They include Bill Boyd, Judith Chapman, Jim Cibulka, Penny Cocker, Roberta Derlin, Pat Duignan, Colin Evers, John Ewington, Neville Grady, Dan Griffiths, Peter Gronn, Christopher Hodgkinson, Jo Howse, Park Sun Hyung, Ron Ikin, Wong Kam-Cheung, Gabrielle Lakomski, Hanne Mawhinney, Cheng Kai Ming, David Monk, Bill Mulford, Peter Ribbins, Chris Roellke, Joyce Upex, Barbara Vann, Donald Willower, Norman Wicks and Ken Wong.

This text also makes use of previously published material. The authors and editors of the publications include in the bibliography are thanked most warmly for their contributions to scholarship.

Each of these audiences, support systems and individuals have helped a sustain a context of critical support and interactive accountability. It is no accident that this text examines the policy proposal that 'educative accountability' might be both desired and desirable in learning communities concerned with the quality of education and society.

Above all I want to thank the primary policy community that made this text and my research commitments possible; my family, Nicki, Kirsty, Shiona, Ewan and Angus, your love sustains me. Guy Macpherson, you inspire me.

1

The International Challenge of Accountability

Introduction

Accountability is a challenge for team, institutional and system leaders around the world. The term can be evoked as a principle of organization, mobilize the normative order, link evaluation to strategic planning, act as a political metaphor of propriety, lodge an appeal to obligations, adjust priorities and purposes or set the stage for the allocation of blame or praise. In this chapter some of the more practical aspects of the challenge of accountability are traced in the United States, Canada, Australia and the UK to the use of restructuring and devolutionary strategies; strategies typically intended to reform organizations and generate economies in systems. Policy and research problems are briefly related to conceptions of democracy and to three theories of accountability often in competition. It is shown that the concept of accountability is in urgent need of rehabilitation.

Restructures and Devolution

A relatively comprehensive and scholarly review of the school restructuring movement in the United States (Murphy and Hallinger 1993) recommended co-operative learning, collaborative planning, teacher empowerment and participative policy making. The review made no mention of public accountability or the use of external criteria to evaluate and improve the quality of school life. Does this mean that the issue of accountability is irrelevant to the reform of schooling? The blunt and contrary answer was President Clinton's *Goals 2000: Educate America Act*, passed in March 1994. The Act virtually drew bipartisan support, reinforced President Bush's agenda, set new national goals, provided a context for state curriculum frameworks, gave more powers to school governors and intensified the 'high stakes' testing of student learning (Boyd and Halpin 1994). And this was all achieved in the face of the US traditions of locally governed and locally managed schooling.

There is a similar contrast evident in the UK. The British government's control of curriculum, assessment of student learning and budgets appears to be as tight as it has ever been, despite the Local Management of Schools (LMS) initiative. The Chair of the Schools Curriculum and Assessment Authority (SCAA), Sir Ron Dearing, recently released another revision of national curriculum 'orders' and assessment processes for Local Education Authority (LEA)-maintained schools. There was some irony in the claim that this meant 'handing professional responsibility back to the teachers' (*Times Education Supplement* (*TES*) 1994, p. 14). Such symbolic willingness to invest absolute responsibility for policy implementation in this one interest group would suspend client, technical and governmental responsibilities, if ever genuinely executed.

In Australia, despite constitutional responsibility being vested in the states, new national curriculum and student achievement profiles have been determined, national textbooks have been commissioned, new testing regimes are being devised, increased monitorial powers have gone to schools councils, and principals are being encouraged to use school-based budgeting to manage the gradually shrinking net resource base available for public education (Beare 1995).

This increasingly common combination of centralist control and devolved responsibility for policy implementation could prove unwise. It could target teachers, polarize or alienate stakeholders, prevent formative evaluation and undercut collaborative planning. A great deal depends on purposes and definitions. To define accountability as a matter of personal, local and private professional honour could place teachers in a vicious circle of arbitrary processes and feral criteria, that is, in the bull's-eye of political blameworthiness. To define accountability in terms of either 'neo-centralism', 'provider-capture' or 'consumerism' could encourage the development of policy options based on the simplistic slogans of the 1980s. These slogans were born out of systems theory and politicized production functions, and are not likely to help school communities explore the potential for collective educational policies and practices in broader historical, socio-economic and political contexts (Macpherson 1993a, b). There are other equally unwise possibilities.

Defining accountability solely as control or measurement mechanisms could provide short and quick answers to blunt questions from extra-education sources. Such an approach could, however, do little to help discharge obligations to the wide range of stakeholders in each school community or to assist with the continuous improvement of schools and their support systems. At a more personal level, professionals might need more sophisticated feedback than blame if they are to be encouraged to constantly recreate the relevance and legitimacy of their services.

Similarly, defining curriculum structures, governance and budgetary powers and the 'high stakes' testing of student learning as the most appropriate forms of accountability could also encourage the artificial partitioning of knowledge about learning, teaching, leadership or governance. Major studies (Fullan 1991, Fuhrman 1993) have shown that schools and systems require coherence between these realms as much as exemplary performances within each.

In sum, I want to suggest that education systems and institutions might need forms of accountability that affect all parts of their organization and link formative evaluation to the governance of policy and strategy, as well as show how much learners have learned. I will argue in later chapters that there is no sound reason for decoupling learning from teaching or leadership with regard to accountability. The more immediate task, however, is to relate the issue of accountability to the quality of government.

Accountability and Democracy

My pragmatic view is that 'accountability' should defined by the purposes it serves. I agree with Simey (1995, p. 20) when she argued that

> accountability is not a mechanism or a routine but a principle. More than that it is a principle which serves a purpose. In a democracy, that purpose is to provide the basis for the relationship between society and its members, between those who govern and those who consent to be governed. The word consent provides the significant clue, implying as it does the striking of a bargain or the drawing up of a contract between people who are partners in some joint enterprise.

Since this makes the governors of education themselves accountable for the quality of accountability policies, we find ourselves asking the age-old question: *quis custodiet ipsos custodes?* Who guards the guardians? Clearly, accountability policies should give educative substance to democratic principles whose spirit was elegantly summarized by Thomas Jefferson (cited in Lange 1988, p. 1):

> I know of no safe depository of the ultimate power of society but the people themselves and if we think them not enlightened enough to exercise their control with a wholesome discretion, the remedy is not to take it from them, but to inform their discretion.

In a modern complex democracy, the principle of accountability is, therefore, to be honoured by generating legitimacy in the minds of the governed. This focuses attention on the origins and indispensable conditions of legitimacy. First, it is commonly assumed in a democracy that the delegation of the authority to govern proceeds only with the consent of the governed. Second, this consent is conditional on the stewards of that delegated authority remaining accountable. 'This holds good whether they are elected members or officials. Accountability is thus the solid plank on which our whole political system rests' (Simey 1995, p. 17).

This 'plank' under democracy uses four main values when it invests political authority in 'the people' (Plano and Greenberg 1972, p. 7):

- individualism—governance is to maximize the achievement of potential,
- liberty—governance is to maximize freedoms consistent with order,
- equality—governance is to maximize equity of opportunity, outcomes and access to power, and
- fraternity—governance is to maximize co-operation in building a wholesome society.

Democratic governance, therefore, provides a moral order for human relationships and social institutions. It requires a decision-making system based on majority rule that also protects minority rights. Particular guarantees are also commonly held to be indispensable to democratic governance; freedom of speech, press, religion, assembly and petition, equality before the law and 'accountability under law' (Plano and Greenberg 1972, p. 1).

In these special and precious ways, constitutional accountability distinguishes democratic governance from absolutism, or government unrestrained by law; examples include totalitarianism and fascism. The basis of democratic accountability is that those who govern remain responsible to the governed for the quality of their stewardship. The guardians of education in a democracy are, therefore, primarily responsible for the quality of accountability policies, and responsible to the stakeholders of public education; the issue considered in the next section.

The Agenda

Recent international case studies of restructures of education systems (Beare and Boyd 1993, Jacobson and Berne 1993, Martin and Macpherson 1993) have shown that the decentralization of pedagogical, administrative and some governance powers to locally managed schools, with a simultaneous recentralization of control functions, has led to a consensus of cynicism among professionals, low policy legitimacy among other stakeholders and uncertainty in the hearts and minds of many education system managers. There is mounting evidence that accountability policies based on three common mechanisms, specifically private professional perspectives and self-evaluation, on local market and political mechanisms, and using the assessment of student achievement as a universal proxy for quality in education, cannot sustain the legitimacy of institutions and systems. A new approach is therefore required if the stewards of public education are to sustain democratic accountability and the legitimacy of their services and the policies they offer.

In this book I try to develop alternatives. I have started with the democratic ideal that accepting responsibility for the quality of public schooling brings with it the moral duty to be publicly accountable. As Thomas Jefferson put it: 'When a man [sic] assumes a public trust, he should consider himself a public property' (Oxford University Press 1990, p. 272). I also take a holistic view of knowledge production in organizations to argue that appropriate forms of accountability in education will need to use processes and criteria that gather data, report on, evaluate and enhance the quality of learning, teaching, leadership and governance, often simultaneously. I argue that leaders of education systems and institutions need to organize the creation of educative accountability policies quite deliberately. I try and demonstrate that an educative policy-research process is required if the new accountability policies produced are to have technical, consumerist and professional merit. I assume from the outset, quite reasonably as it turns out, that such educative accountability criteria and processes are already evident in what are often seen in or regarded as 'best practices'.

In this chapter I try to show why the term 'accountability' has become 'politically incorrect' in the hearts and minds of many educators. Paths toward the rehabilitation of the term are examined. For example, I suggest that inclusionary, normative and re-educative accountability processes that help school communities reflect on the relationship between practices and the broad purposes of education could help create more satisfactory levels of legitimacy. But it is more than simply a question of providing better local policy-making processes.

I propose that incoherent accountability policies derive principally from ambiguous responses to competing theories about what constitutes knowledge and how learning can be demonstrated. As historian Openshaw (1995) and policy analyst Rae (1996) have demonstrated with regard to New Zealand, such policy incoherence is often traceable to an unresolved struggle over basic purposes; is public education to be concerned primarily with the betterment of the individual or society? To begin, nowhere are the ambiguities of purposes to be served by accountability policies better demonstrated than in the literature on the local management of schools.

Restructuring and School-based Management in the United States

A sophisticated analysis of the school restructuring movement in the United States (Murphy 1993) identified the four most common strategies for reorganizing education as providing choice and voice for parents, school-based management, teacher empowerment, and teaching for understanding. These strategies were related to two themes in restructuring schools; the 'marketization' of education and the redefinition of the roles of educational stakeholders. Since each theme carried different sets of assumptions about accountability, and about their relationship to reform, they are now discussed.

The marketization theme was evident in attempts to privatize schooling, introduce market forces, deregulate education and provide greater accountability. The calls for greater accountability, usually specified as outcome-based measures of student and school performance, were traced to three beliefs; that educational bureaucracies were hopelessly dysfunctional, that competition will enhance teaching and learning, and that a good deal of educational expenditure should be transferred from the public sector to the private sector. How these beliefs led logically to the desired components of accountability was left unclarified.

The redefinition of stakeholder roles reportedly involved significant conceptual changes. For example, learning was reputedly being conceived less as 'cognitive behaviour' and more as 'constructivist action in a social context'. Teaching was being defined less as 'technical instruction' and more as 'learning enablement' and as 'professional co-management'. Administration was being seen less as concerned with 'exercising positional authority' and more to do with 'transforming relationships, services and governance'. Parents appeared to be seen less as passive advisers and more as policy partners, co-teachers and community builders. Murphy and Hallinger (1993) concluded that schools should 'backward map' from students' learning, see school restructuring as an ongoing and constructivist process, work systematically at all levels and recognize the centrality of local stakeholders.

These conceptual ideals and politically critical conditions were echoed when Murphy and Hallinger went on to identify the key elements of supportive infrastructure; the backing of vital external constituencies, time, sufficient material resources, professional development, cross-fertilization between schools, trust, working structures and effective policy legitimization. The concept of accountability itself, however, was left undeveloped, as it is in other classic texts (e.g. Fullan 1991). To illustrate, a critical review of school-based management (SBM) (Wohlstetter and Odden 1992) found that, while the concept of SBM is pervasive, there are many forms in existence without clear goals or systematic accountability structures. Sadly, it was also found (p. 537):

> that little substantive decision-making authority actually has been delegated to SBM programs. Where there is substance, the outcome concern is teacher morale and satisfaction; the impact on student learning is usually ignored ... The result is that connections between student learning—the real objective of education policy—and SBM are not probed and thus not discovered.

Similarly critical views of SBM are also available in another literature. Research into the origins of administrative policy during restructures in Australian and New Zealand state school systems (Macpherson 1991a, b), earlier research in the UK and in the United States (Bush *et al.* 1989, p. 86, Elmore and Associates 1990), and a fine longitudinal study of governance in the Scottish education system since 1945, all recorded the extent to which the assumptive world of educational administrators and educators was 'deeply persuasive to those who shared it' (McPherson and Raab 1988, p. 99). When people had been socialized into this culture of professionalism, they became 'busy, but blind' (p. 99). Further, those who held this 'professional' perspective were typically found to regard client and technical views on accountability as largely incomprehensible, even offensive.

Finally, these studies also showed that the client perspective concerning accountability was the least known and the least influential of all three and that the persistent tensions between state officials, professionals and clients were due to fundamentally different theories being used. I will argue in Chapter 10 that these plural perspectives need to be understood as competing theories of accountability if the potential for change is to be realized. They are described in detail in the next section.

Theories of Accountability in Competition

The triplex typology of the perspectives noted above was elaborated by Elmore and Associates (1990). The typology contrasts technical, client and professional perspectives. The *technical* perspective holds that schools will only improve if the teaching, learning and leadership practices that are introduced are based on scientifically validated knowledge. Reforming these 'core technologies' of schooling, it follows, requires a steady stream of new knowledge, effective implementation mechanisms and regular structural adjustments of power relationships. This technical approach highlights the role of mandated systemic priorities, a sound knowledge of what schools can achieve and reliable means of identifying actual achievements. Accountability is, therefore, to be accomplished by being clear on purposes, defining performance indicators, and then collecting objective performance data and giving them prominence in the next planning round. As management consultants typically assert, this extends quality control and quality assurance into Total Quality Management (TQM) (e.g. West-Burnham 1992, West-Burnham and Davies 1994). An interesting blend of traditional bureaucratic, modern management and Islamic values is evident in Malaysia (Ibrahim, 1995).

The *client* perspective, in the Elmore typology, suggests that schools will improve when educators account directly to their clients; parents, students and the community. Reforming the relationship between providers and clients, it is held, requires greater client choice, flexible resource management and policy responsiveness. This approach celebrates consumerism. Accountability is, therefore, to be accomplished through political, market and managerial mechanisms such as clients governing school policy, competition, external audits and responsive human resource management and development. An exemplar of this approach was developed in New Zealand after the Picot Report (Taskforce to Review Education Administration 1988). School charters of purposes and objectives, that expressed the three-way contract between school professionals, community and government, were expected to drive school development planning, management practices and evaluation.

The *professional* perspective, according to Elmore and Associates, takes the view that schools will improve when educators and their immediate leaders are given greater opportunity to develop skills, exercise judgement and have greater control over their work. Reforming the professionalism of teachers and school leaders, it is argued from this view, requires special occupational conditions featuring autonomy, respect, resources and expertise. This approach questions the validity of non-professional views and the reliability of external evaluation. It promotes collegiality as the most appropriate base for accountability criteria and processes. Accountability is, therefore, to be accomplished by deconstructing and reconstructing the meaning of schooling, collaborative planning, and co-operative teaching and learning. One example is teachers' socially critical action research (Kemmis and McTaggart 1988). Another is stakeholders' co-operative action research into 'educational productivity' and learning (Levin 1994). Table 1.1 summarizes the three perspectives in competition.

Table 1.1 also makes it clear why, from the 'professional' perspective, the more traditional technical approach and consumerist perspectives can easily be seen to be driven by bureaucratic, economically rationalist and New Right thinking. They need not be taken seriously, from a professional perspective, when they are so patently 'politically incorrect'. The incidence of this phenomenon of competing perspectives is now briefly reviewed in four international settings to better map the challenges involved.

Australia, New Zealand, Britain and Canada

In Australia, the covert encouragement of privatization by governments (Anderson 1993) and the general adoption of corporate managerialism have been well mapped

Table 1.1: Three competing perspectives on accountability in education

Dimensions of belief	Technical perspective	Client perspective	Professional perspective
Conditions crucial for improvement	Teaching, learning and leadership practices based on scientifically validated knowledge.	Educators account directly to their clients; parents, students and the community.	Educators and their immediate leaders develop skills, exercise judgement and control their work.
Strategic target of reform	Core technology of schooling.	Power relationship between providers and clients.	Professionalism of teachers and school leaders.
Most appropriate reform strategy	New knowledge, effective implementation and regular structural adjustments of powers and relationships.	Greater client choice, flexible resource management and policy responsiveness.	Provide occupational conditions featuring autonomy, respect, resources and expertise.
Most valued knowledge-base	Objective measures of school achievement and learning outcomes.	Inter-subjective views of political and market realities and stakeholder interests.	Collegial norms concerning effective learning.
Source of legitimacy	Managerialism.	Consumerism.	Professionalism.
Most appropriate accountability processes	Define purposes and performance indicators. Collect objective performance and outcomes data for use in the next planning round.	Political, market and managerial mechanisms. Clients help govern school policy, competition, external audits, and responsive HRD.	Professionals empowered to deconstruct and reconstruct 'schooling'. Collaborative planning and co-operative teaching and learning.
Examples	TQM. School improvement and effectiveness strategies.	Charters underpin school planning, management practices and evaluation.	Collaborative staff action research.

(Beare 1995). Some educative antidotes have been proposed (Duignan and Macpherson 1991, Macpherson 1992a, b). The decentralization of administrative power to schools has elicited syntheses of local management skills (e.g. Caldwell and Spinks 1988, 1992) based on the early US school effectiveness literature (Ewington 1996) and on more recent Australian research and scholarship (e.g. Beare *et al.* 1989, Chapman 1990). On the other hand, some observers have noted the displacement of other educational metavalues such as quality pedagogy, democracy and social equity in policy discourse and in organizational structures (Chapman and Dunstan 1991, Angus 1992). Similar to overseas trends, there has been a widespread consideration of social action theories of learning (Walker *et al.* 1987) and constructivist theories of teaching and professional development (Northfield *et al.* 1987). Unlike most international patterns, however, Australia has witnessed efforts to develop holistic, pragmatic and consequencialist theories of leadership (Evers 1987a, b) and more inclusionary (Rizvi 1986) and more educative models of leadership service (Duignan and Macpherson 1993).

Despite these later initiatives, the politics of education are yet to reconcile the divergent interests of federal and state authorities or to accommodate the plural perspectives of administrators, teachers and parents with effective policy-making processes (Crowther and Ogilvie 1992). It is shown in Chapter 2 that the remnants of technical accountability policies are in disarray, even where they have been reasserted with uneven success. Recent evidence (Boston 1995) is beginning to suggest that technical mechanisms are being limited more and more to monitoring and key control functions as the 'business' of the state is rolled back, contrary to a rhetoric of devolution.

In New Zealand, as noted briefly above, the basis of accountability in the post-Picot era has been the School Charter. Negotiated between professionals, the state and the community, it provides the basis for evaluation, reporting and development planning. Unfortunately, just as the school communities were gaining confidence and expertise in collaborative planning, the role of the state was changed by ministerial edict without due regard to the balance of technical, professional, consumerist and democratic principles embodied in the original design (Codd and Gordon 1991). Schools' Boards of Trustees found themselves engaged in bitter political contests with the state over funding responsibilities, relative policy powers and the appropriateness of some market mechanisms (Cusack 1993).

Interpretations of the current situation in New Zealand vary. One view is that the agency theory in use did not anticipate fully the way that market forces would affect the decisions and priorities of school communities (Gordon 1995). A second is that the policy contradictions are so severe that school leaders should make a virtue of 'finessing' complex balancing acts (Wylie 1995, p. 54). My view is that while successive governments were keen to link education to economic prosperity, they have found it hard to reconcile the need to fund the improvement of public education with the need to reduce public expenditure and the deficit. I also suggest that a minimalist and elitist theory of state has became increasingly incompatible with a localist form of neo-institutional governance introduced by the Picot Report. Whatever, the general upshot appears to be that the legitimacy of the state has been damaged and that most schools have had to find ways of operating within the changed terms of the new school–community–government contracting process. Plural policy settlements are being negotiated both locally and vertically (with the Ministry) in order to reconcile temporary and competing views of appropriate accountability and theories of the state.

In the UK, the LMS initiative attempted to shift powers to governing bodies and parents in order to improve accountability procedures and to make the distribution of public funds more equitable and efficient (Sockett 1980, Kogan 1986, Levacic 1992). It is fair to say that school and system administrators initially found LMS complex and challenging. The proposed accountability mechanisms for teaching and school performance implied a major redistribution of power in favour of clients. Implementation then appeared to partly stall amid political controversy (Morris 1991) despite a widespread realization that formative evaluation is vital to professionals learning about how to teach better and develop their school (Mortimore and Mortimore 1991). Vann's (1993) study of how accountability policies can be related to school improvement also indicated that substantial impediments were created by the national reforms of administrative practices.

A change of minister in the UK and two reviews of curriculum and student assessment practices by Dearing (1993) have eased pressures in schools. Dearing's £6 million consultation process and the extensive opportunities for parental participation in school governance, for the moment, appeared to satisfy demand for client representation in policy making. The major tensions that remain are between the government's requirement for information for the purposes of summative accountability, and educators' need for information to enhance learning and teaching. Put another way, technical and professional theories vie for supremacy, with client views temporarily satiated.

While LMS was soon regarded as a relatively congenial aspect of leadership, criticism mounted when the government pressed ahead with its agenda concerning assessment and curriculum. Fears were raised that the revised national testing at 'Key Stage 2' would reintroduce features of the long discredited and abandoned 11-plus examinations (Sweetman 1994, p. 16):

> the assessment process means that credit or criticism is going to be placed fairly and squarely on individual class teachers and subject specialist. Put at its simplest, if things go wrong, the governors will know who to blame.

Galton (1995, p. 152) identified an even deeper problem; the gross limitations of the systems theory of policy making and implementation used by the UK government and its agencies. The first principle of democratic accountability noted above was revisited when he called for curriculum policy production processes that gave a central role to educative relationships:

> the crisis in the primary classroom will not disappear simply by reducing the demands of the national curriculum and providing a moratorium for teachers to master the various programs of study and revised assessment procedures. Curriculum building needs to start from the opposite direction, where relationships

between learning and teaching are clearly articulated and the nature of the knowledge required by various task demands, as these relate to how children think and learn, made explicit.

In Canada, while the dynamics and focus of system restructurings in recent years have varied by province, all jurisdictions have considered market mechanisms and realigned the influence of stakeholders (Martin and Macpherson 1993, Mawhinney 1995). Research into the nature of transformative leadership at school level (Leithwood 1992) has been particularly helpful. A worrying counterpoint, however, is the mounting evidence that teacher empowerment and site-based management have not led to demonstrable improvements in teaching (Sackney and Dibski 1994). This led to Fullan's (1993) call for the co-ordinated co-development of schools, district-by-district. He identified the crucial need for more effective forms of supportive infrastructure, including, presumably, appropriate forms of accountability.

Summary

It was argued in this chapter that the essence of democratic accountability is that governors remain responsible to the governed for the quality of their stewardship. It was assumed that the delegation of the authority to govern proceeds only with the consent of the governed and that this consent is conditional on the stewards of that delegated authority remaining accountable. Democratic accountability also attempts to maximize four values when it invests political authority in 'the people'. This first is individualism defined as the achievement of potential. The second is liberty, that is, freedoms consistent with order. The third is equality conceived as equity of opportunity, outcomes and access to power. The fourth is fraternity or co-operation in building a wholesome society. In sum, democratic accountability and governance provide a moral order for human relationships and the organization of social institutions.

The general international evidence introduced here suggests that the political structures of education are not articulating moral order or policy compromises to an extent where professionals, administrators and clients feel that their views are being honoured, and honoured as reasonable expectations that can and are being discharged publicly. Basic standards of public accountability are not being provided. A common feature is that there are no readily available answers to three questions:

- Are accountability processes open and fair?
- Are accountability criteria explicit, used in decision making and documented?
- Are those who make decisions being held accountable for them?

Policy vacuums concerning accountability are, it appears, felt most keenly wherever lapsed technical structures are no longer regulating performance evaluation, where reporting and improvement processes are not integrated, and where formal alternatives to bureaucratic relationships have not been developed. The paradox is that so long as the accountability issue is not 'politically correct' among professionals, they will continue to be exposed to the corrosive effects of feral criteria, arbitrary process and low legitimacy, all of which will tend to make worse rather than ameliorate the current situation. The danger is that both education and democratic society rely on the principle of accountability to recreate the legitimacy of structures, policies and practices.

An international policy paralysis concerning accountability appears to be induced by the seemingly intractable differences between technical, professional and client perspectives. The operational consequence is to make some measurable aspects of learning a universal proxy for the quality of learning, teaching, leadership and governance and the basis for accountability. It is time to develop processes and criteria that can simultaneously monitor, report on, evaluate and improve teaching, learning, leadership and governance. To begin, the term 'accountability' itself needs rehabilitation. This issue is taken up in the next chapter.

2
Rehabilitating a Politically Incorrect Policy Issue

Introduction

The first step towards rehabilitation involves understanding more fully how a concern for accountability came to be 'politically incorrect'. A feature of the international examples of restructures and devolution briefly examined above is evidence of non-existent, doubtful, impotent, controversial and incoherent accountability policies. There is no case known of where the responsibilities of stakeholders are as understood or as codified as they once were when the myths of bureaucracy prevailed. While this might be a good thing, in many senses, the new policy vacuums and ambiguities appear to have brought a different and no less intense set of challenges.

This chapter maps some of the major antecedents to the current situation, examines some of the policy dilemmas being experienced in Australia, identifies the extent to which a simplistic dualism, specifically hyper-individualism and communitarianism, has constrained accountability policy making in the United States. It is shown that accountability has all the hallmarks of a poorly structured policy problem; it lacks obvious criteria for an adequate policy, a means of making appropriate policies is yet to be developed and there is uncertainty about the nature of appropriate accountability processes.

Antecedents

There are at least five general antecedents to the current situation. As noted in Chapter 1, the first antecedent is that the more traditional and technical accountability criteria and processes were seriously disturbed when centralist bureaucracies were radically down-scaled and when different combinations of pedagogical, administrative and governance powers were decentralized. However, when most governments simultaneously recentralized key control functions, and generated a new consensus of cynicism (Kirst 1990, Beare and Boyd 1993), the legitimacy of accountability structures fell further into technical disarray and disrepute among professionals and clients. I investigate in a later chapter the possibility that this partial anomie triggered local attempts to generate stable occupational ethics in an increasingly communitarian context.

The second antecedent is the general change in policy processes. Up until the late 1970s, the policy cycle in education tended to move sedately through the Westminster process; election mandate, Green Paper, White Paper, Parliamentary debate and decision, and then the mobilization of support structures and implementation (e.g. Frazer *et al.* 1988). The process generated relatively high levels of consensus between stakeholders, permitted reasonable degrees of expert and professional participation, and had recognizable stages and cycles, albeit sometimes accomplished at a pachydermous pace.

By the late 1980s, however, governments were tending to respond much more in haste to issues whenever they became politicized or whenever a minister chose to intervene. Task forces of outsiders, usually contracted personally by ministers or their equivalents, generated proposals within months, often without the significant involvement of educators or researchers. Once proposals were approved by cabinets, and quickly launched as 'policy', schools were expected to implement the changes, with predictable degrees of hilarity, disbelief, resistance and alienation amongst professionals. Accountability proposals were seen as political claims, the vicarious preferences of the temporarily powerful. They competed poorly against 'professional' definitions of value, norms and practices. And given the extent to which professional development is locally organized, professional perspectives tended to be become more plural and more site-specific than technical and systemic perspectives had ever permitted.

The third general antecedent is the changed use of power. With the international collapse of command economies and systems of governance, the use of power-coercive change strategies in education (after Chin and Benne 1976) became far less tenable, especially in ambiguously decentralized contexts. And since rational-empirical approaches continued to have limited impact on the relatively conservative cultures of education, normative and re-educative change strategies became the principal remaining means available to would-be reformers by the early 1990s.

In such conditions it seems reasonable to assume that ideology could easily displace consequences and logic as the basis of policy justification. For example, some of the massive structural changes in some states of Australia were simply announced using ideological rhetoric in a glossy brochure issued by the minister's office. These documents were reputedly greeted with derision in many staffrooms. In schools and school districts, where the emphasis was increasingly on co-operative learning, collegial professionalism, school-based staff development, collaborative consortia, teacher empowerment and participative policy making, the traditional meanings of 'accountability', quite understandably, became ridiculous.

The fourth general antecedent is that policy research in education has focused more and more on measurable inputs, processes and outcomes and on 'value-adding' in economic terms. This atomization only made sense to those who saw schools as a reified social system held together by a production function and the use of power. It alienated those who saw learning, teaching, leadership and governance as symbiotic and reciprocal community relationships. One index of this alienation is that comparatively few recent articles or texts on school or system restructuring use or index the term 'accountability'. One notable exception is scholarship and research in the politics of education, a field whose publications are largely unread by teachers. Major US texts have consistently noted the low levels of consensus between major stakeholder groups on appropriate forms of accountability, and the correlation of this phenomenon with low levels of internal and external legitimacy in systemic administrative policies (Hannaway and Crowson 1989, Mitchell and Goertz 1990, Fuhrman and Malen 1991, Cibulka et al. 1992, Marshall 1993).

The fifth antecedent is that the rationales used to justify some recent 'reforms' have become implausible and regarded as naive. The school restructuring movement has yet to produce an empirical base to add to its normative and subjective claims. Conversely, the movement regularly recommends that education communities look to the quality of their accountability policies and practices if they wish to retain the respect and support of their constituencies (Timar 1990, Fullan 1993). Unmentioned is the dual effect of performance or service data. By this I mean that while performance or service data are usually considered essential to the recreation of legitimacy in education policies and practices, they also change the nature and use of power in education. As Cibulka (1991, pp. 198–199) pointed out:

> Embedded in the controversy is the deeper, often unstated issue, of whether educators, elected officials, or individual parents should have the primary power over access to and uses of performance information. This is a political problem at the heart of democratic theory ... One thing seems certain. Performance information is reshaping the character of educational politics.

To summarize to this point, as centralist policy making retreated into curriculum content, assessment and programme budgeting in many education systems, local accountability criteria and processes were generated more and more through normative, re-educative and socio-political processes. Performance or service data became central both to the reform strategies and politics of accountability. Dilemmas in Australia help clarify these events.

Current Dilemmas in Australia

In the 1980s the Australian state governments moved to dismantle the centralist patterns of policy making and implementation in school systems. Schools were asked to assume new responsibilities, and increasingly, to self-manage their affairs. As noted above, most states devolved responsibilities to do with pedagogy and school services while retaining high degrees of steerage over curriculum content, budgets and industrial relations. As the 'downscaling' of bureaucratic structures unfolded, the traditional criteria and processes of accountability gradually evaporated. The Cresap Report (1990), for example, ensured that Tasmania became one of the most devolved education systems in Australia by drastically reducing the system's central office, eliminating regional structures and providing only a slim presence at eight (later seven) district offices.

Some of the fundamental assumptions about such devolution have encountered stern criticism (Chapman 1990, Chapman and Dunstan 1991). For example, the devolutionary strategies assumed that the leaders of 'self-managing' schools would develop performance criteria and accountability processes that would be acceptable to school communities, remain responsive to state and national policies, and yet, cohere with a corporate managerialist approach to performance management (Beare *et al.* 1989, Caldwell and Spinks 1992, pp. 139–157). Any ambiguity over accountabilities, it was vaguely acknowledged, would detract from the legitimacy of public education, question the expertise presumed to underpin educational and administrative policies, and cast doubt on the professionalism of educators and on the leadership services provided by school executive teams.

The accruing evidence (Macpherson 1991a, Beare 1995), however, suggests that ambiguity over accountabilities is still widespread in Australia's devolved state systems, despite centralist controls, and that this ambiguity is closely associated with concerns over the productivity of state schools. Five indicators are now briefly described.

First, there is considerable doubt over the appropriateness and effectiveness of post-devolution accountability structures in state education systems. For example, the Executive Director of the Ministry of Education, Youth and Women's Affairs in New South Wales (NSW) (Grimshaw 1990, p.13) admitted long-standing difficulties in NSW over how to develop a

> monitoring system that addresses the substance of educational issues, rather than simply structures and processes and simplistic, measurable 'performance indicators'.

In the post-Scott Report (1990) era, NSW reintroduced a technical form of accountability developed first in South Australia: systemic 'Quality Assurance' (QA). Without the benefit of stakeholder preferences or systematic research, Cuttance (1995) claimed that QA was technically sounder than others because it focused not on quality control, but on the management of quality at all phases of the management process, at all levels and in all management systems. Key features of this approach included ubiquitous application and scrutiny, performance indicators, intensive strategic planning between and at each level, school reviews every few years and the production of school development plans. The approach tried to reconcile two competing functionalist theories; control through technical and contractual accountability, and school development through professional advisory functions.

It is my view that there was a limited appreciation of the limits of this form of radical and technical structuralism (Morgan 1980). It was too easily seen as being politically insensitive to plural and local conceptions of professionalism and consumerism. Schools councils, parents and students had no real role. It failed on its own 'Total Quality Management' (TQM) terms in the sense that the teacher performance appraisal and student assessment processes were excluded from the scheme. QA officials in NSW reported not to the line managers with whom they worked in schools and district offices but through separate structures directly to their own Assistant Director General. The QA scheme was, therefore, inevitably associated with schizophrenic purposes, structural ambiguities and conflict, and temporary and negotiated zones of assurance. In a phrase, QA in NSW provided symbolic and political accountability.

The NSW QA scheme finally ran into political sand in September and October 1995. As salary round negotiations stalled, the NSW branch of the Australian Teachers Union banned QA inspectors from schools. The 40 QA inspectors had to be assigned other duties. The union gave five reasons for the decision; poor consultation over the use of student learning outcomes, the unprecedented moves from qualitative to quantitative data, the fear that basic skills and Higher School Certificate results would be used to measure and compare school performances, the limited account taken of value-added measures and the cavalier use of comparative test results by some principals in QA inspection teams. In early 1996 Director General Ken Boston announced to his QA staff with characteristic precision that 'QA has crashed' and commissioned a search for an alternative accountability policy (Clark 1996).

Tentative plans have linked four strategies; (a) a two- or three-day review of school effectiveness and development needs every second year by the principal, staff and QA team leader, (b) annual school achievement reports to superintendents, parents and communities, (c) needs-based resourcing of schools, and (d) school performance improvement. However, given the turbulence in the state political context, and the proposed retention of many features and structures of the QA approach, there was little immediate prospect of a model being agreed in the context of a broader industrial policy settlement.

The second indicator is a persistent concern over the productivity of state schools. Learning in core subjects is monitored with light-sampling standardized tests in most Australian state systems. On the other hand, the comparative outcomes of inter-state education systems are systematically shrouded by the States' data collection, processing, storage and application protocols. In some states data does not go back to teachers or schools. The quality of teaching is sometimes only evaluated using non-transparent processes to establish fitness for promotion. Promotion criteria and processes can change dramatically at state boundaries. Selection and transfer processes generally discourage interstate appointments, and, in Tasmania until the mid 1990s, inter-district movement. Most teacher appraisal schemes use private methodologies and are rarely connected to career advancement or rewards. Leadership services are subjected to critical review only in times of crisis. In sum, performance management in state education generally remains at an early and incoherent stage of development. The productivity of Australian state schools cannot be specified, expected or demonstrated.

In sharp contrast, there is a good deal of anecdotal evidence in every state system that suggests that many school communities are developing increasingly effective horizontal accountability relationships. Further, philosophical research in educational administration has derived performance criteria for educative leaders from a holistic and pragmatic moral theory (Walker *et al.* 1987, Evers *et al.* 1987, Duignan and Macpherson 1993). I will come back to these ideas in Chapter 10. At this point it is enough to suggest that the policy vacuum might yet be filled with sensitive and effective approaches and methods.

Third, as Beare *et al.* (1989) have shown, there is falling surety in Australian state schools about the nature, currency and implications of public expectations. Sophisticated demand data are, in general, not collected and the situation does not appear likely to change. Conversely, the public perception is that Australian state schools are not significantly more responsive to public expectations after the devolution of decision-making powers related to teaching, learning and school development. There is also rising concern among experts, clients and professionals over the relationship between public choice, education policies and professional practices (Beare 1995). There is rising doubt that school leaders can provide educative forms and appropriate levels of accountability related to public-choice-making processes or link them to annual planning processes and to day-to-day management practices.

Fourth, there is considerable uncertainty as to whether or not the Liberal-National (conservative) coalition federal government elected in early 1996 will provide the extensive leadership education promised annually by the Labor federal government since 1993. Relationships between federal and state governments are tentative. While the Commonwealth government alone has income taxing powers, and the states retain constitutional authority over education policy and public school spending, the potential for misunderstanding remains considerable. Most state and national governments are using 'dry' economic policies to cope with massive projected deficits. In this context, budgetary support for leadership education has atrophied.

A number of implications follow from these four conditions. Accountability criteria and processes, especially performance data, are not likely to be negotiated easily in such a contested context. The Australian policy stage is crowded by state and federal players commissioned by ministers and other stakeholders, many of whom tend to be distracted by a concern for relative powers and control within their portfolios. Little serious prospect therefore exists of vertical policy coherence concerning educational matters much above school-district level, not unlike the situation in Britain and in the United States, although for different reasons.

The consequence, nevertheless, is that there is no more chance of Wohlstetter and Odden's recommendations concerning SBM being implemented in Australia than in the United States. They concluded (1992, p. 541) that:

school-based management should be developed by (a) joining SBM with curriculum and instructional reform as part of a coordinated effort to improve school productivity; (b) decentralizing to school sites real power, an aggressive staff development process, a comprehensive school database, and new compensation systems; (c) investigating how SBM can create a new organizational culture; and (d) developing district and school leadership that supports SBM.

It is also notable that, in Canada, Sackney and Dibski (1994, p. 111) took a different research route but came to much the same view, adding that national and provincial SBM policies would need to be more broadly reconceptualized around the 'notions of improvement, equity and equality of opportunity'. It can also be said that the general approach used in all three countries provides policy settlements in a limited range of policy domains while trying to satisfy plural purposes. In Australia this disjoint and mutual accommodation of priorities gives the national policy community, comprising mainly federal and state officials, much what they want; a discourse of reform and the symbolism of policy steerage. It also allows each state and school community to make plural interpretations of the national policies. It allocates each state a *pro rata* share of the captured textbook market. It offers parents standard curriculum and assessment policies and gives teachers a free hand on pedagogical matters.

None of these outcomes, notably, requires the definition and measurement of valued outcomes, only the partial accommodation of interest group preferences. Australia appears to lack the degree of clarity over purposes required for the development of fully coherent educational accountability policies. As the next section demonstrates, this situation is not unique to Australia.

The Nub of the Problem

The decades of debates over accountability in the United States have been described with considerable insight as a 'struggle over the soul of American public education' (Theobald and Mills 1995, p. 462). The rhetoric appears to be warranted and to have international application. The turning point of the 'struggle' is clear. Beliefs about (a) what constitutes valuable knowledge, (b) how learning can be demonstrated, and (c) the appropriate role of leaders and governors in public education, determine how educational accountability is to be accomplished in practice.

Such beliefs, however, appear to have been polarized in the United States between the powerful, who have advocated batteries of tests to monitor schools and learning, and those who have argued that there are better methods than traditional testing to hold education accountable. Prominent advocates such as Dewey (1916) have long promoted democracy and community, co-operative learning, holistic curriculum and pedagogy with formative assessment of individuals. Leading educational psychologists (e.g. Thorndike and Hagen 1969) were far more successful at promoting the individualization of education, the disaggregation of curriculum into skills, competencies and concepts, teaching to behavioural objectives and the mass testing of outcomes. How the historic dualism between hyper-individualism and communitarianism relates to the principle of accountability is summarized in Table 2.1.

The first challenge to the dualism has been epistemological in nature. Thorndike's approach to measurement and evaluation, and its handmaiden of scientific management (Taylor 1911, 1947), proceeded using a conservative version of science, positivism. It omitted values, politics, subjective experience and normative dimensions. While the term 'positivism' has been used in many ways (Phillips 1983), it is no longer considered acceptable to base theories solely on observation reports and logic, and to use empirical concepts derived only from observation. Dewey (1938, 1963), for example, relied on a broader form of 'experimentalism' that took many forms of data into account so that he would be guided by the consequences of actions.

As I argue in greater detail in Chapter 10, a post-positivist view of 'good science' suggests that observation is theory laden, plural views of the world are one of many sources of knowledge about the world, and knowledge grows holistically through competition and refutation between theories (Evers and Lakomski 1991, Berrell and Macpherson 1995). Hence, while Table 2.1 usefully contrasts the stark implications of

Table 2.1: Accountability in education and society

Dimensions of accountability	Hyper-individualism	Communitarianism
Definition of 'learner'	Individuals requiring education to achieve their potential and predictable destinies.	Members of society learning to be citizens while contributing to the character of local community.
Definition of 'educated person'	Individuals who satisfy national standards or competencies in selected skills, attitudes and concepts.	A public project requiring local deliberation in matters of ethics, compassion, justice and democracy.
Learning theory	Linear process comprising input, information processing in ways explained by cognitive science, and measurable learning outcomes.	Building understanding through interaction between cultural beliefs, assumptions, media messages, stories, myths and formal instruction.
Best pedagogy	Individualized instruction using behavioural objectives.	Facilitating co-operative learning in conditions of trust, self-direction and supportive social groups.
Formal curriculum	Valuable skills, attitudes, and concepts; authoritative knowledge.	Partitioned webs of belief governed by those with access to power.
Appropriate policy-making agency	National curriculum and testing services.	School communities.
Disciplinary bases	Psychology, Scientific Management.	Social philosophy.
Epistemology	Positivism; science creates truth.	Post-positivism: good science suggests the extent to which we should trust plural theories in competition.
Primary commitment	Maximizing the potential of the individual.	Developing democracy and community.
Purpose of accountability	Measure learning and school achievements.	School–community evaluation of learning, teaching and leadership.
Appropriate accountability processes	Objective assessment to measure students' grasp of facts, skills and attitudes.	Formative evaluation to assess and develop the depth and quality of understandings and plans.

an ideologically driven choice, educative accountability policy making might be wiser to explore the basis and alternatives to the taking of either side. It might, for example, seek means to reconcile the theories that lie behind empirical, normative and subjective data concerning accountability claims. One approach used and evaluated in the research reported in later chapters is to identify the common ground between competing theories, or touchstone, with the active engagement of stakeholders. Another idea is to use touchstone and a consequencialist moral theory to arbitrate dilemmas and a coherence test of evidence rather that rely overmuch on empirical data.

The second challenge to the dualism above is axiological in nature. Instead of valuing a learner solely in terms of test scores and predicted destinies, and letting science both define and deliver a good life, Dewey (1963) argued that what constitutes an educated person must be answered by communities, if they wish to remain communities. Accountability criteria and processes are required to connect deliberation about ethics, compassion, social justice and democracy to strategic evaluation and planning in educational systems and institutions so that learners and educators can refresh their understanding of civilized living. It follows that a policy community using communitarian principles to debate and select the most appropriate mix of values will have to consider *both* hyper-individualistic and communitarian perspectives. The converse would not need to apply.

It is pertinent here to note that, in recent times, the national debate in the United States over educational choice has been conducted largely using two systems of metaphors; (a) *laissez-faire* visions of free individuals whose actions are harmonized by a beneficent market, and (b) communitarian appeals to the civic role that schools have in creating unified communities and responsible citizens (Margonis and Parker 1995). The serious problem here is that national assessment and curriculum policy making could have pre-empted a community debate over what knowledge should be learned, how it might be demonstrated and the role of wise leadership and governance (Moffat 1994). It seems that more educative accountability processes and criteria are needed to help communities recapitulate education, society and democracy, as well as show how well learners have learned substantive content. Both agendas are to be valued.

The third challenge to the dualism is political and conceptual in nature. Some states in the United States have been providing legislation for 'charter schools' in the belief that more intensive local politics will break the accountability gridlock of technical, client and professional interests in public education and boost the development of schools and classrooms. Elmore and Associates' (1990) analysis of the competing perspectives was summarized in Table 1.1.

There are also at least four reasons for doubting that more intensive local politics of educational choice will have the intended effect of improving US schools. First, as noted above, when seen from a 'professional' perspective, the consumerist, economically rationalist and New Right thinking involved in the politics of choice need not be taken too seriously when it is so obviously 'politically incorrect' (Macpherson 1996a). Second, none of the chartering laws passed in eleven states by the end of 1994 took a broad enough view to provide a suite of accountability mechanisms linking district support, school improvement and classroom development (Wohlstetter *et al.* 1995). Third, penetrating educational productivity research (Odden 1990a, b, Clune 1994, Odden and Clune 1995) suggested that current accountability structures in the United States are not challenging poor leadership services, such as financial management, or encouraging 'best practice' teaching. Worse, the evidence was that equity was being displaced by adequacy as a policy aim. Fourth, the nature of systemic structure itself needs to be far better understood if reforms are to gain greater coherence and to improve teaching practices (Cohen 1995, Kirst 1995, Elmore 1995). Together, these research findings suggest that the dualism of hyper-individualism versus communitarianism needs to be set aside and that the problem of accountability itself needs to be reframed.

Summary

Accountability has become both a 'politically incorrect' concern and a personal and private professional matter at a time when there is international concern in education over fundamental purposes, what counts as knowledge, how learning is to be demonstrated, school productivity and policy legitimacy. There is little evidence, in this era of SBM, LMS, school management initiatives (Wong, 1995) and 'self-management', of coherent or educative accountability policies. Sustaining such policy ambiguities is not wise in institutions and systems where technical, client and professional theories of accountability are in active competition. There is too much at risk.

Antecedents include the technical disarray that followed the devolution of functions and responsibilities, the collapse of Westminster style policy making, the cavalier use of Ministerial reserve powers, the atomization and commodification of education, and the politicization of performance and service data. Unintended consequences are indicated by industrial action, concerns over school productivity and ambiguous public expectations. The situation is likely to continue until policy research in each system breaks the gridlock.

On the other hand, professionals, administrators and clients need to know that their views are being honoured in policy settlements, being expressed as reasonable obligations and that such duties are discharged publicly. Reasonable standards of public accountability should also be evident in open and fair processes, explicit criteria used in decision making and documentation, and those who make decisions being held accountable for them. Their absence exposes professionals to the corrosive effects of feral criteria, arbitrary process and low legitimacy. It was also suggested from a preliminary review of restructures that educative accountability criteria and processes, especially performance data about learning, teaching, leadership and governance services, are essential to the creation of legitimacy in education policies and in practices.

It also appears that beliefs in policy communities about what constitutes knowledge and how learning can be demonstrated largely determine how educational accountability is to be accomplished. Beliefs in the United States tend to be polarized between the powerful, who prefer empirical performance and outcomes data, and those who argue that there are more educative methods than testing to hold education accountable. While the former hold the political and policy high ground, their position is being eroded by epistemological and moral questions. The latter appear to hold better philosophical cards but have yet to produce demonstrably better purposes and methods.

The struggle over accountability policy can too easily divide on hyper-individualist versus communitarianism lines. It has all the hallmarks of an ill-structured problem (Robinson 1993, p. 26). It lacks obvious criteria for an adequate policy. The means for reaching an appropriate policy are not clear. There is uncertainty about the nature and availability of the information required to develop an appropriate policy.

A three-chapter case study of accountability policies and practices in England and Wales is now presented with four intentions in mind. These chapters were developed to illustrate how the accountability policy cycle has moved from one policy settlement to another. They were expected to reveal advances made in conceptual analysis. They were to identify the extent to which changes relate to the broader social, economic and political contexts. They were also to indicate the potential there might be for the reconstruction of public policies by policy research.

3
From Consensus to Conflict in England and Wales

Introduction

Practitioners, policy makers, researchers and theorists in England and Wales have employed the concept of accountability in public education in very different ways since 1945. Events and their interpretations are divided into three broad political eras for convenience, although antecedents and effects traverse the boundaries. The first period, from 1945 until 1979, covers the post-war reconstruction period and the so-called Great Debate. The second period, from 1979 until 1992, saw the imposition of new values in public education: the New Right intervention. The third and current phase is described as a contested policy settlement. This chapter is concerned with implications of the first period which, in essence, ended the post-war consensus.

Norms and Definitions

There are definitional, normative and political features concerning accountability practices and policies in England and Wales that need to be taken into account from the outset. The definitions in general use accept that to be accountable is to be responsible, to be explicit about obligations and to be answerable. In practical terms, an accountability relationship is universally defined as being between parties or stakeholders. It is also widely accepted that this relationship needs to employ criteria and parsimonious processes through which each of the parties accounts for its actions to the other(s).

However, while this relationship exists in the knowledge that outcomes could be used by one group to modify the actions of others, the principle of public accountability does not command automatic assent in public education. Educators in England and Wales appear to have appropriated reserve powers concerning directions and judgements to do with their professionalism.

The normative base of a professional concept of accountability today appears to embody (a) knowing and discharging one's duty in a partially confidential relationship with learners, (b) iterative and self-managed rather than directed and radical improvement, and (c) taking reasonable rather than absolute responsibility. Stenhouse's (1977) justification for these norms both ensured that improvement was contingent on the self-perceived fairness of means and a limited acknowledgement of the principle of public accountability:

> Accountability must be associated with feelings of responsibility. When people feel accountable they attempt unconsciously to improve their performance. When people feel unfairly called to account they devise ways of beating the accountancy without actually improving the balance sheet.

There are, of course, many reasons for this cautious norm. One is the deeply embedded habit of appealing to professionalism in public education when asked to account. It is interesting that the characteristic patterns and implicit norms of professional service in England and Wales, summarized in Table 3.1, have been developing in England since the 1500s (Charlton 1965).

A second reason is that the development of accountability policies in England and Wales over the last five decades has been more heavily influenced by political ideology than by educational philosophy or strategy, and especially by the way in which policy settlements have been reached. It will be shown that educators have been confronted regularly with policy processes and proposals that have offended professional norms, which in turn helps explain why they have developed a consensus of resistance to political forms of accountability, whatever their source of moral, contractual or professional obligations.

The examination of each period now begins with policy content and process, reviews the broader contexts of practices and ideas and maps out important consequences. This

Table 3.1: The nature of professionalism derived from Renaissance England

Dimension	Related characteristics of professional service
Relationship with client	Trust and confidence.
Code of ethics	Public service and duty to the community.
Code of expertise	Based on well-defined knowledge, including theoretical knowledge.
Preparation	Period of systematic training is provided, often in an institution.
Code of practice	Organized and institutionalized to test the competence of members and to maintain expertise.
Socialization	Defined social group with its own hierarchy, social life and commitment.
Status	Claims and is accorded status based on salary, learning acquired at highest level, organization and solidarity, code of conduct and professional ethics, and independence.

approach is intended to link practices, policies, research and theories to each other and to their interdependent development over time.

From Consensus to Ideological Conflict

The transformation of educational decision-making structures and practices in England and Wales has been portrayed as a move from the post-war reconstruction consensus to ideological conflict (Chitty 1994). The post-war consensus grew out of bipartisan commitments to full employment, the welfare state and the co-existence of public and private sectors in the economy. A national consensus-building policy-making style was the norm.

The 1944 *Education Act*, for example, was prefaced by three years of consultations. It established a 'national system, locally administered' that diffused power. Educational services were delivered by a 'tripartite partnership' of central and local governments and schools and colleges. The expansionary economic contexts, the limited number of powerful interests groups and the consultative style of liberal Ministers of Education sustained the consensus well into the idealistic 1960s.

A series of events and interacting factors helped dissolve the consensus. Sir Edward Boyle, the Conservative Minister of Education from 1962 to 1964, favoured turning a two-tiered secondary school system of grammar and secondary modern schools into comprehensive schools that served communities. He also wanted to scrap the rigid 11-plus selection process for academic secondary education referred to in Chapter 1. He supported these policies despite the views of some of his more conservative colleagues with the result that, when the Conservatives lost office in 1974, his influence eroded steadily. A series of Black Papers from the Right made the public very aware of the polarization in the Conservative Party concerning progressive education and comprehensivization. Simultaneously, the recession of 1973–1975 brought on by the OPEC oil shocks shattered secular beliefs in the post-war welfare capitalism and Keynesian social democracy, and revealed the deeper and persistent reality of British class conflict.

If the shattering of belief was so profound, why has education in England and Wales been peculiarly and yet infrequently vulnerable to political intervention on the matter of accountability? A major reason is that the traditional patterns of British representative democracy have made it very difficult to graft new forms of effective participation onto educational governance, especially by interest groups located outside of those structures (Fowler 1978). A recurrent consequence has been that when politicians have been put under pressure from plural constituencies in education they have tended to define complex issues in traditionally recognized ways, that is, as problems of quality, equity or choice, rather than as a problem of 'voice'. And yet, when matters have come to a head and action has been required, they have intervened in dramatic ways and changed structures, policies and forms of governance.

A second, related and major reason for education's vulnerability to infrequent political intervention is the technical incapacity of policy makers and structures to convert complex and plural public values into effective policy settlements. As Taylor (1978, p. 26) pointed out:

Discussions of accountability are peculiarly subject to technical regression: arguments about values soon become arguments about techniques. The weakest link in the goals–objectives–assessment–feedback–improvement model of accountability is its first. It is not difficult to see why this is so. Pluralism ... raises awkward questions about truth, relativism, power, normative order and a host of other issues about which philosophers and social scientists have wrangled for generations.

An example of poor articulation of purposes and objectives is the Assessment Performance Unit (APU) established in 1976. While it produced helpful evaluation prompts and materials for teachers, it lacked clarity and consistency of purpose and proceeded to develop strategies that were eventually considered to be serving little more than public relations functions.

Symbolic Accountability

It is important to note that the context in the mid-1970s predisposed many active in the policy arena to focus on technical matters rather than on the values to be served by accountability processes and criteria. For instance, the first recommendation of The Bullock Report, *A Language for Life* (Department of Education and Science (DES) 1975), significantly increased demand for more effective testing of learning. It called for a new national system of monitoring and for the creation of new instruments. The DES responded in the spring of 1976 by commissioning the APU. The agency was mandated to promote assessment, monitor student achievement and identify the incidence of under achievement, although it soon assumed the role of tracing trends in achievement over time.

It quickly came under attack. Its charter was traced to a desire by the national government to control, rationalize and evaluate the use of resources. The Director (Kay 1976, p.1) conceded that the APU existed as 'part of the response of the DES to demands for greater accountability by the educational service for the resources it consumes' and pressed on. And yet, despite this difficult beginning, the APU took a comprehensive and still influential view of achievement across six aspects of student development (mathematical, language, scientific, physical, aesthetic, and personal and social). It used light sampling of cohorts to monitor mathematics, language and science achievement, arguing that national assessment did not require the testing of each child. It also used matrix sampling so that each student sampled was given only part of the battery of tests used with each cohort. This approach generated generalizable information, limited the backwash effects on school curriculum and classroom pedagogy, and created a good deal of trust amongst professionals, who had feared, initially, that the APU would be used to narrow the scope of curriculum and control teaching.

The agency then ran into technical, educational and political trouble. It was realized (Maclure 1978, p. 16) that national tests using light samples were technically unable to identify the incidence of underachievement accurately, relate achievement to the personal circumstances of learners or to relate achievement to the quality of teaching or leadership in schools. Serious doubts were raised about the capacity of test item banks to fully represent the breadth of curriculum content. The APU, to its chagrin, was also stimulating inappropriate testing practices in some LEAs.

Even more formidable challenges followed. The Rasch statistical procedures employed by the APU were questioned (Goldstein 1979, 1980) to the extent that the National Foundation for Educational Research (NFER) stopped using them. Educationally plausible criteria for aesthetics and physical development proved elusive and the APU's work on criteria for personal and social development became politically unsustainable when some interest groups objected to them as 'invasions' into areas of 'private morality'. Broad political support for the APU began to erode.

In time it became clear that the role of the APU had failed to satisfy basic criteria as an accountability scheme. It was shown to be unfair and non-remedial; it could not look beyond student performance to the processes under the control of teachers or to the broader purposes of education (Nuttall 1982, pp. 42–45). The high intelligibility of published outcomes relied on deceptively simple data and concealed the lack of comparability. Since reasons could not be given for the APU's findings, and remedies could not therefore be suggested, it was concluded that its provision of information primarily served a public relations purpose. The APU showed that the Government and the DES *cared* about standards.

Hence, while the APU served as a source of interesting information, Her Majesty's Inspectorate's (HMI) surveys of schooling continued to have far greater influence. They triggered conferences, encouraged professional development and promoted school improvement activities. One inescapable conclusion in retrospect is that new policies and agencies sponsored by political intervention have to compete for influence with continuing structures, assumptions and practices on technical, educational *and* political grounds. Even when earlier policy and practices have lost their legitimacy in the public eye and are discontinued, and have apparently been replaced by newly legislated accountability policy and mandated agencies, their purposes and objectives will still have to be negotiated with all stakeholders. Further, only provisional and conditional acceptance are achieved at each phase, prior to trials being conducted, and the policy changes being finally accepted and fully embedded in practice. It appears that stakeholders in education in England and Wales long held a veto on quick change. Until more recent times, the implementation of public policy in education could only be achieved very slowly, principally through consultation with all of these veto-holders.

The Great Debate

A national reconstruction of accountability policies in England and Wales was deliberately triggered on the 18 October 1976, by the Labour Prime Minister, the Rt Hon James Callaghan (1987, p. 405). While he never used the term 'accountability' in his Ruskin College speech, he argued 'that where there is a legitimate public concern it will be to the advantage of all involved in the educational field if these concerns are aired and short-comings righted or fears put to rest' (*Times Educational Supplement* (*TES*) 22 October 1976). His call for a 'Great Debate' was 'immediately recognized as the public signal for a change in public attitude' (Maclure 1978, p. 11). It precipitated a debate focusing on standards of achievement, curriculum content, participation by parents and managerial responsibilities, and for much of the next decade, defined accountability in education as a salient political issue.

Callaghan proposed a national core curriculum, the improved monitoring of standards and a better relationship between schools and industry. The organized regional debates that followed tended to promote discussion among professionals and in public life about the purposes and content of a core curriculum rather than engage the citizenry. The process appeared to serve a number of purposes. Elliott et al. (1981, p. x) took the view that it was intended to manufacture 'the appearance of a democratically derived consensus in order to legitimate greater state control over educational policy'. Indeed the evidence reviewed in this chapter does suggest that concept of 'public accountability in education' was co-opted by Prime Minister Callaghan and the DES to justify subordinating the goals of education to the goals of the state, transferring power from professionals to the state, and to focus attention onto the means rather than the ends of accountability.

The Great Debate in education was driven by defensive politics occasioned by problems elsewhere rather than by a proactive and educational strategy. It might have been expected that the incoming Labour Government led by Harold Wilson would have used education to attend to the deepening divisions in society as it restructured the economy. Instead the Wilson–Callaghan administration of 1974–1979 became associated with rising inflation, trading deficits, industrial turmoil, currency depreciation, unemployment and indecision. The turbulent climate generated new political alignments. In search of electoral appeal, a series of 'think-tanks' moved the Conservative Party towards a mix of classical market liberalism and anti-statism. While the Labour Government floundered, the Labour Movement used symbolic revisionism to mask its covert commitment to a mix of neo-Marxism and fundamentalist socialism (Marquand 1988, p. 3). The result was that, as Chitty (1994, p. 15) noted, the:

> Labour party was thrown on the defensive by the ferocity and scale of the right-wing attack on its policies. The leadership appeared to be deeply embarrassed by the association of the Party in the eyes of the public with so-called progressive education, characterized as it often was by a child-centred approach to teaching, informal pedagogic and assessment methods and a general antipathy to hierarchy and inequality.

Hence, not only was Callaghan's Great Debate on accountability created by a Labour government in a deeply embarrassed and aggressive mood, it built and bequeathed policy machinery to the next government that replaced the Westminster consensus-building processes with high-speed presidential politics. The Downing Street Policy Unit provided an executive arm for the Prime Minister, Harold Wilson. It attacked the DES as the brain of the Education Establishment, and later, put education up as a 'leading candidate' for reform when Callaghan took over from Wilson in March–April 1976 (Donoughue 1987, p. 111). The analysis that Callaghan accepted was that teachers' unions had been captured by militants and progressives who paid scant attention to the needs of the clients at a time when children needed to acquire employable skills and attitudes.

Callaghan, therefore, launched the Great Debate on two strategic assumptions and with two primary aims. First, the government should force teachers to be more accountable to politicians, employers and parents, even by centralizing curriculum development. Second, the DES itself was a major part of the problem. The two aims of the Ruskin speech were, therefore, to replace the metaphor of 'partnership' with 'accountability' in policy making and to shift the locus of educational policy making power from the DES to Downing Street. These aims were declared, achieved, and then later trumped when the government changed. But this is to get ahead of events.

After the post-Ruskin consultations with regional and other stakeholders, Callaghan's government set out its position in a Green Paper. Probably the first time that the word 'accountability' appeared in a British government document on education, it was given a very partial definition. Accountability was deemed to be about evaluating learning and schools and the LEAs remedying poor performance (DES 1977, pp. 16–17):

> Growing recognition of the need for schools to demonstrate their accountability to the society which they serve requires a coherent and soundly based means of assessment for the educational system as a whole, for schools, and for individual pupils ... [and] ... it is an essential facet of ... [LEAs'] ... accountability for educational standards that they must be able to identify schools which consistently perform poorly, so that appropriate remedial action can be taken. Such assessment will take account of examination and test results, but will also depend heavily on detailed knowledge of the circumstances of the schools by the authorities' officers, their inspectors and advisers, and such self assessment as may be undertaken by the schools.

The Green Paper, and the following Taylor Report (DES/Welsh Office 1977), both identified the school (not the system or the classroom) as the unit of change, and proposed three broad reform strategies; testing and assessment, inspections and parental involvement. Professional educators soon made it abundantly clear that they preferred forms of self-evaluation (East Sussex Accountability Project (ESAP) 1980, Nuttall 1981) and accountability processes that recreated trust in partnerships. Becher and Eraut (1977, p. 11), for example, argued that a morally accountable educator is:

> accountable to all those who have placed one in a position of trust, and that accountability should be expressed in terms intended to secure the continued renewal of that trust. In practice, the broader definition may well be the more appropriate, because moves to strengthen formal accountability gain support from those who have simply ceased to trust.

As the Great Debate rolled on it gradually became accepted (Becher 1979) that the five forms of accountability set out in Table 3.2 were intimately related.

Table 3.2: Five related forms of accountability in education

Forms	Definitions
Moral	Answerability to clients.
Professional	Responsibility to self and colleagues.
Contractual	Accounting in terms of an employment contract.
Political	Accounting to political masters.
Public	Accounting publicly in terms of the public interest.

These terms and definitions will be used below. Since the public interest is the least discussed of these five forms in education policy research in England and Wales, it will now be considered.

The Public Interest and Accountability

The public interest in education refers to the capacity that schools have to serve society. The public interest in England and Wales tends to be defined in terms of the interests and benefits to stakeholders and by their differing demands for involvement in institutional decision making or policy making. Accountability for curriculum policy is a case in point.

The Taylor Report (DES/Welsh Office 1977) recognized the controversial nature of the curriculum and proposed school governing bodies as the appropriate mechanism for the articulation of the public interest in education. The National Union of Teachers reacted quickly and put it to the Taylor Committee that the education of students should be protected from the intrusion of non-professionals and that powers given to governors were against the best interests of children, parents and society. The point of reconciliation came with the realization that the need for schools to be responsive to the needs of society did not necessarily entail society or its representatives deciding how best to meet those needs. How policy was to be implemented was held to be a matter of professional judgement, albeit in consultation with others. This can be seen as significantly diluting the principle of public accountability in public education.

Similarly, the policy settlement process preferred by the House of Commons Education, Science and Arts Committee (1981, para. 4.21) reflected this desire for local reconciliation dynamics when it emphasized politically critical conditions: 'the responsibility should be a shared one with the fullest involvement of teachers, parents, employers, local community and local education authority'. In so doing the Committee understood the public interest concerning curriculum to be 'a matter of professional judgement tempered and modified by local consultation' (para. 4.22) and a 'contract and as a consensus in which individual schools and their governing bodies play a major part' (para. 4.21) within the policies provided by the LEAs and the DES.

While this approach has the advantage of reconciling potentially plural views and containing political noise, the problem is that 'the public interest' comes to be defined as a socially constructed artifact determined by the intersection of self-interest however defined by one occupational group. This can set aside other criteria such as students' needs, perhaps derived from a careful weighing of principles, consequences or outcomes, and pre-empt public critique of policies and policy making. The idea of a 'public interest' itself can be degraded. This can lead to the erosion of societal benchmarks, permit public policies to advantage sectional interests and allow justifications that actually back up into mere 'consensus' however ingratiated by reference to the 'general interest' of society.

Instead of accepting such political relativism, the relationship between 'the public interest' and accountability in public education is better expressed as standards of service in public life. Since this matter has become a matter of major significance in England and Wales in recent times, it will be dealt with below. What counts as the 'public interest' raises vital matters since, in a pluralistic society, it has to be decided:

> whose right it is to set the value framework within which judgments are to be made. The fact that there are few culturally accepted absolutes does not of itself rule out any determination of merit: it merely means that assessment must take place in a relativistic ethic ... [Hence] ... schools can no longer be labeled as good or bad, but instead be judged as good or bad examples of the particular kind of schools they happen (by circumstances, or sometimes by positive choice) to be. (Becher 1979, p. 64)

Thus, in a pluralistic society, evaluation needs to be relative rather than absolute in nature. A culturally privileged set of criteria and processes will be disrespectful, and therefore ultimately, inadequate in political terms. Nisbet (1978, pp. 108–109) captured the broad policy consensus of the late 1970s with regard to accountability:

> tests are most appropriate for monitoring national standards and for calibrating school assessment; examinations must continue to be used for some secondary assessment; inspection, or external assessment by a consultant team, is flexible and therefore suited for special probes; pupil records meet parental requirements for accountability; and self-assessment by school is essentially concerned with communication between school and community. The distinction made is artificial: there is, and there should be, much overlap ... This is a pluralist solution, which acknowledges the multiple values in education, and the wide variety among 'stakeholders'.

Policy Research in the late 1970s

Accountability policy research did not review the effectiveness of the widely shared policy of 'accountability in partnership' in the late 1970s. Instead, it tended to focus on the development of criteria to evaluate the quality of accountability schemes. Becher and Maclure (1978, p. 224) took a minimalist approach (see Table 3.3).

Table 3.3: Becher and Maclure's criteria for evaluating accountability schemes

Criteria	Specification
Maintains confidence	The scheme should maintain confidence in schools. This is the prime concern. It should determine how much information a scheme should provide.
Lay intelligibility	The scheme should provide a lay audience with enough information to base a reasonable judgement on whether the school is discharging its responsibilities.
Fairness	The scheme should not be biased; teachers and the public should have reason to trust the fairness of the procedures.
Remedial	The procedures must be formative, acknowledge real constraints and avoid making destructive or pointless criticisms.

There are a number of limitations to this formulation. First, the emphasis on the conservation of confidence could prevent the emergence of data that might disturb the tranquillity of a school. It is not too difficult to imagine circumstances when a disruptive intervention would be thoroughly warranted. Second, restricting intelligibility to a lay audience assumes that there is a method of dividing the plural audiences that comprise a school community into lay and non-lay groups, and further, that it is wise to define the 'lay audience' as *the* primary audience. If it is accepted that (a) an operating school is seen as an interactive social system located in a public service system embedded in broader political and bureaucratic systems that provide employment to professionals, and (b) an important purpose of accountability in that social system is that it should be formative and educative, then (c) there is a need for feedback that serves the moral, professional, contractual, political and public audiences summarized in Table 3.2. The data required are therefore to be determined by the needs of plural audiences, audiences that also have overlapping memberships. The complexities of process and criteria thus multiply.

A considerable degree of attention was given to this issue in the late 1970s. Nisbet's (1978, p. 100) formulation in Table 3.4 expanded on Becher and Maclure's list and suggested eight criteria for judging the appropriateness of accountability schemes.

Fresh themes in Nisbets' list include natural justice for all stakeholders and the need for a sophisticated methodology. To these criteria and specifications Nuttall (1982, p. 30) added that any scheme should also be (a) economic in its use of resources and (b) an acceptable blend of centralized and delegated control and evaluation. In all three approaches above, however, it was accepted that while accountability implies evaluation, evaluation does not necessarily imply accountability. And differences can also be distinguished between accountability and evaluation methods. To illustrate, Stake (1976) identified nine basic evaluation models each with its own purpose and strengths (see Table 3.5).

Although all of Stakes evaluation models require standards, data and judgements, accountability processes take the information gathered further to provide answerability

Table 3.4: Nisbet's criteria for evaluating accountability schemes

Criteria	Specification
Fair	Processes, criteria, findings and generalization seen to be fair.
Valid	High face validity, items and measures appropriate to relevant concerns.
Provides feedback	Data required for decision making is collected and available to stakeholders.
Objective	Methods and data are objective or the bias due to subjectivity is controlled.
Verifiable	Procedures and evidence are open to checking.
Non-damaging	Learning and teaching should not be distorted by processes or criteria.
Understandable	Processes, criteria and findings are communicable.
Comprehensive	Take full account of aspects and purposes of education.

Table 3.5: Stake's evaluation models

Model	Purpose	Strengths
Student gain by testing	Measure student performance and progress.	Precision and clarity over student progress.
Institution self-study by the staff	Review and develop staff effectiveness.	Motivates staff and raises commitment.
Blue-ribbon panel [committee of inquiry]	Resolve crises, preserve the institution.	Gathers expertise, insights and external legitimacy.
Transaction-observation	Develop understanding of activities and values.	Creates broad appreciation of services and relativity of values.
Management analysis	Increase rationality of day-to-day decisions.	Feedback on decision making.
Instructional research	Generate explanations and methods of instruction.	New principles of teaching and materials development.
Social policy analysis	Develop institutional policies.	Social choices and constraints clarified.
Goal-free evaluation	Assess effects of programs.	Data on effect with little co-option.
Adversary evaluation	Resolve a two-option choice.	Public testing of data and claims.

to clients (moral accountability), discharge responsibility to colleagues and public (professional and public accountability) and demonstrate policy subordinacy (contractual accountability). While both evaluation and accountability models in education usually attempt to move away from punishment towards remediation, the presence of contractual accountability clearly implies that retribution is possible. Similarly, as Hoyle (1980, p. 161) pointed out:

> Evaluation can be part of the process of the accountability of the school to its stakeholders, or it can be part of the management process of the school whereby it acquires feedback on its performance in order to improve it ... if it is to serve both purposes [i.e. external accountability and performance feedback] the strategies of evaluation will have to be thought out very carefully in order to avoid distortion.

When the government changed in 1979, however, many of these subtleties were simply swept aside. It can be assumed that accountability policy research at the time was fundamentally disconnected from institutional, systemic, and especially, the national political processes that reviewed administrative policy and practices.

Summary

The evolution of thinking about accountability in public education in England and Wales has three phases. The post-war consensus began to unravel from 1979 with the Great Debate. It will be shown in the coming chapter that the New Right intervention was pressed hard until about 1992 when a more pragmatic and contested search for a new policy settlement began.

It has long been accepted in England and Wales that to be accountable is to be responsible, to be explicit about obligations and to be answerable. In practical terms, an accountability relationship is defined as being between parties or stakeholders that exists in the knowledge that outcomes could be used by one group to modify the actions of others. On the other hand, the principle of public accountability does not command automatic assent. Educators have appropriated reserve powers concerning directions and judgements to do with their professionalism. There is a consensus of resistance to political forms of accountability, whatever their moral, contractual or professional value.

Two persistent impediments to the development of accountability policies have been the structures of representative democracy that have limited participation in educational governance, and the technical incapacity of policy makers and structures to convert complex and plural public values into effective policy settlements. The hostile reception given to largely symbolic structures of accountability showed that even when earlier policy and practices have failed in the public view, and ostensibly replaced by newly legislated accountability policy and mandated agencies, purposes and objectives still have to be negotiated with all stakeholders to gain provisional and conditional acceptance at each phase, prior to changes being achieved in practice.

The Great Debate raised the saliency of accountability in education as a political issue and defined it in terms of standards of achievement, curriculum content, participation by parents and managerial responsibilities. 'Public accountability in education' was co-opted by the state to justify subordinating the goals of education to the goals of the state, transferring power from professionals to the state, and to focus attention onto the means rather than the ends of accountability. The metaphor of 'partnership' was displaced by 'accountability' in policy making and the locus of educational policy-making power shifted from the DES to Downing Street.

Policy research during the Great Debate noted the degradation of the principle of public accountability in the public service in education. It gave priority to natural justice for all stakeholders and the need for sophisticated yet economic methodologies. It suggested a blend of centralized and delegated control and evaluation. Many argued that accountability should use evaluation information to provide answerability to clients (moral accountability), discharge responsibility to colleagues (professional accountability) and demonstrate policy subordinacy (contractual accountability). Unfortunately, there were relatively few connections between accountability policy research and the processes used to review administrative policy and practices. These few connections were severed soon after Margaret Thatcher's government came to power.

4

The New Right Intervention in England and Wales

Introduction

When Margaret Thatcher came to power in May 1979, she trimmed the size and influence of the Downing Street Policy Unit. Then, bolstered by the results of the 1983 election and determined to boost the pace of change, she gave it the capacity and power to create policy even more quickly than it had under Callaghan. The pace of change was further accelerated later in her second term when she began to rely more and more on her own principles and intuition, rather than on evidence and argument. It soon became routine to announce decisions and hand them over to ministries for implementation, with inevitable problems (Rosenhead 1992, pp. 297–8).

Two serious casualties during this period were the contestability of advice, since proposals were seldom referred to Cabinet for debate, and the quality and credibility of accountability policies in education. The involvement of educational experts, researchers and stakeholders was severely curtailed. A legacy of cynicism, simplistic policy dualisms and resistance by professionals dates from this period. One of the most persistent dualisms is now examined.

Productivity versus Responsiveness

The teaching profession in England and Wales was under siege concerning accountabilities in the early 1980s. There were two competing perspectives on how accountability processes in education would best encourage school improvement. The 'productivity' model, favoured by the state, promoted greater public control over decisions about school organization, teaching methods and the curriculum. The 'responsive' model, favoured by most educators, held that schooling would improve when professionals retained control over these decisions and become more responsive to those whose interests are affected by those decisions (Elliott *et al.* 1981, p. xiii).

The latter approach was actively promoted within the profession. The Cambridge Accountability Project (CAP) used case-study and action research methods in 1979 and 1980 to explore the extent to which schools could be improved using the responsive accountability processes. Six secondary schools in three LEAs who claimed to be well advanced in relating to and responding to local interest groups and in developing self-accounting procedures were selected. After a year of research, case study reports were presented to each school. The second year involved a second round of action research, another report to each school and then cross-site analysis to develop 'grounded theory' about 'responsive' accountability practices.

The CAP's methodology can be criticized for its determined promotion of the 'responsive' model, for failing to acknowledge any educative virtue in 'productivity' approaches, for remaining uncritical of the simplistic dualism involved, for disavowing any political stance and for providing some generalizations that outreach the data. On the other hand, the CAP provided fresh evidence and swept aside many myths. The maps of interest group perspectives were graphic and informative and can be summarized.

For accountability purposes, teachers divided their audiences into two groups; professional peers and clients. Many teachers felt accountable to clients on an individual basis rather than collectively. Intra-professional accountability was segmented into super-ordinates, peers and sub-ordinates. The more hierarchical the school, the less individual teachers felt accountable to each other or collectively to client groups. The more teachers felt collectively responsible for the school as a whole and to peers, the more they felt collectively accountable to client groups.

The majority of teachers did not feel individually or collectively accountable to governors and local government officers. Many teachers saw their accountability and their obligations as being determined by face-to-face interaction and to be valued in terms of proximity to their classroom work. Accountability was defined by teachers in two ways; 'fitting in' with role expectations pre-determined by others, and, explaining and justifying the decisions and actions taken to others. The first definition implied contractual accountability while the second implied the need for effective 'answerability' processes in order to be morally and professionally accountable.

The parents' perspectives were far less clear in the CAP data and seemed to be based on smaller and more opportunistic samples. Nevertheless, the parents interviewed, and surveyed in one school, appeared to value the humanistic qualities of the six schools above their capacities to enhance examination successes. They judged schools using particular criteria; their child's personal happiness, how caring teachers were, how well teachers 'stretch' students without stressing them, how approachable and open teachers were, particularly their willingness to discuss their teaching practices, and the quality of discipline in the school. The challenges posed by non-communicating, non-involved or non-attending parents were usefully traced to practical difficulties, undue deference being given to teachers, the learned cynicism of parents, over dominant parents, alienating social events at school and some parent's long standing dread of school. It was later confirmed by other research (ESAP 1980) that the majority of parents did *not* seek greater control over what went on in schools, or greater opportunities to exercise choice between schools, they instead wanted more information about what *was* going on in their children's schools.

There were major differences found by the CAP between employers' and teachers' views. The first was the gap between the aspirations that teachers had for their pupils and the job opportunities then provided by employers. The second concerned the attitudes they each wished students to develop, especially towards work application and authority. The third gap was over priorities in the curriculum and the teaching of particular subjects, most notably basic skills and intellectual disciplines. The difference between teachers' and employers' attitudes towards the values and importance of the world of industry and commerce was the fourth gap identified. The fifth concerned the way that industry, commerce and capitalism were portrayed and criticized in schools, and the promotion of occupations associated with spending rather than creating public wealth.

It is particularly important to note that the professional's description of the situation did not cohere with the remedial strategies they held to be appropriate responses. Instead, the CAP found that the teachers in the six schools tended to treat the policy challenge of accountability as being synonymous with the problems of communication. This had the most unfortunate effect of setting aside the need to (a) review the comparative power of teachers and parents, (b) respond to complaints about the failure of schools to deliver preferred attitudes and skills, (c) broaden participation in school governance, (d) modernize the curriculum and (e) give greater consideration to legitimate external perspectives. The CAP team came to the view that what was being proposed by most professional groups, namely 'self-accounting', usually meant fuller accounting or more of the same, and was 'associated with a desire to control more effectively the kind of picture of the school which is communicated to outside audiences' (Elliott *et al*. 1981 p. 228). This self-selected sample of professionals, who were very aware of the need for accountability, were, nevertheless, reluctant to respond to the opinions and arguments of their audiences and to change their practices. Their preferred strategy was better image management.

The overall conclusions of the CAP concerning optional forms of professionalism that relate to accountability have been summed up in Table 4.1.

Table 4.1: The CAP typology of school responsiveness and accountability

Forms of professionalism	Response to external groups	Response to external perspectives
Isolationist	Given minimal information, not encouraged to seek information and discouraged from questioning teaching practices.	Unconcerned with external perspectives and decoupled them from the improvement of practice.
Rational	Given maximal information and encouraged to seek information and to provide suggestions.	Consulted but encouraged to consider the limits of their professional expertise and to respect the professional autonomy of teachers when they make educational decisions.
Participatory	Given maximal information and encouraged to seek further information and to question practices.	Consulted, encouraged to suggest improvements and invited to participate in the reformulation of policy and professional practices.

The teachers in the six schools were found by the CAP team to favour the second option of rational professionalism. It appeared to best represent their concepts of professional authority and professional accountability. This was in sharp contrast with their own analysis of the situation and with demands for direct 'lay participation' in educational decision making increasingly being articulated by parent activists (e.g. Sallis 1977, 1979a, b). The CAP research team itself favoured the second option, with some reservations, accepting in principle that teachers' professional expertise justified privileged status in the control of practice. The CAP team had three reservations suggesting that they saw the third option as a longer term ideal.

First, the boundary between professional practice and other broader curricular and organizational policy questions was not as clear cut as teachers' calls for professional autonomy suggested. Second, in situations where other parties were not participating in decision making in the six schools, the teachers commonly referred to, anticipated and tried to accommodate their supposed views, i.e. they did not act autonomously. Thirdly, and finally, since the teachers never flouted or even risked public disapproval, it suggested that they preferred to have policies and to use practices that were able to withstand public tests of reason, evidence and argument.

It can be assumed, therefore, that CAP teachers recognized limits to peer-based legitimization and professional autonomy, whatever the overt claims. And, as Burgess (1992) verified a decade later, they also accepted that a key purpose of revising accountability arrangements was to restore confidence in public education. Put simply, despite the rhetoric and the methodological limitations, the CAP showed that productivity and responsiveness were not regarded as mutually exclusive options. The CAP findings also implied that the normative base could tolerate quite high levels of centrally driven change concerning accountability.

Indeed, within five years of the announcement of the Great Debate, the DES published its view, in *The School Curriculum* (DES 1981), that a new national core curriculum of Mathematics, English, Science, Modern Languages and Technology should be introduced. This core was immediately characterized by its critics (e.g. Elliott *et al.* 1981, p. x) as 'the industrialists' curriculum,' since the DES was giving unprecedented priority to the revival of Britain's industrial base. There was, nevertheless, a similar sea change soon under way in the evaluation of schools.

School Evaluation

Inspections involving external participants had long been intended to add greater credibility to evaluations of each schools' performance than could be achieved by the LEAs or the professionals in schools alone. In the early 1980s, LEA-initiated school evaluations tended to be loosely structured processes guided by documents that suggested 'starting points' for evaluation. Some LEAs provided tighter prescriptions but encountered reactions from teachers and head teachers who disputed the degree of centralization involved. Other LEAs merely offered guidelines and support services, such as specialist consultants, accepted the processes and criteria that schools developed and made a virtue of the degree of decentralization involved. This latter approach, institutional self-evaluation, was usually defended using the comfortable metaphor of 'professionalism'. 'School' was obligingly defined as a professional learning community that was constantly concerned with evaluation and improvement. When the level of trust required of clients and the political realities of community life periodically outran the capacity of these accountability mechanisms, this approach came to be regarded as a little too incestuous to be plausible. The political litmus test of capacity, in times of crises, was usually how 'the school' or 'the system' handled the issue of ineffective teachers.

The approach most favoured in most LEAs by the mid-1980s was termed 'audited self-evaluation'. Head teachers and teachers prepared internal evaluations of school performance which were then audited by outsiders such as LEA staff, some governors and occasionally by the HMI. The assessment of achievement was determined by external examinations that were set by a range of examining bodies, or by moderation panels comprising insider and outsider professionals. Visitations by panels of moderators tended to accredit curriculum already developed by insiders. HMI inspections were by the invitation of professionals or as part of commissioned national reviews, that is, not directly related to annual school development processes. Stakeholder participation was, typically, limited. Involvement in the evaluation of school performance and in the assessment of student achievement remained a privilege confined to the profession. Since there are many and varied interpretations of the situation that prevailed in the 1980s, it is important to set aside the myths of a 'golden age' of LEA audited self-evaluation.

The quality of LEA inspection processes was closely examined by research mounted in 1991 and 1992, prior to the mechanisms being swept aside by the Office for Standards in Education (OFSTED) model. Maychell and Keys (1994) were asked to clarify the range of LEA monitoring and evaluation strategies, to gather experiences of these strategies nationally, and then to advise LEAs on 'best practice'. Six LEA case studies provided an initial database prior to the project being overtaken by the publication of the *Schools Bill* (1991). The national surveys of schools and LEA personnel were then adapted to clarify the nature and likely outcomes of 'full inspections'. The highly charged context guaranteed both high response rates and degrees of bias.

The findings show quite clearly that LEA schools at the time were not being subjected to regular or rigorous inspection and that most heads were antagonistic to this form of accountability. Only 60% of LEAs carried out any 'full inspections', with the numbers of schools actually inspected annually well below the 25% proposed by the new OFSTED. Only half of the 850 schools surveyed had *ever* been inspected. While head teachers and LEA inspectors favoured 'full inspections' every four years, 90% were against privatized inspections. No more than 10% of LEA personnel and 20% of heads believed that it was in the best interests of schools to separate advisory and inspection functions. The general view was that the summative and formative evaluation of schools should be simultaneous and co-operative.

Hence, Maychell and Keys (1994) accurately predicted that the introduction of privatized inspection would be controversial, as would the publication of examination results and inspection reports. They also reported that head teachers were particularly concerned that they would no longer have access to the information, advice and ideas that they traditionally associated with LEA inspection. This concern, prophetically, was magnified by the fear that the delegation of responsibility for advisory and professional development services to schools would not be matched by a commensurate allocation of funds. In sum, this reasonably definitive statement about the way things were in the early 1990s in UK public schools left precious little room for romantic recollections about the 'golden era' of LEA-managed school accountability.

Persistently Ambiguous Purposes

The New Right intervention proceeded on the assumption that the ends to be served by fresh accountability mechanisms were given. Logically, specifying accountability relationships presumes that there is clarity and agreement over aims and methods. Herein lay a major and persistent challenge. As public property, the ends and means of public education are inevitably contested, and in a democracy, must remain contestable. Accountability models and frameworks must, therefore, reflect a range of theories in competition about purposes and processes. Kogan (1986) provided a seminal analysis of the plethora of frameworks actually in use, with complex reasons as to why they coexisted, a vital point often lost on those who saw them as exclusive options. He identified three major models, as summarized in Table 4.2.

Kogan's analysis also made it clear that each model had its own way of determining appropriate partners for accountability relationships, the most appropriate processes that should be used by a partner to exercise control over another, and the appropriate source and nature of criteria that should be used to make judgements. It also identified the related issues of powers, responsibilities, rights, professionalism and entitlements and the extent to which models carried with them discrete theories of state and knowledge (Kogan 1986, p. 16):

Table 4.2: Kogan's three models of accountability in education

Dimensions	Public or state control	Professional control	Consumerist control
Purposes of accountability	Given and legitimized by democratic processes.	Arbitrary, therefore to be determined by experts.	Arbitrary, therefore to be determined by clients.
Appropriate accountability processes	Bureaucratic structures and lines of authority. Hierarchical and one-way relationships and top-down external reviews.	Team based structures and expertise-based authority. Interactive relationships, internal and external reviews.	Temporary functional structures. Contracted partnerships. Political relationships and external reviews.
Source of criteria	Superordinates.	Professional peers.	Elected representatives and the market.

For example, the belief in the professional control of schools carries with it assumptions, often implicit and unclarified, about the nature of democracy and participation. It entails assumptions about the purposes of education and schooling and the feelings, or affect, which are aroused by particular forms of power relationships between teachers, the larger political system, and client groups such as parents and pupils. The professional model of accountability rests on assumptions too, about how knowledge is generated, how it is communicated, and the ways in which different forms of curriculum lend themselves to external control.

On the other hand, Kogan's analysis can be criticized for defining accountability too simply as the rendering of accounts between partners, and for underplaying formative and moral dimensions, especially the capacity of accountability processes and criteria to recreate moral economies. Taylor (1978, p. 55) had emphasized this point a decade earlier when he predicted that:

> The complexity of the variables involved means that attempts to make individual schools and teachers accountable for pre-specified outcomes in relation to given resources are unlikely to succeed. Worse, by stressing product rather than process, such attempts may erode the moral basis of a curriculum and pedagogy which should properly exemplify, sustain and develop understanding and skills in relation to such values as tolerance, respect for persons, liberty within the law, democratic pluralism and community participation ... A worthwhile and workable concept of accountability embraces much more than achievement in the basics or the overall monitoring of performance. It includes systematic self-monitoring by individuals and institutions, regular reviews of curriculum and methods, a variety of types of evaluation, wider public discussion of educational objectives, and a more considered attempt than hitherto to bring existing and new research efforts to bear on relationships between schooling and other social processes.
>
> Too narrow an approach to accountability will weaken the contribution that teachers and schools can make to the sustenance of those personal and social values ... identified as important. A broader base is needed if damage is to be avoided and improvement secured.

Taylor's inclusionary accommodation of pluralism was further justified by appeal to an appropriate range of values that translate into forms of educational accountability. Pateman (1978) demonstrated that schooling should, simultaneously, respond to parental preferences, use public resources efficiently, allow teachers professional freedom, meet the requirements of society, and satisfy children's needs. The troubling current problem, he noted (p. 69), however, was that:

> the collapse of agreed values—the legitimization crisis again—has put teachers on the spot, and put 'professional freedom' in danger of being relegated to the fourth division of values. It is not surprising, therefore, to find teachers actively suggesting new forms of accountability which are effectively proposals for more

effective and responsive forms of intra-professional accountability than have hitherto been used, though there are plainly limits beyond which they cannot go. Thus ... politicians are putting the question of getting rid of dud teachers on the agenda ... something that teachers could scarcely do themselves.

Compounding this values vacuum and legitimacy crisis was the confusion created by politicians, principals, teachers and parents holding sets of intentions and expectations that were not so much in conflict as being confused priority sets in complex structures. As Kogan (1978, p. 114) put it:

while there is no clear correspondence between levels and values being propounded, different types of assessments will be appropriate as we move from the relationship between individual teacher and the pupil through the different levels of the institutional system ... [leading to] ... the problem of multiple legitimacies ... Institutional systems have to provide mechanisms for establishing priorities and resolving conflicts between different purposes and the tasks arising therefrom. So multiple purposes lead, logically, to multiple assessments.

Developing multiple forms of accountability to produce multiple zones of legitimacy requires networks of answers to at least three prior questions. Who is to be accountable to whom? Accountable for what? What purposes are to be served by accountability criteria and processes? There was no follow up of this sophisticated line of inquiry at the time. I will come back to this issue in Chapter 10.

Attempts to Reconcile Competing Perspectives

Policy research tended instead to search for ways of reconciling a professional perspective with other priorities such as industrial and social justice, validity and reliability, comprehensive pragmatism and educative processes. Nisbet's (1978) proposals are summarized in Table 4.3.

The dangers Nisbet referred to were then seen as very real. With regard to school accountability policies, MacDonald (1978, p. 147) claimed that the current initiatives commanding attention, resources and support were:

technocratic in form, deterministic in educational values, and precariously dependent on a costly and defect-ridden technology. If pressed to fulfilment they will constrain and standardize the curriculum, penalize the non-conformist teacher, eliminate experiment and decisively reinforce the schools' already well attested proneness to conservatism.

He recommended an alternative; a 'process model of accountability' which would have teachers move from the collection of descriptive data to anecdotal evidence that revealed educational values, and then to the 'organization of self-reflection in the school ... creating the conditions for self-criticism (where the school rather than the

Table 4.3: Nisbet's criteria and dangers of four perspectives on accountability

Dimension of accountability	Industrial and social justice	Validity and reliability	Comprehensive pragmatism	Educative processes
Criteria for selecting audiences, criteria and purposes	Methods need to be fair as well as seen to be fair by all concerned. Criteria and processes must be understandable and communicable.	Data must have face and content validity and be objective or make their subjectivity explicit. Processes must be replicable, reliable and use verifiable data.	Methods must avoid distorting or damaging learning or teaching. Criteria and processes must be comprehensive in scope.	Methods must provide feedback for decision making.
Dangers to be avoided	Any approach that appears to offer payment by results.	Reductionist effects that are generated by adopting pre-specified objectives and batteries of tests.	Any approach that assumes the existence of a simplistic production function and uses relatively few and quantifiable input and output measures.	Any national system that centralizes power and control, decentralizes responsibility and blame, and diminishes the capacities for organizational learning.

individual is the 'self')' (p. 145). While an advance at the time, it lacked the external-internal dynamic and a program of planned change and evaluation later considered essential in action research (Burns 1994, pp. 293–311).

Dualisms also persisted and retarded the production of policy options. Many educators and policy researchers shared a culture that conducted situational analysis in stark terms; bureaucratic versus professional, centralism versus localism. Eraut (1978, p. 155), for example, argued that:

> we ultimately have to settle either for a centralized education system based on central accountability with teachers as closely controlled employees or on a decentralized system based on a concept of delegation with teachers as autonomous but accountable professionals.

To be fair, there was also evidence that Eraut was among the first to recognize the benefits that might follow a mix of change management strategies (p. 187):

> The power-coercive approach may provide a framework which forces a school to consult widely and to respond to those who have delegated authority to it, without imposing externally determined policies; and the empirical-rational approach may contribute an important element to internal evaluations; but the normative-reeductive dimension is crucial if evaluation is ever going to lead to action which brings about desired improvements in pupil learning.

In this context characterized by plural values and prescriptions, and increasingly strident demands for intervention from their right wing, the government decided to use circuit breakers. By 1987 a complex array of right-wing education interest groups with overlapping memberships (Griggs 1989, pp. 116–117) were providing Mrs Thatcher with two common views; undermine the power of the LEAs and 'ginger up' schools by 'repositioning' them in a free market. What they were less able to agree on easily was the need for a centrally imposed national curriculum.

Nationalizing the Curriculum and Establishing a New Moral Order

The division in views in Conservative ranks over the need for a national curriculum revealed a basic paradox of Thatcherism; free economy versus strong state (Belsey 1986, Jones 1989, Chitty 1994). While neo-liberalism celebrates freedom of choice, individualism, market mechanisms and minimal government, neo-conservatism favours social authoritarianism, a disciplined and subordinated society, hierarchy and strong government. The dilemma of imposing a national curriculum while creating a 'free' market in public education was eventually resolved by the new Education Secretary Kenneth Baker and influential DES officials. They persuaded neo-liberals in Cabinet that a national curriculum would actually *help* establish a 'free' market. Their winning proposition was that since testing was required at important stages to provide parents with the information they needed to exercise choice, a national curriculum was actually needed to specify appropriate benchmarks (Coordingley and Wilby 1987, Chitty 1989, p. 218). This proposition was the first circuit-breaker.

The second involved redefining accountability in solely market and political terms. This dramatically simplified the problem of plural purposes in public education. Coopers & Lybrand's (1988) deficit analysis promoted moral accountability to clients and helped ensure that the *Education Reform Act* (1988) was interpreted as a mandate for increasing the accountability and the responsiveness of schools and LEAs. Coopers & Lybrand defined parents and business as the consumers, schools and LEAs as the providers and recommended market and political accountability mechanisms. The official interpretation of moral accountability was, thereafter, defined by consumerist metaphors. There were serious problems with this logic and its consequences; the latter became particularly evident through comparative research conducted in Scottish policy settings.

First, the definition of parents as consumers devalued partnership responsibilities and alienated students in secondary and primary schools (Aitchison 1995, Sullivan 1995). Second, reducing 'public schools' to the status of a market that 'cleared' did not acknowledge the more complex, long standing and functional social, political and economic interdependencies (Gallacher 1995, Macbeth 1995). Coopers & Lybrand's exchange theory had no category for co-operative community partnerships. Third, defining 'a school' as a polity potentially politicized all roles, redefining them as representing competing interests. When things went wrong, community problem solving capacities were devalued in favour of political accountability; evoking a destructive politics of blame and adversarialism (Brighouse 1995). Fourth, moves towards more consumerist and political relationships in school communities did not improve the quality of teaching or learning (Glatter 1995) or relationships between schools and businesses (Warwick 1995). Emphasizing consumer and political rights increased confrontation rather than improved co-operative problem solving capacities. It also reduced attention to the reciprocal aspects of moral accountability, such as obligations and responsibilities, by both professionals and citizens (Thody 1995).

In sum, the political and market strategies selected were unable to achieve the declared aim of the new policy; greater moral accountability. Coopers & Lybrand's consumerism and managerialism swamped moral and professional accountability with contractual accountability. They did not seem to realize the limited technical capacity that political and market mechanisms have to provide reasonable levels of moral and professional accountability in education (Macbeth *et al.* 1995). As Petch (1992, pp. 91-92) put it:

> It is easy to demand accountability, to write documents promising accountability, even to draft legislation requiring accountability. It is not so easy to achieve effective accountability in practice. Success or failure depends on attitudes ... If people feel threatened, if they feel that their views do not matter or that it is not worthwhile, any process to make them more accountable will be a token. Successful accountability can enhance the quality of relationships; token accountability may damage them, as people become resigned, disappointed or hostile. The consumer route to accountability in education is dangerous. Fortunately we can

reject confrontation and develop through co-operation. A more constructive route lies through partnership and mutual accountability.

There is a third reason why consumerist and political mechanisms cannot generate more than a limited degree of moral accountability. When 'a school' is defined as a market or a polity it is no longer held to be a moral community. In such circumstances it no longer needs to account in a moral manner and, therefore, is no longer required to account according to a moral code. It only needs to account according to the political and economic theory of the moment. This can lead to the dangerous relativism of 'situational ethics': where current political and market perceptions of 'the situation' may be used to justify ends and means. These political and economic contingencies will inevitably distort organizational norms. Since schools tend to be bounded social systems, with major penalties attached to the disturbance of the moral order, the relativism involved can be rendered beyond question until aberrant consequences trigger external intervention.

This, of course, begs another question; why should a 'school' account according to a public and improvable moral theory? The short answer is that because decisions taken and services provided by schools affect others and their life chances, schools need to get better at decision making and justification. The longer answer is that, as Hodgkinson (1991) demonstrated, public schools in civilized societies have always had a blend of ideological, economic and aesthetic purposes to serve. It is reasonable that the balance of these services should be governed and justified by a moral order. The moral order must also remain contestable and contested as change is a necessary though insufficient condition for improvement.

Since the moral order itself needs to be capable of improvement, four conditions follow. Policy settlements should reflect the best and shared wisdom of partners, since moral knowledge is intrinsic to collective webs of belief about purposes, and thus, how best to be organized. The revision of this wisdom also needs to be guaranteed by an openness to new ideas and approaches. This wisdom also needs to be regularly tested with fresh information and critical and collaborative dialogue (Robinson 1989, Fullan 1991). Logically, these processes need to feature relatively undistorted communication and be managed so that each partner may satisfy statutory and negotiated roles.

LMS and Legislation

Many aspects of the ideals developed in the section above had long been evident in professional norms and practices. It can be speculated that their continuing presence in public education meant that the Conservative party-political reconciliation of neo-liberalism with neo-conservatism then had to 'run the gauntlet' of professional, administrative and societal perspectives before ambiguous policy settlements were eventually achieved. The evidence below is that each settlement required many complex agreements by sector, by level of government, by institution and by issue. Evidence of differential rates and patterns of settlement has also accrued.

To illustrate, the *Education Reform Act* (1988) mandated new student assessment and management policies. The general changes to accountability prescribed by the 1988 Act are set out in Table 4.4.

Table 4.4: The *Education Reform Act* (1988) and accountability

Area	Changes	Accountability mechanisms
Curriculum	Introduced National Curriculum of core subjects (mathematics, English and/or Welsh, religious education, science), foundational subjects (history, geography, technology, music, art, physical education, modern foreign languages), and cross-curricular themes (e.g. economic awareness).	Learning attainment targets at the end of four key stages. Students are tested at ages 7, 11, 14 and 16. 'League tables' of results published.
Finance	Secretary of State to approve LEAs distribution of finances to county and voluntary schools, budget management powers delegated from LEAs to schools' Boards of Governors. Governors may delegate powers to the Head.	State steerage of LEA distribution formulae. Heads prepare program budgets and account to Governors.
Competition	Schools induced to opt out of LEA administration and to become 'Grant Maintained Schools' (GMS) and thus gain property rights, acquire property, contract independently, accept gifts and invest. Grants to establish City Technology Colleges in urban areas may cater for 11–18 year olds, provide broad curricular with an emphasis on science and technology. Schools market themselves.	The enrolment market clears as demand is matched by supply.
User-pays	Legalization of the power of schools to raise charges for extra services, providing parents have agreed.	Local fund raising and competitive tendering.

Implementation strategies included boosting the local management of schools (LMS) and nationalizing and privatizing school inspection processes. Consultants Coopers & Lybrand (1988) also recommended that schools budgets be determined by a three-part formula (*pro rata* grant plus current operating costs plus special needs) and that LMS should include financial and personnel management. How it worked out has been the focus of some very sophisticated policy research.

The LMS policy set high priority on local responsiveness and gave power to governors and head teachers to plan accordingly. Its relatively short-term effects were systematically evaluated (Levacic 1992) using three categories; inputs (organizational structures, national and community policies, expectations of government, parents, community and industry; resources and finance), school processes (school goals, resource limits, resource allocation) and outputs and accountability (performance indicators, market accountability through enrollments, financial controls and inspection). While this evaluation was limited by the systems theory used, it recognized the importance of stakeholders (central government, LEAs, parents, community and industry) in determining the criteria and processes of accountability and respected the interdependence of inputs, processes and outputs. It is crucial to note that the specification of 'appropriate inputs' and the reliability of determining 'educational outputs' were found to be weak. The quality of 'rational goal setting and budgeting' was also discovered to be doubtful, and concern was expressed about the use of market mechanisms as a proxy for the discharging of public obligations.

The impact of the LMS policy was found to be both rapid and considerable. The six traditional roles of LEAs (educational leadership, facilities planning, partner-to-schools, banker-to-schools, information-brokers to service-users and quality control) had been changed. The first two roles had been downgraded, the third redefined and the last three enhanced. LMS had helped redistribute resources, with primary schools in general gaining at the expense of secondary schools, especially small secondary schools. Differential impacts on schools depended on resource expansion or contraction, on the LEAs using budgets as a tool for control or for planning, on the attitudes and experience of school governors and on the entrepreneurialism of the head teacher. Levacic (1992, p. 16) ranked the possibility of the intended aims for LMS being achieved as follows:

- enhanced powers of parents and governing bodies at the expense of LEAs,
- improved accountability for the use of resources,
- improved efficiency in resource allocation,
- improved equity between schools of resource distribution, and
- improved quality of teaching and learning.

Levacic's evidence relating LMS to the quality of teaching and learning was particularly interesting. School effectiveness factors had an ambivalent relationship with LMS policy and practices. Some aspects of LMS (e.g. resource allocation using school priorities) enhanced school effectiveness while others (e.g. head teachers' reduced capacity to offer curriculum leadership) acted as inhibitors. This led to her overall conclusion that an improvement in the quality of teaching was unlikely to be a direct consequence of LMS.

While the rapid adoption of LMS can be explained by the considerable extent to which the scheme promised to service a professional and community impulse among school governors and educators to develop the local moral economy and the effectiveness of their schools, it also brought a new suite of problems. Levacic (1992) showed that LMS gave lower priority to inter-school co-operation, industrial equity, social justice and to regional and national strategic perspectives. In essence, under LMS, accountability became a local process involving financial, managerial, political and market mechanisms.

The Continuing Search for Alternative Policies

The rapid implementation of the *Education Reform Act* (1988) accelerated the search for more appropriate accountability policies. Some sought to develop more politically sensitive models in polarized political climates. Becher *et al.*'s typology above, for example, was expanded by Burgess (1992, pp. 5–8) to identify seven forms of accountability differentiated by stakeholder, all concerned with securing quality performances by others (see Table 4.5).

Burgess (1992, pp. 10–11) also argued that accountability policies need to provide checks on tyranny by locating policy making and political responsibility with governors while expecting experts to determine, organize and deliver services:

> This division of function (which is similar to that in schools between governors and heads and in government between Ministers and civil servants) rests on a constitutional scepticism about politicians and experts. In English local government there are no directly elected officials. The politicians are accountable to the

Table 4.5: Burgess' differentiated forms of stakeholder accountability

Form	Examples of differentials in stakeholder accountability
Personal	Students are accountable to teachers for their learning. Parents are morally accountable to society for their children's education.
Professional	Teachers are accountable to peers and inspectors for practice, standards, qualifications, training and conduct.
Political	Governors are accountable to the electorate for institutional policy. Elected leaders of teaching teams account to peer followers.
Financial	Financial managers account to institutional managers for practice and standards. Institutional managers account to auditors and governors for policies and probity.
Managerial	Institutional managers are accountable to the governing body for institutional performance.
Legal	Parents legally accountable for their children's education. Teachers and school accountable *in loco parentis*.
Contractual	Teachers accountable to employers for the terms of their contracted service. Institutional managers accountable to employers for the quality of organization and leadership.

public, but do not run the service; the officers who do this do not have the last word. Each is a check on the other. The object is to guard against the weaknesses of democracy—that politicians may act ignorantly or corruptly—and of bureaucracy—that officials may be over mighty. The strengths of the system are too often taken for granted: they are appreciated when it breaks down. In the late 1940s and 1950s the danger was from authoritarian chief officers who were largely out of members' control. More recently, the danger had been from meddling members who have bullied through their personal and political fads.

Burgess' reiteration of the traditional Westminster division of functions concluded with the advice that a plurality of processes should be managed (pp. 13–14) 'with confidence. They are better exercised, not in isolation, still less in confrontation, but in partnership. It is here, not in prescription, instruction and regulation that quality is to be found'. The partnership he proposed, however, was limited to two stakeholders; the elected and the appointed. The former was presumed to have an absolute mandate, and the latter, faultless professional expertise. It was, in sum, a proposal to return to a normative and elitist theory of centralist accountability.

The practical tensions for other stakeholders at this time were considerable. For a head teacher in London (Pringle 1992, p. 68), the exercising of expertise meant 'managing the middle ground between conflicting interests ... trying to give due consideration to three sometimes conflicting principles: considerate treatment of individual staff, equitable treatment of all staff and our accountability to our pupils.' Looking in on schools, the Chief Adviser of a London Borough (Goddard 1992, pp. 83–84) agreed and discerned the need to embed four evaluative elements in systemic accountability processes and criteria:

> First is a sound institutional evaluation, located within a professional framework of integrated support and external moderation. It links with professional accountability and enhances responsibility and understanding.
> Second is openness to parents, governors and the public which brings together professional and contractual accountability. Again this should lead to understanding and responsibility for all those involved.
> The third element is the means to deal satisfactorily with the unacceptable, either within a confidential or public arena.
> The fourth element is information about the quality of the school. There is a difficulty here: raw data rarely constitute information ... information has to be presented for a purpose and with an audience in mind.

Goddard then identified four types of systemic review mechanisms that together met these criteria. They are summarized in Table 4.6.

There are also limits to this integrated, hyper-professional and normative model. It assumed that all stakeholders and participants would be both teacher and learner in a context of differentiated power. It assumed that the LEAs and school managers could establish and sustain a positive environment in a hostile broader political context. It

Table 4.6: Goddard's systemic accountability mechanisms for schools

Mechanism	Purpose	Links created	Significant features
School self-evaluation	Build school capacity for improvement through rigorous internal scrutiny.	School development and planning, complimentary to staff appraisal.	Fosters professional development and action research.
Specific review	Provide judgements on specific areas of work or functions.	Self-evaluation and school's development planning.	Agenda and timing set by the school using both internal and external critiques.
Periodic whole-school inspection	Evaluate management, school achievements, and evaluation practices using stakeholders' criteria.	Review of policy making and implementation processes.	Report presented to governors, LEA and the public.
LEA survey	Inform LEA on quality of services.	LEA statutory functions for resource management, policy and teacher development.	Data used to inform LEA decision making is also made available to the school.

denied governors and market forces any role in professional accountability. It defined students as powerless and made very limited provision for contractual accountability. It thus reinforced a simplistic dualism; summative evaluation serving contractual accountability versus formative evaluation serving professional accountability, promoted the latter at the expense of the former, and set aside moral accountability to students and parents.

The machinery of accountability needs to have the capacity to create practical meaning in staffrooms and classrooms. From a director's vantage point, Stoten (1992, p.105) argued that it needs to have four characteristics; 'it must be understood by the parties involved; achievable for them and acceptable to them; it must have an end product which can be judged'. Achieving this, he argued, implied locating accountabilities with individuals as well as with groups, effective and improvable communication, responsible and participative leadership, subsidiarity and responsibility at the point of action, setting and achieving objectives, developing appropriate performance indicators and rigorous evaluation.

There was, however, some distance between such ideals and practice at the time. In Cambridgeshire (Lister 1992, p. 111), the effects of LMS, consumerism and politicized governance were baldly explicit in the LEA's accountability policy being limited to:

> *Planning:* providing or reducing provision in relation to demographic changes and parental choice, and deciding on the most effective pattern of school and college organization;
> *Quality control and improvement of standards:* information to parents and students about all the choices of establishments and courses available to them;
> *Funding of locally-determined services:* like community education, nursery education and certain kinds of home-to-school transport which, at no cost, add value to the basic level of education.

It should not be assumed that the centrally imposed changes were rejected by all educators. There were attempts to ingratiate the changes. Earlier, Lello (1979, p. 10) had used selected aspects of the Westminster principles of government to argue that accountability:

> involves reporting to other people voluntarily or compulsorily. It means having a moral conscience or a moral responsibility about what you are doing. It means being answerable to other people both junior and senior to yourself. It implies a dependence both on ideas, and on others. It is part of the essential cement in a democratic society.

His later work, *Accountability in Practice* (Lello 1993), posed as descriptive and non-political yet conceded that 'fitting in with government policy might be the subtitle of the book' (p. 6). He supported the Prime Minister's policies of dismantling LEAs (p.11), attacking socialism (p.19) and introducing competition and open management (p.19). It should be noted here that Mrs Thatcher had been replaced by John Major in November 1990 to little discernible effect. He reportedly (Bolton 1993, pp. 14–15) continued her practice of turning 'a deaf ear to the professionals in education'. Nevertheless, writing in praise of the government's political acumen, Lello concluded (pp. 49–50):

> What is accountability expected to bring? ... The benefits are total control, total knowledge of what is going on, and a greater chance of a reduction in costs. If it is thought important to compare the progress of a child on Merseyside with a child in Sevenoaks then there is now the machinery to enable this to be done. It also means that every teacher in the land can have his or her teaching monitored. It means that the teacher on Merseyside who wants to spend all his time teaching Greek will not be allowed to do, nor will the fanatical sociology teacher in Sevenoaks be allowed to be lop-sided.

While the absolutism involved is inconsistent with the principle of democratic accountability, and his conclusions totally at odds with the research reviewed in the next chapter, the existence of this view alongside the perspectives reviewed above tends to confirm Simkin's (1992, pp. 6–12) view that four distinct models of accountability were in active competition by April 1992. The four models are summarized in Table 4.7.

Summary

Two serious casualties of the New Right intervention were the contestability of policy advice and the quality and credibility of accountability policies in education. A legacy of cynicism, simplistic policy dualisms and resistance by educators was created, a legacy which tended to reinforce rather than reconstruct practices and norms. Benchmark research found that many teachers saw accountability as a communication issue. This position set aside the need to review the comparative powers, respond to complaints, broaden participation in school governance, modernize the curriculum and give greater consideration to legitimate external perspectives. Image management was the preferred response.

There was no 'golden era' of LEA-managed school accountability. While the evidence indicates an enduring commitment to democratic pluralism and community

Table 4.7: Simkin's four models of accountability

Models	Professional	Managerial	Political	Market
Key actors.	Professionals.	Managers.	Representatives.	Consumers.
Influence factors.	Peer review.	Hierarchy.	Governance.	Choice.
Success criteria.	Good practice.	Effectiveness and efficiency.	Policy conformance.	Competitive success.

participation, LEA accountability practices lacked coherence and comprehensiveness. There was a values vacuum and a growing legitimacy crisis. Stakeholders in complex structures had confused priority sets. Policy research, especially by Kogan, made it clear that a sophisticated accountability policy would have to acknowledge the complexity of structures in public education with levels of organizational purposes and tasks, and employ multiple assessments to establish multiple legitimacies

The incoming Conservative government took a different view. It saw a confluence of contested values and a policy paralysis expressed as simplistic dualisms. The Right campaigned for intervention. Two political circuit-breakers were devised. The government (a) nationalized curriculum and assessment, so that parents could exercise choice between schools, and (b) undermined the power of LEAs and introduced LMS to localize accountability through financial, managerial, political and market mechanisms. The changes were legislated using the *Education Reform Act* (1988) and the *Education (Schools) Act* (1992).

Policy research reviewed showed that these Acts and mechanisms did not recognize schools as communities or the limited extent to which consumerist and political characteristics could actually achieve a major policy aim: provide moral accountability to clients. They had little appreciation of the sense of security, significance and solidarity intrinsic to school communities. The research also noted the immediate and considerable impact of LMS and its ambiguous relationship with the quality of teaching and learning. LMS had limited policy coherence with other initiatives controlled from the center; changes to the assessment of student learning, school governance and school evaluation.

The general conclusion that can be drawn at this point is that the changes that followed the *Education Reform Act* (1988), and which were reinforced by the *Education (Schools) Act* (1992) in England and Wales, were (a) intended to boost managerial, political and market forms of accountability practices, but (b) by late 1992, their implementation was increasingly being contested in school communities. The next chapter examines the process that led to an uneasy accountability policy settlement.

5

The Contested Policy Settlement in England and Wales

Introduction

The period since 1992 has been characterized by complex attempts to reach a fresh working consensus in public education, policy issue by policy issue, within and between levels and stakeholders, often clarifying purposes after the fact. Student assessment, school governance and school evaluations have been three major cases in point and are discussed below. It is also important to note at the outset that the accruing damage done to community was becoming very evident in the broader context. The related field of social welfare services in England and Wales is a useful example (Allen and Martin 1992).

The 'new consensus of disenchantment' with the post-war welfare settlement in social welfare (Papadakis and Taylor-Goodby 1987) took three main forms that had direct equivalents in education; a 'legitimization crisis' (a general loss of confidence in the efficacy of public sector services), a 'distribution failure' (a general acceptance that state interventions had not progressively redistributed opportunities), and 'paternalism' (the least advantaged coming to see education and welfare services as oppressive and manipulative). The targeting of simplistic indicators, such as 'standards' and 'discipline' in education, and 'dependency' in social welfare, helped redefine long-standing communal values related to accountability. An instance in education was David's (1989) analysis that showed how a consumerist notion of individuality was employed to redefine 'parent–teacher partnerships'. The concepts of 'provider capture' and 'parent power' were used to convert a 'sense of community' into 'competitive self-interest', and to recast 'equal opportunities' as a matter of 'individual choice'.

The fundamental challenge this New Right agenda posed to the purposes of accountability in education also became apparent in the field of community adult education. Community adult education in Britain had long been based on three articles of belief (Ward and Taylor 1986); an egalitarian ethic that saw adult education as a means of addressing dangerous inequalities within and between communities, an empowerment ethic that saw adult education as giving people capacities needed to change community life, and a localist ethic that insisted on sophisticated and locally governed forms of intervention rather than those that reflect the priorities of central and professional groups. As noted above, Margaret Thatcher's unique blend of neo-liberalism attempted to establish a free market, at direct cost to egalitarianism, by using a strong state to drive 'reforms'. It neither realized the need for empowerment or localism in capacity building at all levels, nor the need for social and political dimensions to compliment the economic aspects of relationships between people. Her brand of economic privatism valued property above citizenship, saw little purpose in co-operative social systems and denied any legitimacy to the deliberate and democratic accrual of power by interest groups with collective identities.

While this context encouraged community educators to explore new opportunities in private, voluntary and informal sectors, the policy initiatives were never widely 'regarded as an adequate substitute for democratically controlled and accountable public policy at the local level' (Martin 1992, p. 28). Indeed, by the end of the 1980s, the outcomes in community adult education were still more ideological and hegemonic than structural and institutional in nature.

It appears that economic privatism lacked a sophisticated theory of change. The seventeenth century roots of economic privatism or 'possessive individualism' (MacPherson 1962) defined humanity as 'freedom from dependence on the will of others'. Freedom was seen as a function of possession. Society was held to be no more than interaction between free and equal individuals who own and use their capacities to make exchanges. Political activity was seen as a calculated device to protect this property and to maintain the order of exchanges.

The crucial implication here is that such a theory could present publicly financed institutions of education, health and welfare as unwanted and unnatural national liabilities, rather than as national assets. It also lacked a wider philosophical, sociological and political imagination. In essence, Margaret Thatcher's ideology of economic privatism reduced the role of education to economic and ideological purposes, and attempted to set aside the need for aesthetic and social dimensions to accountability. This became particularly apparent when her government's accountability proposals concerned with student learning were forced through several iterations before being reluctantly and provisionally implemented.

The Assessment of Student Learning

The current assessment policy had to go through four distinct modifications and partial implementations over six years (Scott 1994) before a barely operable policy settlement was finally achieved. It will be recalled that the *Education Reform Act* (1988) had legislated for 'learning attainment targets' at the end of four 'key stages', which meant that students were to be tested at ages 7, 11, 14 and 16 and the 'league tables' of results were to be published.

During the first phase of implementation, the DES (1988a, b) Task Group on Assessment and Testing (TGAT) commissioned by the Secretary of State proposed two major strategies; teachers assessing students as part of the teaching process, and terminal and summative assessments at the end of each key stage that would use a criterion-referenced framework to establish each teacher's reliability. It is important at this point to identify some of the major differences between such criterion-referenced and norm-referenced assessment, as set out in Table 5.1 (Willms 1992, p. 17).

These differences explained why although norm-referenced tests 'generally serve the administrative and accountability purposes of administrators better' than do criterion-referenced tests, they are 'inevitably met with resistance from teachers' (Willms 1992, p. 17) who favour other purposes. The TGAT recognized this inherent tension by suggesting an integrated formative and summative model that linked professional judgement making to the systematic development of judgement capacities in a national context. They argued that assessment outcomes should be adjusted to take account of socio-economic disadvantages before being published in the public domain. A crucial strategy proposed was that moderation be used to bring teachers' judgement into line with national standards and lead to the development of 'standard assessment tasks' (SATs). The Government's views prevailed, however, and, for example, it quickly became practice to publish unadjusted 'raw' data in 'league tables'.

The TGAT model also tried to attend to school improvement by putting the assessment of learning in context and encouraging co-operative teacher development. Again, this agenda did not match the priorities of the Government which tended to emphasize comparability, competition and the need for consumerist data. The second phase, therefore, downgraded teacher judgement (much to their annoyance) and the formative functions of evaluation, and boosted 'key stage' summative assessment. Ironically, teacher assessments had to be retained in the areas of 'attainment targets' that could not be covered by SATs. Teachers also had to determine the level at which SATs pupils would be entered. Education Secretary John MacGregor (1990) continued to argue for TGAT principles, in particular that assessment should be firmly linked to curricula and that teachers should play a major role. At this point the Prime Minister intervened and MacGregor was replaced.

The third phase, at the behest of the new Education Secretary, Kenneth Baker, attempted to replace the lengthy, interactive and potentially formative SATs process. The SATs process provided teachers with potential links into curriculum development. Doubts had crept in concerning the reliability and comparability of SATs, especially when the numbers of candidates gaining high grades increased each year. This put teachers in a catch-22 situation. It potentially blamed them for poor teaching if the numbers fell and yet could blame them for poor evaluation if the numbers rose!

Baker intervened. He expressed a preference for quickly administered 'pencil and paper' norm-referenced tests at the end of each key stage. He rationalized SAT's specifications by edict and statute, limited the role of teacher marking and moderation and emphasized the maintenance of standards. And although the *Education (Schools) Act* (1992) then legitimated the publication of 'raw' test scores, teachers became increasingly incensed, many governing bodies of schools rejected his analysis and edicts, and in the summer of 1993, many school governors publicly challenged the authority of the Secretary of State for Education by abandoning the Key Stage 3 tests.

Table 5.1: Major differences between criterion-referenced and norm-referenced assessment

Assessment	Criterion referenced	Norm referenced
Educational purpose	Describe tasks pupil can perform by relating performance to given criteria.	Describe comparative performance of individuals or groups using norms derived from the performances of other pupils.
Approach	Usually several tests covering a limited domain of learning tasks tied closely to the content of the formal curriculum.	One test covers a large domain of learning tasks, with relatively few items covering the skills related to any specific objective.
Design	Often developed locally by teachers or provided in teachers' guides that come with textbooks.	Items of moderate difficulty are selected to discriminate amongst individuals.
Typical use	To identify mastery, understanding or to diagnose the areas requiring further teaching.	To compare students' achievements at the end of a course of study.
Outcomes	Focused, formative and local evaluation.	Broad, comparative and summative evaluation.
Subsidiary outcomes	Motivation of learning and better pedagogy.	Standards of achievement and national assessment.
Limitations	Norms vary between professional groups, school cultures, intake cohorts and authorities, thus requiring calibration (moderation) through discussions between professionals.	Insensitive to parents' and pupils' aspirations, variances in cultural capital, different access to resources and opportunity, and require careful professional interpretation.

The escalating stand-off between politicians and governing bodies became politically intolerable and was eventually defused when Dearing (1993, pp. 65–66) was commissioned to slim down the curriculum, review the ten-level scale for rating students' achievements, simplify the testing arrangements and improve the central administration of the National Curriculum and testing.

The fourth phase finally produced a workable policy settlement concerning assessment, essentially through a string of compromises. Dearing separated formative and summative assessments, recommended that the ten-level system introduced by TGAT be modified or abolished, gave equal status to teacher judgement and SATs through co-reporting, and dramatically cut the prescription of curriculum and assessment. Most stakeholders appeared to greet the policy settlement with a deep and uneasy sigh of pragmatic relief.

The Dearing model of accountability was ideologically pluralistic and politically pragmatic. There was something for everyone. It combined greater professional discretion over curriculum and teaching in a context of continuing external review. Published performance data permitted some comparison of students and schools for consumerist purposes. A similarly troubled settlement was reached over school governance and inspections

School Governance

The *Education Reform Act* (1988) introduced a model of local management that gave school governors the power to act as a board of directors with the head teacher serving as the chief executive officer. School governors were empowered to set school policy, goals and objectives, allocate resources, monitor the performance of the school and to hold the managers accountable. Specific duties of the governors included senior appointments, salaries, discipline and dismissal of staff, suspension of pupils and ensuring that both national curriculum and collective worship policies were implemented. The governing body itself was deemed by the legislation to be responsible to the LEA, which retained the power to withdraw delegated powers in times of crisis, and to the parents. This legislated role located governorship in the first of four categories of types of governing bodies as defined by Kogan *et al.* (1984); an issue I come back to.

Whatever the legislation, a great deal clearly depended on the quality of the relationships between governors and members of the school executive. These interactions were found by research to be affected by many factors such as capacities, values and the nature of particular issues. The most persistently difficult factor was role clarity (Kogan *et al.* 1984, p. 164, Sallis 1991, p. 217, Golensky 1993).

The extent to which school governing bodies were actually providing educative forms of accountability by the mid-1990s was clarified by Earley (1994). He provided well-researched answers to four key questions. How are school governors coping with their responsibilities? Has the delegation of powers led to new partnerships between clients and administrators? How is business being done, especially decisions being made? Is there an emerging consensus over the role of governors? Survey data were received in 1993–1994 from 499 schools in England and Wales. Response rates from heads were high (63%) whereas only 35% of chairs of boards and 36% of governors responded. Data were, however, triangulated with detailed case studies in six LEAs.

It was found that governors tended to come from white, male, middle class and well-educated sections of society. Eighty-six percent of chairs and 76% of all governors had received, and 80% valued, the training they had received for their roles. Eighty percent of heads had opted to become governors. Recruitment difficulties and imbalances of expertise and inexperience were concentrated in inner-city schools that were located in poorer areas. Skewed recruitment was often attributed to governance workload although primary heads were more inclined to report such overload problems which suggested that they were less comfortable with the changes.

Governors tended to meet five times a year for about three hours on each occasion with most also serving on subcommittees. Governors were also experiencing heavy expectations and were increasingly valuing commitment, involvement and a willingness to serve. Head teachers instead tended to value other governors as critical friends. The other governors defined ideal service as providing support, encouragement and guidance. This meant that advisory roles, in Kogan's terms (see Table 5.1), were far more common in practice in 1994 than the legislated accountability role.

Earley (1994, p. 23) also showed that governors were discussing particular matters at length; resources (including LMS), building maintenance, annual parents' meetings, staff appointments, the school development plan, National Curriculum and assessment, staffing structures, school aims and objectives, and special needs provisions. Other marginally less frequently mentioned topics included disciplinary issues, governor training, school prospectus, respective managerial roles, external links, GMS status, sex education, school inspection, collective worship, in-service training and extra-curricular activities. This order of frequency was remarkably similar to that found by Keys and Fernandes (1990), showing that role change in school governance had been very slow.

The features of effective governing bodies most commonly mentioned were co-operative teamwork, sharing tasks and responsibilities, commitment, and a balance of

Table 5.2: Kogan, Johnson, Packwood and Whitaker's types of governing bodies and accountabilities

Governance type	Purposes	Limitations
Accountable	Policy making and conformance.	Head teacher facilitation.
Advisory	Laity provide feedback and advice to professionals.	Legitimacy of lay judgement and responsiveness of professionals.
Supportive	Provision of resources and legitimization.	Unquestioning trust in professionals.
Mediating	Represent diverse interests, create consensus and influence school.	Low participation.

expertise and representation. Over two-thirds of heads believed that the shortage of time available to governing bodies, their limited knowledge and understanding of educational matters and limited LEA support prevented even greater effectiveness. Conversely, the extent to which governors were making a difference to the quality of teaching and learning was believed by governors to be in large part determined by two crucial factors; the attitude of the head towards local governance, and the relationship between the head and the chair of the board. Many governors in the case study schools also noted the crucial role of high-quality staff and the need to improve learning environments and to appoint such staff. They consistently identified two key factors to success; partnership and their links with the school.

In sum, the evidence by 1995 pointed broadly (Levacic 1995, p. 135) to:

> governing bodies performing supportive and advisory roles rather than operating the accountability model of official expectations ... [their] ... working style has been very considerably affected by LMS and is still evolving towards an accountability role. The ability of governors to do this depends very much on their individual expertise, competence and commitment, and on the willingness of the head teacher to recognize the legitimacy of the governors' accountability role and to permit it or encourage it to develop.
>
> Governors exercising their accountability role, and particularly interpreting it to include intervention in day-to-day management, are likely to provoke conflict between themselves and the head teacher. Mutual trust and respect, and clarification of the respective roles of board and executive are essential if governors' efforts are to bring benefits both to their school and to the education system as a whole.

Levacic (1995, p. 197) advised head teachers to couple the 'technical core' of teaching and learning to a formal, documented and rational budget-setting process and evaluation to enable governors to (a) better understand the educational rationale that underpinned the budget and (b) better discharge their mandated accountability role.

School Evaluation

The overall developmental trend has been towards greater consistency and greater contextualization of school evaluations. In many senses these trends date from 1983 when the HMI's sophisticated evaluations of schools were placed in the public domain for the first time. The methodology (Browne 1979) purportedly involved:

> close observation exercised with an open mind by persons with appropriate experience and a framework of relevant principles. HMI's first duty is to record what is and seek to understand why it is as it is. The second step is to try and answer whether or not it is good enough. To do so, HMI uses as a first set of measures the school's—or other institution's—own aims; and, as a second, those which derive from practice across the country and from public demand or aspiration.

The research evidence suggests that practices were long otherwise. Gray and Wilcox (1995) identified four basic approaches used by the HMI in 1983 to evaluate schools; comparisons with national or local averages, that is, using standards, 'pass rates', comparisons of the school's performance over time, and comparisons with schools of similar intakes.

Clearly, the use of a standards framework is at odds with using socio-economic status (SES)-based comparisons, depending on whether or not students' background is to be taken into account. Since the framework for comparisons and determining 'pass rates' rely on statistical norms, and elapsed time comparisons employ trend analysis, it can be concluded that all four methods provided different and potentially useful information, depending on purposes.

Despite the public policy favouring a 'balanced' approach using all four approaches, Gray and Wilcox (1995, pp. 56–60) found that 35 HMI school reports written in 1983 actually favoured the use of 'pass rates' (86%), often made comparisons with the national average (66%), sometimes conducted trend analysis (37%) and irregularly contextualized the data (17%) in various configurations. By 1989, follow up research showed that more of all four methods were being used in HMI reports, with less an emphasis on 'pass rates' and more on context and trends (albeit sometimes only over two years). And while the interpretation of data was gaining more in sensitivity and sophistication, the 'standards' yardstick used was still the 'national average', even with schools deemed to be 'at risk' or 'failing'.

Despite heroic efforts in some LEAs, this reliance on 'national standards' has clearly persisted for a number of practical reasons. Inspectors did not have trustworthy expectations or reasonable benchmarks for schools serving different communities. The balance in purposes between evaluating performance to enhance choice or to encourage improvement kept shifting in the policy context. Even the proposed development of fine-grained 'success criteria' (Hargreaves and Hopkins 1991) or educational 'performance indicators' (Beare *et al.* 1989) implied an advanced capacity in school communities to clarify basic purposes and coherent strategies. Instead, as Gray and Wilcox (1995, pp. 82–83) pointed out:

> Performance indicators are usually something done to schools, rather than for or with them ... it is an exceptionally well organized school that can move quickly towards the kind of coherence of purpose and practice that systems of performance indicators imply. In such circumstances performance indicators can easily come to be seen as unwelcome harbingers of changing regimes of accountability rather than as potential contributors to school's development.

The concept of 'school performance indicators' in England and Wales was redefined by two events in the early 1990s. First, the 'Parents' Charter', first issued in 1991 and subsequently revised, entitled parents to information in three forms; reports on student progress and school development, quantitative indicators that compared school performance to national trends, and reports by independent inspectors. As explained above, this was driven by a concept of parents as consumers of school services.

By 1994, the emphasis (DFE 1994) had shifted even further towards providing summative consumer information for parents so that they could exercise choice *between* schools. The entitlement included regular reports on student progress, regular inspectors' reports, comparative performance tables for all local schools, school brochures and annual reports from school governors to students' parents. All forms of information required school-based evaluation that used external criteria. Student reports, for example, had to use national curriculum levels. On the other hand, the extent to which parents could actually exercise choice between schools remained a contentious issue. As the OECD (1995, p. 68) noted, actual:

> choice of school varies a great deal from area to area, according to how much flexibility there is in the system, and how many surplus places there are in local schools. Some are in fact unable to do more than 'express a preference' for a particular school (their only legal right) because the popular schools are already full.

The second event that altered the concept of 'school performance indicators' was fresh legislation that established a new agency of state; the Office for Standards in Education (OFSTED). The *Education (Schools) Act* (1992) empowered OFSTED to fundamentally alter the ways in which schools were to be held accountable by changing the form of inspection. The changes and implications are summarized in Table 5.3.

From the outset, a great deal of the OFSTED scheme depended on the reliability of contracted inspectors and on school communities coming to terms with the criteria and processes of being inspected. Induction training was made available for those seeking registration as inspectors. Phase One, a residential week, introduced neophytes to the inspection system devised by OFSTED and assessed their suitability. During Phase Two, neophytes participated in a school review and were again assessed for suitability. The OECD (1995, p. 56) evaluation found both merit and flaws in the scheme:

> Launched in 1993, the new inspection system is still in its infancy, although more than 4300 schools have been inspected since it began. So far, its most positive effect has been to encourage schools to improve their own evaluation and development procedures in preparing for inspection, helped by a clear framework published by OFSTED setting out what is expected of them. The most obvious problem with the system—as with most inspection regimes—is that it can put stress on schools and from time to time divert teachers' attention away from teaching pupils to looking good for the inspectors.

The process of inspection proposed by OFSTED also triggered debate in two important areas. First was that the model made the use of quality evidence mandatory. There was a fresh focus on the validity and reliability of data and methods. Second, although perhaps not originally intended, the OFSTED model highlighted the capacity of each school to 'add value' to the relative advantages experienced by their students.

Table 5.3: The Education (Schools) Act (1992) and changes to school evaluation

Agency	Accountability mechanisms	Forms of accountability
Her Majesty's Inspectorate	Inspectors of schools became supervisors of an independent inspection system. Chief Inspector regulates school inspections, trains, selects and registers inspectors, monitors and provides guidelines for inspectors.	National, public and formative evaluation changed to summative, localized and political forms of accountability.
Office for Standards in Education (OFSTED)	LEA advisory and inspection services changed into performance monitoring by OFSTED. OFSTED identifies institutions to be inspected, invites tenders from registered inspectors and commissions inspections.	Political accountability regarding quality of provision, standards achieved and efficiency of financial management.
Schools	Audited self-evaluation changed into inspections every four years. Schools generate required information. Inspection teams are led by a registered inspector. Detailed written report goes to OFSTED. Summary report and each school's subsequent plan of action goes to its governing body and into the public domain.	Local public accountability according to state-defined criteria and processes. Intense, infrequent, external scrutiny.

The latter debate produced four main candidates for collecting, reporting and using data about school performance; external inspection, reporting 'raw' examination results, reporting scores adjusted for socio-economic status, and using adjusted scores to report school improvement strategies (Scott 1994). Table 5.4 is a summary of the major dimensions of each approach. It must be noted that while the OFSTED inspection process and the publication of raw data have become familiar practices in England and Wales, the publication of SES-adjusted data and the improvement of schools being helped by value-added performance data have not proceeded past trials due to major technical difficulties.

The 1993 OFSTED model employed criteria in four areas. Organizational criteria were specified as resource efficiency, management, governance and external links. Curricular criteria included quality, range, equality of opportunity, assessment, recording and reporting matters. Pedagogical criteria focused on standards of achievement, quality of teaching, staffing, learning resources and accommodation. Student development matters traversed spiritual, moral, social, cultural, behavioural, student welfare and guidance areas.

Table 5.4 shows that the 1993 OFSTED model made but modest use of 'value adding' criteria. This is even more evident in Table 5.5 where the OFSTED criteria are compared to the features of schools found to have performed best according to test data adjusted for socio-economic factors (Mortimore 1995, p. 11).

The comparison confirms that the OFSTED model tended to focus on issues other than 'value adding', such as providing information to parents so that they could exercise choice between schools. By the mid-1990s, as indicated above, such an entitlement was being reiterated as a 'right' in the 'Parents' Charter' in an attempt to anchor accountability firmly to standards, performance data and parental choice, rather than relating accountability to the adding of value or school improvement.

Research in the mid-1990s into parents' perceptions (Ouston and Klenowski 1995, p.8) found general support for the OFSTED inspection policy but with some concerns being expressed over the costs of inspection, the quality of parental involvement, disturbed relationships between teachers and parents and governors experiencing role conflict. Parents were also found (Tabberer 1995) to be broadly positive about the new professionalism triggered by inspection and to be aware that improvement was dependent on the quality of follow-up planning, the visibility of improvements, the willingness of educators to attribute improvements to the inspection and to planning in ways that valued parental involvement.

Similarly, teachers were reported (Shaw *et al.* 1995) to see significant potential in the OFSTED process although concerned about inspectors not interacting with professionals, variances in quality of local professional leadership and OFSTED values imposing an orthodoxy concerning evaluation and accountability. These findings were confirmed by Jeffrey and Woods (1995, p. 26) who concluded that a stressful 'climate of surveillance now exists.'

On the other hand, about 94% of head teachers were reported (Ouston *et al.* 1995) to be positive about the OFSTED inspection process. The head teachers of all schools inspected between September and December 1993 were surveyed. The 60% responding head teachers believed that their inspection made a significant contribution to school improvement, especially when the report was integrated into the follow-up

Table 5.4: Four approaches to school performance evaluation and reporting

	OFSTED (1993) inspection	'Raw' performance data published	Data adjusted for SES and published	Value added results used to improve schools
Data collection processes	Staff prepare basic data on the school, its intake and area served, and performance indicators using OFSTED Handbook. One week external contracted inspection. Lessons are observed, pupils' work is inspected, discussions are held with staff, pupils and parents.	Public examination results in key subjects summed.	Public exam results adjusted for socio-economic status of students to indicate the value added by schooling.	Value added data used to select school improvement strategies.
Data reporting processes	Inspectors report to senior staff and parents. Report placed in the public domain.	League tables of schools' exam results.	League tables of schools' value adding.	Value adding strategies reported.
Data form and use	Summative report about standards of achievement, quality of curriculum, efficiency of resource management, students personal development and behaviour. School Governors prepare Action Plan.	Raw scores used by parents to compare schools in the market.	Adjusted scores used by parents to compare schools' value adding capacities in the market.	Adjusted scores used by parents to compare and employ value adding strategies in the market.
Criteria	Organizational, curricular, pedagogical and student development (defined below).	Comparative assessment of student achievement.	Comparative assessment of school achievement.	Comparative assessment of value adding strategies.
Strengths	Code of Conduct for Inspectors. Explicit and detailed standards of evidence and indicators. Participation of lay inspectors and stakeholders. Seeks to reconcile central and local perspectives.	Most accurate comparative measure of each student's achievement.	Most accurate comparative measure of students' progress.	Most accurate predictor to date of effective school improvement strategies.
Limitations	Samples and data can be manipulated. Data can be accepted uncritically. Non-negotiable criteria create resistance. Unsubtle processes not integrated with annual school planning cycle. More summative than formative. Explicitly consumerist and tends to reproduce traditions, order, differentiation and control.	Poor indicator of teacher or school performance. Often unreliable and give false positive and false negative measures of each student's performances.	Poor indicator of when and how in-school and ex-school factors related to progress take effect with individuals.	Poor indicator of context specific strategies or when school features are necessary and sufficient, and can promote technical rather than moral deliberation.

Table 5.5: OFSTED inspection criteria compared to features of 'value-adding' schools

OFSTED criteria	OFSTED specifications	Features of 'value adding' schools
Organizational	Resource efficiency, management, governance, external links.	Purposeful leadership of staff by the head teacher (firm and purposeful, a participative approach, the leading professional). Shared vision and goals (unity of purpose, consistency of practice, collegiality and collaboration). A learning organization (school-based staff development).
Curricular	Quality, range, equality of opportunity, assessment, recording and reporting.	Monitoring progress (monitoring pupil performance, evaluating school performance). Home-school partnership (parental involvement). High expectations (high expectations all round, communication expectations, providing intellectual challenge).
Pedagogical	Standards of achievement, quality of teaching, staffing, learning resources and accommodation.	A learning environment (an orderly atmosphere, an attractive working environment). Concentration on teaching and learning (maximization of learning time, academic emphasis, focus on achievement). Purposeful teaching (efficient organization, clarity of purpose, structured lessons, adaptive practice).
Student development	Spiritual, moral, social, cultural, behavioural, student welfare and guidance areas.	Positive reinforcement (clear and fair discipline, feedback). Pupil rights and responsibilities (raising student self-esteem, positions of responsibility, control of work).

action plan, and when the governing body of the school and external advisers were involved in formulating the action plan. Head teachers were also found (Radnor *et al.* 1995) to be sensing a growing alignment between participative and representative democracy in school governance and a falling level of accountability to LEAs.

Two research reports concerned with the general impact of the OFSTED model on school development (Fidler *et al.* 1995) and how resource management related to school effectiveness (Levacic and Glover 1995) raised doubts about the reliability of the model and the validity of inspector's reports. The former called for more effective formative mechanisms while the latter reported (p. 21):

> A more disturbing finding is that poorer assessments of the quality of teaching and in particular of learning and overall educational quality are associated with the socio-economic background of pupils ... [and this] ... provides additional evidence to that already accumulating from other studies, that schools in socially disadvantaged areas find it more difficult to provide good quality education ... in establishing desired educational outcomes.

It must be stressed that the OFSTED model was being developed as it was being used. It has been the subject of ongoing evaluation and development (OFSTED 1994). Another example, Barber and Fuller's (1995) review, highlighted the limitations of satisfaction ratings and participants' perceptions as data when evaluating quality, the limited use of triangulation, the low differentiation being achieved between schools, and modest levels of coherence between the OFSTED and other review and accountability mechanisms. This is an important point given the limited extent to which the OFSTED model was shown above to be concerned with value adding or school improvement. It is also an appropriate point to review recent analytic research on accountability in England and Wales.

Values and Accountability

An interesting conceptual advance on Kogan's seminal work came with Halstead's (1994) analytical treatment relating accountability to values. He noted that the concept of 'educational accountability' had to be defined in terms of both 'giving an account' while 'taking into account' broader requirements; the law, the values of broader society, local and national government guidelines, the professional code of conduct, the rights of the parents, the interests of children and the views of legitimate stakeholders.

It is interesting that this definition rules out equating responsibility with accountability, a position which would permit professionals justifying their actions in terms of their own informed conception of their role. Responsibility is a broader concept that entails capacity of rational conduct and freedom to choose between courses of action but it need not imply answerability or the need for an audience. To illustrate, a professional model of accountability which insists on teacher autonomy may be morally responsible but it would fail to meet two basic facets of educational accountability; answerability and the need for an audience.

Hence, by avoiding the absolutes of professional control and professional autonomy, Halstead (1994, p. 148) developed six conditions for accountability policy proposals in education;

- the person who is accountable is the holder of a defined role,
- the role-holder's accountability relates to actions carried out in connection with the requirements of the role, actions for which the role-holder carries responsibility,
- the role-holder's accountability is to one or more specific audiences – those who have delegated the responsibilities of his role to him, and/or those who are on the receiving end of his actions [sic],
- the audience has legitimate expectations which the role-holder should take into account, and has grounds for insisting that these expectations be satisfied,
- the role-holder should be willing to accept that some account, some explanation or justification of how the expectations are being satisfied should be prepared if the audience requires it, or at least evidence should be made available to the audience so that some assessment of how the expectations are being satisfied can be made, and
- sanctions or other forms of appropriate action (including professional advice, remedial help, further feedback) are available if the account or assessment indicates that the legitimate expectations are not being satisfied.

While this schema has the capacity to help create relatively straightforward technical answers to questions about who is accountable to whom, for what, in what manner and in what circumstances, it lacks the moral philosophy to handle other related issues such as negotiating accountabilities in circumstances where educational purposes are changing or being reviewed. The scheme also provides no guidance on how obligations should be discharged. The process could be limited to responding to expectations and requirements. Responsive processes need to be capable of being extended to meet moral obligations, to use 'due process', to provide opportunities for policy-making participation or alternative control, and to integrate accountability processes and criteria with longer term formative evaluation and planning functions to achieve degrees of moral steerage in moral economies and society.

There are other limits. Halstead's scheme does not attend to the problems of many hands or multiple roles diffusing responsibility. The not unusual tripartite accountability relationship in education between employers, professionals and consumers tends to rely on simplistic categories. Employers can include school governors, local and national government as well as taxpayers. Consumers can include students, parents, the school community, other local stakeholders and society. Another problem is the absence of clear definitions between governance and management services and between policy making and policy implementation. As illustrated above, complex sets of relationships between stakeholders can be too easily reduced to simplistic and adversarial distinctions such as 'providers' and 'clients'.

The strength of Halstead's (1994) position is the fresh distinction he drew between contractual and responsive forms of accountability (p. 149):

on one hand the answerability of educators, their responsibility to demonstrate that they are satisfying the expectations of the audience and that, for example, pupils are in fact learning what they are supposed to learn (which I call 'contractual accountability'), and on the other the process of taking into account the requirements of all interested parties when making educational decisions (which I shall call 'responsive accountability').

The distinction is further developed in Table 5.6.

Halstead (1994, pp. 149–162) then contrasted this distinction between contractual and responsive forms of accountability with varying power in the tripartite relationship between employers, professionals and consumers. This produced six models of accountability, summarized in Table 5.7.

The overall framework presumes a liberal set of values, trustful interaction and generally shared views on the basic purposes of education. There is, therefore, some doubt that these liberal values would be accepted in some of the more fundamentalist sub cultures of England and Wales or that accountability criteria and processes based on them would satisfy disaffected minorities (Whitfield 1976, pp. 24–25). Another implication is that policy making will have to proceed nationally on the assumption that a plurality of irreconcilable views prevail and will persist, that only general guidelines will be politically viable, and that more politically sensitive operational policies will need to be articulated at more immediate levels.

Halstead's framework also presumes to aggregate diverse positions. The contractual models celebrate controlling the causal relationship between teaching and learning outcomes, a relationship that is deemed to be known or ultimately knowable. The responsive models honour stakeholder constructivism which favours consensus over consequences and principles as the basis for the moral justification of policy claims.

Simultaneously, the models imply very different change strategies. The Central Control Model stresses managerialism. The Self-accounting Model recommends

Table 5.6: Halstead's contractual and responsive forms of accountability

Dimension	Contractual accountability	Responsive accountability
Primary concern	Educational outcomes and results.	Outcomes and results while emphasizing processes and decision making.
Key issue	Value for money from educational services.	Stakeholders' interests and values.
Purpose	Improve control of decision-makers.	Improve interaction between decision-makers and clients.
Accountability processes	Educators give a descriptive account of their actions and outcomes.	Educators account for outcomes, explaining and justifying actions.

Table 5.7: Halstead's six models of accountability

Dominant stakeholder	Contractual accountability	Responsive accountability
Employer	Central control model	Chain of responsibility model
	Teachers (employees) contracted to provide measurable learning. Testing and inspection considered appropriate methods. Can have low internal ownership or formative dynamics.	Decision-makers at each level in a hierarchy also responsive to legitimate stakeholders at their level. Can stimulate growth of bureaucracy, power struggles and structural ambiguities.
Professional	Self-accounting model	Professional model
	Teachers (autonomous professionals) self-monitor learning and teaching using internal and subjective methods. Can have low external credibility.	Contractual matters delegated to the governors. Matters of responsiveness delegated to the head and teachers. Can lead to localism and 'provider capture'.
Consumer	Consumerist model	Partnership model
	Teachers (providers) exposed to market and political mechanisms such as league tables, parental choice and LMS. Can intensify work and inequalities.	Legitimate stakeholders pool options, interact critically, decide, plan and evaluate. Can lack external legitimacy and be undermined by local politics.

absolute professional autonomy. The Consumer Model trusts in the market. The Chain of Responsibility Model puts its faith in distributive justice while the Professional Model has confidence in 'the learning community'. The Partnership Model values democratic participation. Each of these claims are contestable.

On the one hand, there is also some reason to believe that changing societal values in England and Wales could be affecting preferences concerning school governance, especially to do with lay participation in educational administration. From ten case studies, Deem (1994) found that a collective concern ideology (that valued democracy, public accountability and the public interest) was fading though still evident among some parent and governor pressure groups. In contrast, a consumer interest ideology (that favoured markets, competition, consumer rights and private interests) was growing, particularly in school communities which had 'opted out' for grant-maintained status (Gold 1996).

On the other, contrary indicative data from head teachers (Radnor et al. 1995) suggested that school governance practices could yet prove to be a cradle for democratic accountability. It is salutary that by early 1996, only 1100 out of 25,000 schools had exercised the option to 'opt out'. In essence, the governance and accountability policy settlement preferred by John Major's Government was being contested by the vast majority of school communities.

This takes us back to the two-part argument in Chapter 1 concerning policy legitimacy in a democratic community. Delegation of the authority to govern requires the consent of the governed. Such consent depends on the stewards remaining accountable.

Sources of Legitimacy

Perceptions of appropriate governance in education are apparently informed by broader judgements about the quality of government at the national level. The capacity of poor accountability mechanisms to undermine the legitimacy of governance services became particularly evident in Britain by the mid-1990s. National government was regarded by the populace as being seriously corrupted by slack personal standards, secrecy and venality. A series of personal and administrative scandals rocked the government and defeats in by-elections and local government elections were increasingly attributed to the 'sleaze' factor. The situation was exacerbated by the government having insulated itself from public accountability by having quasi-autonomous non-government organizations (QUANGOs) handle controversial funding decisions. The authority to rule was presumed without the obligation to account. The effects concerning the legitimacy of public policy were singularly instructive.

By 1995, polling showed that a large majority of the British public had lost faith in the Westminster system of government (Linton 1995). There was growing support for a bill of rights (79%), a written constitution (79%) and a freedom of information act (81%). Support for the proposition that the Westminster system of government was working well dropped from 43% to 32% to 22% in successive polls. Over two-thirds of those polled took the view that the MP's code of conduct should be enforced by the civil and criminal courts, and further, that the investigation of alleged misconduct should not be conducted by politicians.

Prime Minister John Major responded, eventually, by establishing a parliamentary *Committee on Standards in Public Life* chaired by Lord Nolan. It was asked to inquire into the quality of practices of parliamentarians and public servants. The Nolan Report (1995) found that the public's concerns were warranted and recommended the adoption of seven cardinal principles of public service summarized in Table 5.8.

The aggressively self-interested response by Government backbenchers took three forms; to reject the need for reform, to vilify the messenger and to attack the Prime Minister for inviting the message (Wintour *et al.* 1995). The British public were affronted and delivered crushing judgements in following by-elections (Wintour and Clouston 1995).

Taken together, this string of national political events suggest that effective accountability processes and criteria are strongly associated in England and Wales with, and considered to be causally related to, perceptions of appropriate government. They also affirm secular support for the idea that the consent of the governed and the legitimacy of policies should be freshened regularly by the stewards of authority providing responsive forms of accountability.

Table 5.8: Nolan's principles of public life

Principle	Specifications
Selflessness	Holders of public office should take decisions solely in terms of the public interest. They should not do so in order to gain financial or other material benefits for themselves, their family or their friends.
Integrity	Holders of public office should not place themselves under any financial or other obligation to outside individuals or organizations that might influence them in the performance of their official duties.
Objectivity	In carrying out public business, including making public appointments, awarding contracts, or recommending individuals for rewards or benefits, holders of public office should make choices on merit.
Accountability	Holders of pubic office are accountable for their decisions and actions to the public and must submit themselves to whatever scrutiny is appropriate to their office.
Openness	Holders of public office should be as open as possible about all the decisions and actions they take. They should give reasons for their decisions and restrict information only when the wider public interests clearly demands.
Honesty	Holders of public office have a duty to declare any private interests relating to their public duties and to take steps to resolve any conflicts arising in a way that protects the public interest.
Leadership	Holders of public office should promote and support these principles by leadership and example.

At school level, the practical challenges of policy legitimacy continued to be considerable. The drive 'away from professional accountability for performance—previously based on self-evaluation and peer review—to more public forms of accountability, based on forms of evaluation such as external inspection or performance indicators' (Riley and Nuttall 1994, p.124) was encountering formidable technical and philosophical problems. The technology of evaluation and accountability were increasingly regarded as socially constructed, and therefore arbitrary and contestable. The indices of change selected were more and more seen as unreliable. The values base of inchoate national accountability policies were coming to be seen as confused (Gray and Wilcox 1994, p. 9). Even more remarkable, it was reported (Vann 1995, p. 190) that:

> schools are beginning to regain the political high ground through the very stakeholders that the government has empowered in the legislation: principally parents and governors ... parents have repeatedly voted against their schools opting out of the control of their LEAs ... schools themselves have refused to allow themselves to be drawn into cut-throat competition with their neighbouring schools ... the government is going too far. It is losing the popular mandate for change that it enjoyed ... governors are supporting the teachers and threatening to set deficit budgets—which is forbidden by the legislation.

Hence the view (Vann 1995, p. 191) that:

> schools are establishing their own [in]formal networks of accountability that run alongside and contextualize the blunt, formal processes of OFSTED inspections and SATs. The assessment of schools is still an inexact science ... the most successful measures will be those which are local and link the stakeholders to the school's policies and processes in a very real way so that line management from client to the provider ... [is] ... short and effective.

The evidence accruing at the LEA level indicated that the education system was undergoing an ongoing cultural transformation. Depth case studies in four LEAs (Radnor *et al.* 1996) identified three distinct yet emergent interpretations of how LEA accountability policies and practices were unfolding, as summarized in Table 5.9.

It is important to note that this modelling both clarified yet overstated the differences between the four LEAs examined. As Radnor *et al.* (1996, pp. 10–11) put it:

> neither in practice nor in the language of legitimation are things as clear cut as they might first appear. Officers and councillors have a broad discursive repertoire upon which they can draw to legitimate a variety of practices and relationships. At different points in time, for different audiences and for different purposes different languages and practices are employed. Ideology, practicality and constraint intertwine in complex ways in the current functioning of LEAs ... apart from their various residual statutory responsibilities, LEAs are now operating in a kind of organisational and political twilight world where issues of accountability and democratic politics are often obscured in deep shadow.

Table 5.9: Radnor et al.'s three models of LEA accountability

Dimensions of accountability	Local management model	Regional service model	Community government model
Values base of governance and politics	Expertise providing efficiency contracts; politics of the Right.	Representative politics and democracy accommodating pluralism.	Community stewardship; politics of the Left and Centre Left.
Priorities in day-to-day management	Cost reduction, performance and monitoring.	Maintenance of service provision and support for schools.	Service provision as defined by community needs and local political interests.
Origins of policies	Exchange relationships between consumers and providers.	Satisfaction and management of diverse interests.	Negotiated with school communities.
Focus of policies	Comparative and measured performance of students and schools.	Service provision and relationships with schools.	Relationships with schools.
School performance criteria	Fixed performance indicators and consumerist data.	Stable consensus about the structure and adequacy of provision.	Negotiated social and management consequences.
System performance criteria	Unproblematic since politics and education are commodified.	Not an issue since the functions of education are regularly mandated.	A public project, hence issues such as equity and problem solving capacities.
Source of legitimacy	Current legislation and the market.	Representative democracy.	Communitarianism.
Dominant form	Market and moral accountability.	Professional accountability.	Local political accountability.

Equally importantly, they also noted examples of 'surrogate accountability' (p. 11) and informal 'residual practices' (p. 12) through which public and systemic accountability was being sustained, enabling schools to be accountable to each other and to the regional society of which they were part, in a situation where the national legislation was incoherent or hostile to democratic accountability. Head teachers and governors reportedly recognized the importance of being answerable to their communities and preferred forms of accountability based on democratic participation above the consumerist models of choice and market enshrined in current legislation. By then international research (OECD 1994) had confirmed the substantial limitations of market strategies in Australia, The Netherlands, New Zealand, Sweden and the United States. Simkins (1995) argued that the concern for excellence and choice during the reform had served mostly procedural forms of equity to replicate historical allocation patterns, rather than use distributive equity principles to reconstruct expectations and to distribute resources according to student need. There was also some evidence in England and Wales that (Radnor *et al.* 1996, pp. 21–22):

> A sense of 'empowerment' is seeping into the consciousness of knowledgeable and articulate governors and parents which may create the basis for a new participatory culture. While it appears that many governing bodies have 'de-politicised' themselves in eshewing the traditional representative model of party democracy, some at least may be prepared to 're-politicise' themselves in a democratic model of associational life.

Summary

By the mid-1990s, the situation in England and Wales is best characterized as a contested policy settlement, however ideologically pluralistic and politically pragmatic Dearing's compromises were. School governance remained mostly advisory in nature, well short of the legislated role of direct, full, local and political accountability. The OFSTED model of school inspection had improved steadily as an accountability methodology but was yet to discount SES effects, guarantee the reliability of inspectors, use 'value added' analysis, boost the improvement of schools, or triangulate with other review and accountability mechanisms in public education.

Policy research has also advanced recent thinking in England and Wales. New theory is available concerned with role-holding. Variance in stakeholder dominance appears to interact with contractual and responsive dimensions of accountability and explain six different models; central control, chain of responsibility, self-accounting, professional, consumerist and partnership. There were also signs that the legitimacy of governance in education is being threatened by the drop in standards in public life at national level.

The purposes, coherence between and technical reliability of accountability mechanisms and practices continues to be questioned by those empowered by recent legislation at different levels to exercise authority. The most appropriate balance between contractual and responsive models is yet be expressed or accepted as public policy. Adequate and multiple legitimacies in the complex structures that comprise public education in England and Wales are still some way off. Ironically, the general trend seems to be away from greater systemic coherence and towards greater local responsiveness and to more differentiated zones of legitimacy, especially as school communities gain their voice regarding distributional equity and assert their legislated role in accountability.

Another irony is that despite the damage done by consumerist policies and practices in recent decades to the networks of rights, reciprocal obligations, responsibilities and duties in school communities, and to the norms of, and the value once placed in, representational democracy, there is reason to believe that a communitarian ethic is finding favour. It is a central theme to the New Labour rhetoric that sets aside Marxist economic determinism and class analysis to emphasize (Mandelson and Liddle 1996, p. 30) the so-called:

> other brand of socialism—the ethical approach—that has unsurprisingly stood the test of time. This is a socialism based on a set of beliefs and values, and similar to the social democracy found in other European countries. It is founded on the simple notion that human beings are socially interdependent and cannot be divorced from the society they live in ... We stand for a strong society and an efficient economy because we need both, and each needs the other. That is the essence of our belief in One Nation policies and the principle of the stakeholder economy.

Communitarian liberalism is held by Gray (1995, p. 16) to be an inevitable successor of neo-liberalism and social democracy as it:

> departs from individualist liberalism in that it conceives of choosing individuals as themselves creations of forms of common life ... It differs from conservative and neo-traditionalist communitarianisms by acknowledging the strength and urgency of the need for individual autonomy. People voice and act upon that demand, to make their own choices and to be at part authors of their own lives, in all of the institutions and practices of contemporary liberal societies. It recognises the pluralism of such cultures. Few of us are defined by membership of a single, all-embracing community and there is no going back to any simple 'organic' way of life.

The possibility of such degrees of responsible autonomy, however, remains dependent on broader structures that both create, recreate and guarantee the recreation of relative autonomies. Such broader structures, perhaps inevitably given the nature of humankind's propensities and maldistributions of power and resources, are more likely to reproduce than transform the social capacities needed by peoples to be relatively autonomous. On the other hand, a communitarian ethic defines the challenge of accountability as both an idealistic and pragmatic public project.

Hence, although the focus moves to a case study of the United States in the next three chapters, the theme of how the principle of accountability relates to purposes, governance and policy legitimacy persists.

6
Accountability Policy Evolution in US Public Education

Introduction

The people of the United States have long been aware of the intimate relationships between the legitimacy of public policy, the consent of the governed and need for governors to remain accountable. In 1776, the Congress of the Thirteen United States of America declared independence from British constitutional authority on the grounds that certain truths were self-evident:

> that all men are created equal, that they are endowed by their Creator with certain unalienable Rights, that among these are Life, Liberty and the pursuit of Happiness. That to secure these rights, Governments are instituted among Men, deriving their just powers from the consent of the governed. That whenever any form of government becomes destructive of those ends, it is the right of the People to alter or to abolish it, and to institute new Government, laying its foundations on such principles and organizing its powers in such a form, as to them shall seem most likely to effect their Safety and Happiness. Prudence, indeed, will dictate that Governments long established should not be changed for light and transient causes; and accordingly all experience hath shewn, that mankind are more disposed to suffer, while evils are sufferable, than to right themselves by abolishing the forms to which they are most accustomed. But when a long train of abuses and usurpations, pursuing invariably the same Object evinces a design to reduce them under absolute Despotism, it is their right, it is their duty, to throw off such Government, and to provide new Guards for their future security.

In this chapter I show that the evolution of accountability policies in public education in the United States in the 1970s and 1980s was characterized by a 'long train of abuses and usurpations' to the body corporate of public education. In this and the next two chapters I demonstrate that there has been recurrent and good reason for policy research in public education systems to provide principled alternatives to the unintended 'despotism' sustained by inappropriate forms of accountability.

To illustrate, a major feature of accountability practices in US public education over the last three decades has been a relentless testing of students using standardized, multiple-choice instruments, instruments usually unrelated to the taught curriculum. School districts have used these tests to monitor the learning of large numbers of students, in order to identify standards being achieved, or to determine those ready for a college education. School examinations set by educators have had far less impact on students' prospects than the standardized tests due to the generally comprehensive nature of secondary schools, the tradition of offering many 'second chances' to students, and later, providing multiple routes into higher education. Test results became a universal proxy for the quality of learning, teaching, leadership and governance. This will be shown to be technically implausible, morally indefensible and anti-educative. It has placed a 'despotic' burden on the shoulders of students.

The reform of accountability seemed unlikely throughout most of the 1970s and 1980s. The states exercised their constitutional authority to define curricula, to varying degrees, and federal governments historically showed interest only in the outcomes of the initiatives they funded. Responsibility and accountability for public education was almost completely decentralized to local school districts for the better part of the two decades. And yet, ironically, when the states were driven into action by the presidential coalition of governors in the late 1980s and early 1990s, they tended to replicate and intensify many of the mistakes of the past, rather than develop demonstrably educative alternatives.

In this chapter I focus on the ways in which practitioners, policy makers, researchers and theorists in US public education used the concept of accountability in the 1970s and in the early 1980s. In sum, a number of international, economic and social trends helped embed the concept in a receptive culture of managerial efficiency. Accountability was then variously redefined during three distinct 'waves' of reform in the 1980s; the naive impulse to legislate, the artless empowerment of professionals, and then a presidential politics of rhetoric that defined 'national performance standards'. By the late 1980s, as noted above in England and Wales, accountability policy making had become a central feature of state-driven school reform initiatives (McDonnell and Fuhrman 1986, Mitchell 1988). Despite this turbulent context, some things remained largely unchanged.

It is shown that policy research concerned with accountability was conceptually limited by a closed systems theory of schooling, locked into a behaviourist technology of student testing and student performance indicators, and that it remained unresponsive to unfortunate educational outcomes.

The Onset of Doubt

A combination of international demographic and economic factors impacted on US public education in the 1970s and altered the focus and role of accountability policies. The 1960s and early 1970s were decades of coping with the post-war demographics of the 'baby boomers'. It was a period of largely uncritical expansionism.

Accounting for the quality of education in this context, understandably, focused on the quantitative aspects of public investment in education, the nature of curriculum development required to cope with extending the period of compulsory education, and the degree of infrastructure implied by the comprehensivization of secondary schooling.

From the mid-1970s, however, accountability policy discourse began to shift remorselessly from quantity to quality. Many commentators trace the beginnings of the shift to the OPEC oil crisis. It began to undermine the secular belief in unlimited progress. It triggered an economic recession, which in turn, eroded public confidence.

The development of anxiety in many developed countries in the late 1970s was accelerated by ten interdependent factors (OECD 1995, pp. 15–16):

- the growing belief that national prosperity would be driven not by industrial productivity but by the production and sale of knowledge and the management of information, and that both would require highly educated populations,
- comparative tests showing that some students, especially American and British, were performing poorly in mathematics and science,
- fiscal restraint required lower unit-costs and prompted value-for-money and cost-benefit analyses of inputs, processes, outputs and outcomes,
- the growing belief that privatization, consumer choice and decentralization could improve the cost-effectiveness of service delivery across the caring portfolios,
- unusually high youth unemployment increasingly associated with unproductive spending on benefits, wasted lives, alienation and social disruption,
- the function of public education becoming controversial—was it to be a form of social welfare, community investment, human-resource development or vocational training,
- with fewer traditional families in developed countries, schools had to assume a greater role in reproducing mainstream community values,
- the decentralization of administrative functions with policy guidelines and targeted funding increased the emphasis on monitoring, evaluation and accountability,
- the transparency of government was seen increasingly as crucial for informed and participative democracy and to raise the effectiveness of services, and
- while massified secondary and post-secondary services made students increasingly heterogeneous, quality and equity were no more evident in the outcomes.

The policy makers and administrators in US public education responded to these circumstances much as the rest of public administration have always done: with the latest conceptual and practical tools available at the time. Their response was fashioned by the precedent that had been set earlier by Robert F. Kennedy; the 'father of program accountability in education' who had attached 'a mysterious black box' known as 'Federal education evaluation' (Haney and Raczek 1994, p. 10) to the 1965 ESEA Title 1 legislation. Evaluation had become mandatory. It was conceived as the quantitative measurement of outcomes. When these imperatives were combined with the broader press for economic rationalism, it helped ensure that managerial functionalism became the order of the day.

Accountability as Legal and Technical Obligations

The path that accountability policy development took in the 1970s was paved almost exclusively with legislation and structural functionalism. Thirty-one states enacted legislation dealing with accountability. They put their faith in closed management systems, including 'assessment of student achievement, evaluation of programmes, setting goals for education, specifying objectives for learners, PPBS (planning, programming, budgeting system), MBO (management by objective), MIS (management information systems), uniform accounting systems and performance accreditation systems' (Pipho 1989, p. 1). Accountability was intended to gain 'more bang for the buck ... [through] ... the application of the tools of business management to education ... [in order to achieve] ... a new era of efficiency' (p. 1).

This cult of efficiency (Callahan 1962) was sustained by a hyper-rationalist form of managerialism that set out to define goals in behavioural terms so that they could be measured and used for 'scientific' resource and personnel management (Marland 1972). I come back to the nature of this 'science' in Chapter 10. At the time Leon Lessinger was regarded as 'probably the foremost advocate of accountability' (Levit 1973, p. 41). His influential text (Lessinger 1970), *Every Kid a Winner: Accountability in Education*, was still in print in 1993. Unfortunately it drew on the now doubted Coleman Report (Coleman *et al.* 1966, p. 315) in which it was reported that 'schools bring little influence to bear on a child's achievement that is independent of his background and general social context' [*sic*].

The point here is that the Coleman Report and Lessinger's text helped switch policy makers attention away from resource inputs to outcomes accountability or performance evaluation of schools. Further, Lessinger's attempt to establish a modernist technology of management defined accountability as a suite of behavioural obligations for individuals which, if properly discharged, would promote better outcomes, for less cost, especially for the educationally disadvantaged.

As noted above, a substantial vote of confidence in this relatively new behavioural 'science' of educational management was apparent in a broad raft of new state and federal legislation. Seventy-three state accountability laws were passed between 1963 and 1974, with Florida passing a new accountability law each year between 1969 and 1976 (Wise 1979, pp. 12–14). Also noted above was the passage of the Federal Elementary and Secondary Education Act in 1965. Its amendments created a new market by requiring evaluations in all Title 1-funded programmes. This new market dramatically expanded demand for standardized tests and related evaluation services. These services cohered with the relatively standard behaviourist and summative definition of accountability in use (Glass 1972); disclosure concerning the product or service being provided, product or performance testing, and redress for false representation or poor performance. The latter more pejorative dimension was confirmed by James March in 1972 when he reported that a demand for 'accountability is [now taken to be] a sign of pathology in the social system' (cited in Cronbach *et al.* 1980, p. 139).

By the late 1970s, a 'basic skills movement' had gained minimum competency testing for advancement or graduation in 33 states (Wise 1979, p.2). Michigan developed one of the earliest and more comprehensive basic skills testing programmes (Murphy and Cohen 1974). Many of the methods used, such as testing, behavioural objectives and competency-based education, were criticized as embodying overly rationalist assumptions about teaching, learning, schools and educational systems (Wise 1979). An example of a questionable assumption is that goals can be converted into measurable relationships. Another is that the actions of educators and students can be governed by policy or a science of education.

There were also criticisms heard at the broader policy level concerning the general direction of the closed systems, objectivist and functionalist 'accountability movement'. House (1972), for example, used a radical humanist perspective to attribute the movement to economic decline and the growth of value pluralism in US society concerning the purposes of education. He argued that the movement's rhetoric masked a covert attempt to maximize the utility of education for the most powerful interest groups in society. He claimed that it did not reflect an authentic social consensus and that the movement failed to do justice to the legitimate interests of all clients, particularly those of underprivileged minority parents. The pressure exerted on schools by system bureaucrats to raise standardized test scores, he suggested, comprised a simplistic 'productivity accountability policy' rather than one that encouraged a 'responsively accountable school'. Where the former approach is characterized by coercive power satisfying external criteria, he argued that the latter policy should be used to generate and communicate data on schooling in the light of concerns expressed by local audiences. In arguing this way, House assumed that accountability policy was a social construction and that knowledge about appropriate forms was relative and often subjective and normative in nature: a comparatively rare view at the time.

Research in the late 1970s (Boyd 1979) and history have tended to support rather than undermine the House analysis. A more recent examination of the broader forces that lay behind the genesis of the 'outcomes based accountability movement' (Haney et al. 1993, Ch. 5) found that they cohered with the sources that powered the massive increase in standardized testing from the 1970s, namely:

- recurring public dissatisfaction with the quality of education, and efforts to reform education,
- an array of legislation, both at federal and state levels, promoting or explicitly mandating standardized testing programmes,
- a broad shift in attention from a focus on inputs or resources devoted to education towards outputs or results produced by our educational institutions, and
- the increased bureaucratization of schooling and society.

Accountability policies continued to develop on explicitly rationalist and behaviourist assumptions into the 1980s. An example is the 'outcome-based education movement' promoted by William Spady. It encouraged many districts and states to develop 'learner outcomes', sometimes defined as 'content and process standards'. They were increasingly used to assess student learning in ways that had serious consequences for students, teachers and school communities. These increasingly 'high stakes' methods shared and extended the rationalistic assumptions about teaching, learning and school management noted above.

To summarize this section, accountability policy makers in the 1970s generally shared four assumptions (Martin *et al.* 1976); that the purposes of schooling could be expressed as instrumental ends and means, that obtaining the outcomes of learning could be both expressed and measured in terms of a production function, that students and teachers are pliable recipients of policies, and that a 'science' of education both existed and justified this approach to accountability. The limits of these assumptions became evident in the 1980s.

Three Waves of Reform and Learned Incapacities

The evolution of thought concerning accountability in US public education in the 1980s was shaped by the three 'waves' of the 'educational reform movement' (Passow 1988). The first wave started in August 1981, when the Secretary of Education, T. H. Bell, established a National Commission on Excellence in Education. It was in response to growing public discontent over the quality of public schooling. The provocative language of the Commission's report, *A Nation at Risk*, triggered a political furore when it tapped into this discontent. It concluded that 'If an unfriendly foreign power had attempted to impose on America the mediocre educational performance that exists today, we might well have viewed it as an act of war' (National Commission 1983, p. 3). As Griffiths (1993, p. 36) noted:

> The results were highly predictable: mainly state-level legislation, regulations, and mandates to be implemented at the local level. High on the list was the mandate to toughen the route to high school diplomas ... This requirement was enforced by periodic basic competency tests. Standards, accompanied by testing, were likewise raised for teachers ... [and] ... a highly centralized management style was adopted by state education departments mandating curricula, textbooks, tests, and standards.

The massive flurry of activity of the first wave created few educational gains but reinforced the idea of accountability as a political impulse driven by economic conditions and to be serviced by bureaucratic technique. As Goodlad (1984) noted, mandating ways of improving student achievement is futile and dangerous; futile because it cannot work, and dangerous in the sense that it defines appropriate accountability processes as implementing policy mandates, competency testing, imposing standards and bureaucratic managerialism. The mandates imposed conditions that were antithetical to the improvement of teaching and learning.

The second wave then had to challenge the 'iron cage' of bureaucracy promoted by the first wave, essentially by seeking to empower schools and the profession in ways that encouraged improvement. The second wave was, in large part, driven by three reports. The first, the Carnegie Forum Report (1986), *A Nation Prepared: Teachers for the 21st Century*, argued that schools should be restructured to provide a more professional environment. It called for appropriate salaries, technology and support, for the profession itself to be restructured with market incentives for performance and be supported by a national Board of Professional Training Standards, and that site-based management (SBM) should become the norm.

The second report was by the 100 Deans of Education that comprise the Holmes Group (1986). They suggested in *Tomorrow's Teachers* that undergraduate education majors should be abolished, that initial teacher education should become a fifth year following a liberal arts bachelors degree, and that the certification of teachers should mark development through three stages; instructor, professional teacher (with a masters degree), and career professional (the 'lead' teachers of the Carnegie Plan).

The third report came from the National Governors' Association (1986). *A Time for Results* promoted the concept of 'effective schools' and the role of 'effective leadership' in SBM, leadership education and certification, and performance incentives.

These 'second wave' reports all helped redefine the concept of accountability in passing, rather than by commission. While they stressed the empowerment of teachers, leaders and schools, they did not review the methods and criteria that should be used to define responsibilities and publicly discharge obligations. The concept of 'the public interest' in accountability was not treated as a professional matter. And, while they boosted the reconstruction of the profession and its certification processes, they gave little attention to how this might regenerate the legitimacy of policies and services and how schools and professionals might better account to clients.

Hence, moral and political forms of accountability were left undeveloped. While they focused on SBM, effectiveness and incentive regimes, student achievement remained the universal proxy for success and tests retained their methodological monopoly. Without school inspection data and stakeholder accountability, the possibility of responsive accountability remained low.

The 'third wave' began when President George Bush convened the first Nations' Governors education summit at the University of Virginia in October 1989. Its purpose was to legitimate and generate national student performance standards. The President and Governors concluded that specific results-orientated goals were needed along with more direct accountabilities for outcome-related results. Thirty state legislatures then discussed educational accountability (Pipho 1989), and, after consultation with major stakeholder groups, identified seven areas for the formulation of goals:

- the readiness of children to start school,
- performance on international achievement tests, especially Maths and Science,
- reducing the drop-out rate,
- the functional literacy of adults,
- training levels for a 'competitive' workforce,
- the supply of qualified teachers and of up-to-date technology, and
- safe, disciplined, drug-free schools.

Clearly, these seven areas were considered to be the key remedial components of a comprehensive, universal and effective theory of educational systems, and by implication, redefined accountability priorities. While a subsequent Bush budget actually reduced funding to education, these goals were sustained in national policy discourse, and eventually appeared in legislation in 1994.

What announcing these national goals did not do was trigger a critical review of accountability policies and practices. Instead, the announcement served to broaden and reinforce the application of student achievement data. A former official in the Bush administration made it clear that achievement data were central to a key implementation strategy: 'outcomes accountability'. He took the rather narrow view that holding schools accountable for outcomes 'is the only kind of accountability worth having' (Finn 1991, p. 149). This approach also helped to further centralize policy making and to decentralize policy implementation and political blameworthiness.

It might now be seen, in retrospect, that simply handing goals down to local schools to implement, with an overlay of test-based outcomes accountability, was unlikely to promote a sense of moral accountability to clients or responsive forms of accountability to local stakeholders. Griffiths (1993, pp. 40–41) noted other serious strategic problems with the initiatives:

> Practically all of the reforms deal with form, not substance ... With the exception of The Paideia Network this criticism is true of all the major reports. The Paideia Network ... prescribes both a method of teaching and the content to be taught. It, too, is alone in defining what is meant by the term, education. The reform reports all claim to be improve education, yet, do not define it.
>
> Even more devastating than not having a clear idea of what is meant by education is that the reform movement considers the schools to be a closed system unaffected by their environment ... There is nothing in the history of education to suggest that schools can be improved by ignoring the social factors that play on them.

Why did all three 'waves' fail to provide strategies that acknowledged the external factors that help determine how students learn, how teachers teach and why schools operate as they do? Among many possibilities, two options are that the policy actors of the time did not *want* to acknowledge the external factors, or, were not *able* to do so. The latter is regarded as a more reasonable explanation. Griffiths' most acerbic critique was reserved for the quality of school leadership services and the preparation of educational administrators, which had been found to 'range in quality from the embarrassing to the disastrous' (1993, p. 43). The National Commission on Excellence in Educational Administration (Griffiths *et al.* 1988) had earlier revealed that the field lacked what might be seen as prerequisites:

- a definition of educational leadership,
- leader recruitment programmes,
- collaboration between schools and universities,
- appropriate representation of minorities and women,
- systematic professional development,
- relevant and clinical preparatory programmes,
- licensure systems promoting excellence and
- national co-ordination.

This raised the possibility that, until relatively recent times, educational administrators in US public school systems have probably lacked the intellectual capacities required to review accountability as a policy issue in a sophisticated way. Indeed, it was not until October 1989, that the peak academic body of the field, the University Council of Educational Administration, agreed common standards for the preparation of educational administrators, including the need to develop a specialist knowledge base through research. It is, therefore, only since about 1990 that preparatory programmes have systematically exposed neophyte educational administrators to an increasingly comprehensive curricula. It was recommended that curricula include social and cultural influences on schooling, teaching and learning processes sensitive to individual differences, theories of organization and organizational change, methodologies of organizational studies and policy analysis, leadership and management processes and functions, policy studies including issues of law, politics, and economic dimensions of education, and moral and ethical dimensions of schooling in a pluralistic society. The inescapable implication is that those with the positional authority to review, recommend and reconstruct systemic and institutional accountability policies in the 1970s, 1980s and early 1990s were probably not able to challenge the assumptive base of current accountability policies and practices. It can be tentatively concluded that the development of accountability policy reform was probably retarded by learned incapacities. The way that educational administrators persisted with systems theory and learning performance indicators adds credence to this conclusion.

Accountability, Systems Theory and Indicators

Debates over the nature of knowledge implicit in accountability policy making remained, understandably, comparatively rare in the 1980s. Instead, the construct termed 'education indicators' continued to be elaborated iteratively within a theory of organizations as closed systems (Seldon 1994). Definitions hardened to the stage where an 'educational indicator' achieved the status of a cultural artefact that guaranteed empirical facts about the constitution of an organization, an organization conceived as a living system. Oakes (1986, p. 1), for example, held an educational indicator to be 'a statistic about the educational system that reveals something about its performance or health'. Smith (1988) saw indicators depicting 'a desirable outcome or as describing a core feature of that system' (p. 487) but cautioned that 'our track record in developing valid models of education systems is weak at best' (p. 488) and 'once underway, it will be practically impossible to change direction' (p. 491).

Smith's words proved prophetic. The very influential Rand study (Shavelson *et al.* 1989) held that educational indicators were individual or composite statistics about basic constructs useful to policy makers, and proposed that they should include the dimensions summarized in Table 6.1.

The Rand Report also recommended eight criteria for evaluating the effectiveness of an indicator system as summarized in Table 6.2.

Some concerns were expressed about the nature and influence of this approach but

Table 6.1: The Rand model of educational indicator systems

Dimensions	Components
Inputs	Fiscal, material and other resources, teacher quality, student background.
Processes	School context and organization, curriculum, teaching quality, instructional quality.
Outputs	Student achievement, participation, attitudes and aspirations.

Table 6.2: Rand criteria for evaluating educational indicator systems

Criteria	Objectives
Functional	Reflects central or core features of the educational system.
Beneficial	Provides information pertinent to current problems.
Pragmatic	Measures factors that policy can influence.
Behaviourist	Measures observable behaviour rather than perceptions.
Objective	Uses reliable and valid measures.
Analytic	Provides analytic linkages among the indicators.
Uncomplicated	Feasible for implementation and uses indicators that can be collected.
Responsive	Addresses a wide set of audiences.

without much impact. Odden (1990a) later pointed out that while single or grouped indicators could give useful information about students and schools, they could not attend to or convey the full complexity of a school. On the other hand he tended to reiterate in international settings (Odden 1990b) the Rand view that a system of indicators should be designed to reveal how the different indicators interacted to produce each school's level of performance: reiterating the promise of a production function theory of schooling.

In sharp contrast, Tyler (1983) and Haertel et al. (1989) argued that (a) the variables chosen for indicators had to reflect the many purposes of educational systems, (b) the reliance on standardized tests had resulted in a narrow view of schools' purposes, (c) indicator systems solely reliant on standardized test results had limited validity, and (d) that only a comprehensive array of data, methods and indicators were likely to prove satisfactory in public education systems. Such advice, however, had comparatively little influence due to the confluence of factors in the broader context, many of them the ongoing effects of the three 'waves' of reform.

The most evident cumulative effect (noted above) was that accountability policies evolved as theories of indicator systems comprising inputs, processes and outcome factors, and as part of wider production function theories of how variables interacted to generate educational outputs. Proponents relied on universal, logically positive and relatively simple causal stories about effective schooling using systems theory. Their search for objective and reliable result systems in education continued to rely on the measurement of students' learning as a proxy for the interdependent subtleties of public schooling.

The second effect was structural. As the positivist causal story gained in political credibility, it was codified as policy, objectified in law and reified in structure. So sure were the President and the Governors of the veracity of the underlying causal story, they raised the political stakes. They declared their willingness to be held accountable for the achievement of national goals but that they would be, in turn, holding others accountable using yet-to-be devised measures of student performance and yet-to-be defined data. The National Centre for Education Statistics (NCES) in the US Department of Education was then directed to meet this fresh demand for more descriptive, more detailed, more accurate and more quickly produced empirical data. The NCES caught the political temper of the times when it identified (Griffith 1990, p. 4):

> indicators as statistics that have been adjudged as important in that they inform on the health and quality of education and they monitor important developments in education. They include some basis for comparison, for a relationship is built in—frequently against a standard or in terms of time; but at least as importantly, across subpopulations, across States, or across countries. Indicators feed into our understanding of how we are performing not only overall in education, but also how we are reaching all the different children in our diverse society.

The third effect was that the indicators developed were more suitable for uncomplicated, summative and system-based evaluation rather than for deep and formative evaluation in classrooms and in schools. They could be used to monitor improvement without necessarily informing or generating improvement. Ironically, it was their potency as a reporting mechanism that created an unintended political controversy.

Their development was given high priority at the NCES through three projects; indicators publications, an expert indicators panel and international indicators. The first project provided empirical data in three forms; an annual report, the *Condition of Education* (NCES 1989), occasional issue-related reports, and the *Wall Chart*, a table of data from the states displaying comparative test scores, graduation rates, financing and school organization. The first *Wall Chart* was released in 1984, a year after *A Nation at Risk*. It provided input, process and output data on all 50 state education systems.

The resultant uproar, however, indicated that there were still starkly different theories of accountability in competition at the time. The nature of the debate, and how accountability policies and definitions have evolved since, showed yet again the extent to which these diverse theories have been subordinated by *a priori* commitments to systems thinking. As shown in the next section, the policy outcomes were surprisingly incoherent and evidenced in a continuing debate over who controlled schools (Kirst 1988). The professionals retained control within schools. Political or systemic appointees retained control of external accountability criteria and processes. Other stakeholders remained marginalized. The system remained unquestioned and the legitimacy of public schools continued to erode.

Contested Systems Theories of Accountability

Three reasons were suggested (Odden 1990b, pp. 37–38) why the *Wall Chart* generated the furore it did. Firstly, it violated a condition agreed when the nationwide testing programme, the National Assessment of Educational Progress (NAEP), was introduced, specifically, that inter-state comparisons would never be made. This also revealed deep sets of conflicted values. The very existence of the agreement suggested that many anticipated that performance data would provide evidence of;

- the values preferred by those with the power to define them as 'policy' and with the capacity to create or sustain structures needed to implement policies,
- the values actually being served, as evidenced by structured outcomes,
- the gap between values in policies and structured outcomes, and
- what needed to be attended to in order to close the gap.

Putting these matters into the public domain rendered all components contestable, particularly the values embedded in purposes, criteria and processes. It seems reasonable to suppose, therefore, that in the absence of a coherent and educative accountability policy, professionals, officials and politicians preferred to avoid the public controversy that such disclosure can generate.

Second, state political and educational leaders were antagonized when they were not involved centrally in deciding what data categories were to be used. Apart from the part played by political turf, and its handmaiden, ego, these leaders demonstrated how important it is for those who are held accountable in education to be able to play a part in the creation of criteria and processes, and similarly, to participate in the political processes of reconciling contested values evoked by the release of performance and service data. Seen from another angle, those who had delegated authority to govern at the time appeared to understand well the fragility of public consent. They also seemed to understand the link between such consent and accountability policy legitimacy. They also appeared to be aware of why it was important to determine the nature of accountability data.

Third, the output measures used, such as drop-out rates and average Scholastic Aptitude Test (SAT) scores, were beginning to be seen as technically inadequate. There were growing doubts that they were measuring performances and services in a full and trustworthy manner. I will come back to this matter in greater detail below.

On the other hand, all stakeholders had reasons to fear the use of alternative, arbitrary and feral processes or criteria. Most notably, the Coalition for Essential Schools (Sizer 1984) began to campaign hard for two ideals. The first appealed directly to the consent of the governed. 'Models and programmes, to have sustenance and integrity, must arise independently out of their communities and school' (p. 225). Second, schooling must be based on nine general principles; focused curriculum, mastery learning, universal values, individualized education, student-centred learning, diploma by exhibition, high expectations, multi-skilled staff and staff-supportive budgeting.

Shaky policy settlements (masking deep values conflicts, contested data specifications and growing doubts over methodology) also seemed to intimate that while policy actors still held onto systems theories of accountability, they were beginning to sense the need to search for ways of accommodating massive political, professional and technical contradictions. This impression was supported when the debate created by the *Wall Chart* stimulated responses.

The Early Search for Better Theory and Policy

The immediate administrative response to the *Wall Chart* furore was predictably technical in nature: the development of more efficient achievement data collection methods. The most senior officials from the 50 states created a State Education Assessment Center to help in this process, eventually publishing *State Education Indicators*. They also devised early methods of comparing outcome data from different socio-demographic contexts. The second response was the commissioning of a national process whereby political *and* 'expert' touchstone could be discovered or created. An Educational Indicators Panel at the NCES was mandated by the US Congress to recommend national education indicators. A group of 'experts' in education, economic and social indicators, education researchers, policy makers and educators were selected. They were asked to recommend on (Odden 1990, p. 39):

- theoretical models of educational systems,
- indicators that show the health and changes to the health in the nation's education system,
- new data and data collection systems required,
- a plan for developing indicators in the NCES, and
- a new way of reporting indicators in the *Condition of Education* (1989).

At least two unfortunate assumptions were explicit in this agenda. It assumed that a comprehensive causal story of educational systems was available along with the specifications of a remedial accountability sub-system. This reflected faith in the promise offered by systems theory; that a workable production function already existed or, at least, would very soon exist. It also presumed that systems theory could be used to control the evolution of education. The NCES clearly anticipated that the development of the Education Index would eventually describe and proscribe the educational progress of US children as well as generate consensus over the criteria to be used in the *Wall Chart*.

The best that can be said is that this situation began to alert policy actors to the limits of systems and production function theory in social systems. Two prerequisites of a more convincing theory of educative accountability began to emerge; the need for both external coherence and the internal use of students' interests to justify policy touchstone. The former was manifest by participation in the Organization for Economic Cooperation and Development (OECD) project that was intended to develop a broad set of education indicators for comparing developed countries. The latter was seen in new forms of political pragmatism. Where the earlier accountability policies were concerned with the technical effectiveness of management in education, the NCES strategies and national accountability policies 'moved from a system in which an individual, or institution, is accountable for *process* standards to one demanded by parents and other taxpayers where accountability is expressed in terms of *student achievement and outcomes'* (Griffith 1990, p. 7). It is a pity, therefore, that this lofty aim was not matched by an equally sophisticated methodology.

The changes at that time did help to define some threshold attributes of an acceptable theory of accountability; it had to accommodate economic, political, educational, national, regional and personal dimensions. For example, the unprecedented engagement of national and state politicians in accountability policy making was interpreted as 'a measure of the importance the US public attaches to education as we comprehend education's role in our economy' (Griffith 1990, p. 7). The shift from process to outcomes indicators, especially to outcomes deemed desirable by political and system leaders, gave fresh salience to the benefits of education and to broader social values. Hence, and a third feature of the emerging accountability policies, was that policy actors struggled to place more emphasis on curriculum content, learning exposure time, higher-order thinking skills, decision making at the school site and the role of teachers.

The fourth change was that the NCES had been given the role of National Guardian; it was expected to develop and refine indicators, to collect, compare and to report data to Congress, and to help ensure that performance data were not used inappropriately. How this worked out in practice can be illustrated by reference to events in California; a state that has accurately been described (Odden 1990) as having tried as hard as any to develop a comprehensive and educative accountability framework.

California has trialed many mechanisms. The first, Quality Indicator Reports, was introduced in 1983 to report the impact of Senate Bill 813 to the state legislature and the public. The reports provided percentages of students enrolled in academic courses, meeting graduation requirements and enrolled in pre-university courses, standardized tests scores in Reading and Maths, SAT scores, achievement tests scores, Advanced Placement test scores, as well as drop-out and attendance rates. Targets for system improvement in each area were defined. Bill 813 also led to an expansion of the second mechanism; the criterion-referenced Californian Assessment Program (CalAP). The CalAP began sampling children in grades 3, 6, 8 and 12 to identify changes to student learning in maths, language, history, social studies and science. As indicated in the next chapter, the CalAP only collected quantitative data and was replaced in the early 1990s by the California Learning Assessment System.

The third mechanism trialed was the School Accountability Report Card. Schools were directed to assess and report publicly regarding four inputs (expenditure and services, development of staff and curriculum, availability of substitute teachers, and availability of qualified support staff), six process factors (facilities, discipline and climate, class sizes and teaching loads, teaching materials, assigning teachers outside areas of competence, evaluation and development of teachers, quality of teaching and leadership) and two outcomes (student achievement in basic skills and academic subjects and dropout rates). In Odden's view (1990, p. 42), California's Report Cards were 'close to the full complement of data variables that RAND suggests should constitute the core variables for a full-fledged educational indicator system.' The quality of report cards used across the United States in more recent times is evaluated in Chapter 7.

The fourth mechanism was an annual research report on the *Conditions of Education in California* assembled by the Policy Analysis for Californian Education (PACE) centre (e.g. Guthrie, Hayward, Kirst and Odden 1988). It added trend data to detailed discussions of inputs, processes and outputs and to evaluations of major state initiatives. The PACE team released the report each fall to inform the annual state policy and budget cycle. The mechanism became a victim of state politics in the mid-1990s. An unlikely agreement was negotiated between teachers unions and school board associations and an incoming Republican Governor, Pete Wilson. While Wilson had been a long time advocate of educational reform, 'high stakes' testing and consistently critical of the performance of the state's public schools, he aligned himself with those campaigning for greater local control and professional autonomy. On winning office he suspended the state's testing system and cut funds to the state intervention programme that had been supporting poorly performing schools.

Rather than dwell here on the perversities of partisan state politics, it is important to focus on six dilemmas that indicated the continuing presence of alternative policy preferences and theories. They are helpful in that they indicate other prerequisites of an appropriate theory of accountability.

Emerging Prerequisites of an Educative Theory of Accountability

Despite truly heroic efforts to the contrary, the limited technical capacity of systems accountability criteria and processes to cause, explain or control changes in practices continued to confront researchers and politicians (Griffith 1990, p. 9). Instead there was persistent evidence of performance measurement corrupting rather than reporting on and improving teaching. Proxy measures, such as retention rates, were being improved without any perceived change in the quality of education. Some people in schools were manipulating testing procedures and using results in improper ways. The integrity of externally imposed forms of accountability proved very difficult to sustain.

The first dilemma here for policy communities in US public education is that imposed accountability policies can have high external validity but be resisted by those who have not been consulted or who do not believe in them. Conversely, adapted policies with higher internal validity tend to have lower levels of fidelity. The evidence in the late 1980s was that the systems theory in use lacked an adequate account of effective policy making and implementation in professional conditions.

The second dilemma concerns how to make fair comparisons between schools and students when they begin from different points and have different resources. School communities add value in different ways and support learning in a context where the purposes of education and making comparisons are disputed. The dilemma involved is that while simplistic averages had to give way to multiple measures, this made public reporting more complex and significantly less compelling. This problem was aggravated whenever the professionals associated with the production of outcomes were marginalized from the process of evaluation, or worse, alienated. In essence, the elegant simplicity of the closed system theories in use were not modelling adequately the communicative prerequisites of formative accountability processes.

This problem was also illustrated by a third persistent dilemma; how to identify the most important factors to be evaluated, the most appropriate criteria and the best methodology. In most states this question was simply never raised. Multiple-choice tests continued to narrow and partition the curriculum into unrelated fragments (Griffith 1990, pp. 10–11).

The fourth dilemma that Griffiths reported was that the profession was avoiding any real evaluation of its own pedagogical and leadership practices. The problem here is that while autonomy can be held to be a threshold condition of professionalism, ruling the evaluation of teaching out of an accountability policy is implausible in broader contexts and undercuts the external legitimacy of schooling. Kaagan and Coley's (1989) review, for example, found that nearly all states had developed state-specific norm-referenced or criterion-referenced achievement tests without ever evaluating teaching and leadership. They also found that the educationalists involved were using tests that did not permit inter-state comparisons. The same research concluded that the most advanced performance indicator systems in California, Connecticut, New York and South Carolina were based on very different theories of how schooling worked and expressed the unique goals and priorities of each state. Political conditions were apparently antagonistic to the development of a nationally accepted and comprehensive systems theory of accountability. Kaagan and Coley came to the view that three political, economic and educational conditions were widely prevalent; premature pressure to use education indicator results to hold school system systems accountable, investment not commensurate with demands for measurable quality, and a general unwillingness among education's stakeholders to trust indicators to represent the crucial relationships between inputs, processes and outcomes when policy making.

The fifth dilemma is that accountability processes predicated on an objectivist conception of social reality, that assumes empirically stable relationships and factors, is epistemologically incompatible with the rubbery nature of 'political realities' at either system or site levels. McDonnell (1989) demonstrated the willingness of the policy community in the United States to politicize indicator data and interpretations.

A related and sixth dilemma is that while all data have political properties, they are not subjected to an epistemic critique. Guthrie (1989) provided many reasons for the politicization of data, and yet, was so determined to defend his systems theory of Californian education that he used a personal political ideology to recommend the establishment of an apolitical organization apparently to be staffed by philosopher kings. It should, he said, be independent of advocacy groups, possess institutional credibility, be knowledgeable about education and government, be competent in methodology, and be able to synthesize and broker information to specialists, educators, government officials, the media and the public. Odden (1990a, b) suggested what appeared to be a more realistic and liberal democratic alternative: a partially independent policy advisory board comprising expert stakeholder representatives. Both theorists agreed, nevertheless, that the 'missing ingredient' in educational systems was the capacity to explain the 'relations among the inputs, processes and outputs of the education system and make recommendations for change' (Odden 1990, p. 48). In sum, in the absence of such knowledge, systems faith continued to legitimate simplistic and unsatisfactory answers to problems in ways that outreached its epistemic capacities. I return to this matter in Chapter 10.

The Unfortunate Outcomes of Accountability Policies

By the end of the 1980s there was a general acceptance in the United States that schools, rather than districts, were going to have to be the sites of reform. This troubled many policy makers given the persistent research evidence that, while most principals acknowledged the importance of 'instructional leadership', large city principals, elementary and high school principals had been found to focus on issues other than the quality of teaching and student learning (Crowson and Porter-Gehrie 1980, Leithwood and Montgomery 1982, Martin and Willower 1981). They tended to be involved in managerial and political challenges rather than improving the quality of instruction (Cuban 1988). They felt least able to help teachers teach better, and moreover, experienced deep guilt about this (Lortie 1986, Cuban 1986). The guilt apparently derived from the heroic leadership promoted by the standard literature of Educational Administration, and two deeply embedded causal myths of success; that such leadership transformed teachers and learning, and that principals held the levers of change. Fortunately, Pitner's (1986, 1988) research gave excellent grounds for setting these myths aside. He mapped the incredibly intricate, non cyclical and interactive organizational dynamics in schools and pointed out the inappropriateness of a heroic leadership style.

Sykes and Elmore (1989) took a sharply different approach; instead of researching how people perform prescribed roles and relationships in a highly regulated structure, they recommended a search for roles and structures that supported preferred educational practices. Early casualties of their approach were bureaucratic constructions of the relationship between teachers and administrators or any proposal that principals could disengage from the bureaucracy of their system. The key issue they identified (p. 84) was 'How might we go about 'unfreezing' structure, practice, research, and policy in a way that demonstrates concern for the multiple purposes of public education and for public accountability?'. While the methodological implications will be addressed in Chapters 9 and 10, it can be noted here that they advanced four principles to guide 'unfreezing'; core technology to drive structure, uncouple leadership from role, reduce the complexity of the authorizing environment, and deepen public discourse on results.

Greater depth in public discourse as a guiding principle recognized that the US public expects schools to serve broad intellectual, vocational, social and personal goals (Goodlad 1984). Ironically, it appears in retrospect that the use of a narrow range of indicators with low face validity helped keep the public appraisal of schooling at a relatively shallow level. Schools with privileged access to resources and social capital typically measured well on such indicators and, it followed, were rarely subjected to public scrutiny. When the opposite applied, there was little evidence of a policy debate informed by equity considerations. Instead, 'poorly performing' schools were typically subjected to increased regulation of their curriculum and instruction.

More broadly, policy debates focused on the narrowly defined 'success' in some schools and legitimated the managed and incremental alignment of curriculum and teaching methods with tested content in others (Cohen 1987). Both sets of practices ingratiated two demonstrably false conclusions; that the achievement of the goals of schooling *were* being measured by the tests, and, since no fundamental restructuring of schooling was required, schools could be managed by a series of technical adjustments to the regulations and other tools of policy implementation. The evidence was otherwise. Sykes and Elmore (1989, pp. 90–91) captured the sum and substance of research findings of the 1980s and deserve citation at length:

> First, an emphasis on indicators of organisational performance, particularly when linked to rewards and sanctions, directs attention to the indicators not to the underlying purposes, goals, and mission of the organization. Test scores become the result, not learning; children become test takers not students; and counts come to dominate educator consciousness, not the deeper significance of their work.
>
> Second, what is not counted, measured, or tested gets overlooked and de-emphasised ... [as] ... the accountability system focuses on a narrow range of outcomes, leaving a broader range of educational goals and purposes out of the account of successful schools. Educators, parents, and students lose sight of education's liberal, democratic, and humane purposes in the face of accountability pressures ...
>
> third ... teachers must be responsive to student diversity, providing learning experiences that accommodate a wide range of differences in prior knowledge, learning style and ability, interest, and socio-cultural background. A regulatory approach to teaching tends to produce a standardized pedagogy—all students moving at the same pace through the same material, with coverage driving the instruction and the test driving the coverage ... [hence] teachers do not ... become students of children's learning, do not discover and invent ways to encourage and invite engagement. Again, the deeper ideals of education are too easily frustrated by accountability systems as technically sophisticated as they are crude educationally ...
>
> Improperly managed, accountability systems can create disincentives to innovate, because risk taking and entrepreneurship is punished. Healthy organisations create a dynamic balance between accountability and innovation, encouraging all workers to be innovative some of the time ... [because] ... professionals operate under twin imperatives; to produce results and to contribute improvements in producing results ...
>
> The danger of accountability-driven education is that the motivational climate for teaching and learning becomes distorted. Students and teachers become alienated and disengage from real learning. They go through the motions; they

are 'schooled', not educated. Sadly, substantial ethnographic evidence, particularly at the secondary level, exposes such perversions ... Accountability pressures are not solely responsible for the contemporary pathologies of teaching and learning, but they contribute to school climates that are profoundly anti-educational.

Summary

Accountability policies were expressed in qualitative and expansionist terms in the early 1970s. Public doubts after the Oil Shocks, however, saw the metaphors of policy express increasingly anxious expectations and offer assurance in more quantitative dimensions. Legislation began to dominate policy development. Structural functionalism supported a cult of managerial efficiency and standardized testing was boosted by federal legislation requiring summative programme evaluation. The standard behaviourist and summative definition of accountability emphasized disclosure, performance testing and consequences for poor performance.

In the late 1970s a basic skills movement promoted minimum competency testing for advancement or graduation. Standardized tests reinforced 'productivity accountability' rather than 'responsively accountable' schools. By the end of the 1970s, four assumptions underpinned policy making; the purposes of schooling could be expressed as instrumental ends and means, an outcomes production function was nigh, students and teachers would respond to mandated policies, and that a 'science' of education justified a managerial approach to accountability.

The first 'wave' of reform in the 1980s reiterated accountability as a bureaucratic process responding to a political impulse in turn driven by economic conditions. Mandating ways of improving student achievement proved to be futile and dangerous because these methods insisted on conditions that largely prevented the improvement of teaching and learning. The second 'wave' sought to empower schools and the profession in ways that encouraged improvement but did not review the methods and criteria that should be used to define responsibilities and publicly discharge obligations. The possibilities of responsive, moral and political accountability were left undeveloped. The third 'wave' valued 'outcomes accountability' and comparative mechanisms such as 'high stakes' testing. It helped to further centralize policy making and to decentralize policy implementation and political blameworthiness. It was not able to promote moral accountability to clients or responsive forms of accountability to local stakeholders.

All three 'waves' of reform dealt with form rather than substance, never defined education and persisted with the assumption that schools are closed systems. Those best located to identify the need to reconstruct accountability policies, namely educational administrators, had a learned incapacity to challenge the assumptive base of accountability policies and practices. Instead 'education indicators' continued to be elaborated iteratively and claimed to guarantee empirical facts about the constitution of a living system. Accountability sub-systems were expected by the Rand Report to be functional, beneficial, behaviourist, pragmatic, objective, analytic, uncomplicated and responsive.

Serious problems began to emerge in the mid- and late 1980s. The variables chosen for indicators did not reflect the many purposes of educational systems. Relying on standardized tests was narrowing schools' purposes. Indicator systems relying solely on standardized test results had limited validity. There were demands by a broadening array of stakeholders for a comprehensive array of data, methods and indicators.

Some problems were due to cumulative effects. Accountability policies themselves were evolving as theories of indicator systems comprising inputs, processes and outcome factors, and as part of wider production function theories of how variables interacted to generate educational outputs. Students' learning became a universal proxy for the interactive complexities of learning, teaching, leadership, governance and context. Nevertheless, as the simplistic causal story gained in political credibility, it was codified as policy, objectified in law, reified in structure and sanctified as indicator systems.

Performance indicators were, however, consistently found to be more suitable for summative and system-based evaluation than for deep and formative evaluation in classrooms and in schools. The policy outcomes were also shown above to be incoherent; professional control within schools, political or systemic control of external accountability criteria and processes, and low levels of public legitimacy. By the end of the 1980s, many were beginning to realize the need for educative administrative strategies that could help reconcile the norms of professionalism, bureaucracy and community.

However, in the late 1980s, instead of moving to nurture governance, leadership and relationships in schools that would directly encourage and report improvement in the quality of learning and teaching, state governments began using aggressively interventionist styles of policy making and implementation. They created fresh policies in curriculum and accountability. They devised systems of indicators of student performance and published quantitative data. On the other hand, the introduction of some more 'authentic' methods of student assessment showed that state departments of education were beginning to see the need for greater consistency between their curriculum, instruction and assessment policies.

The policies promoted by the Council of Chief State School Officers (CCSSO) eliminated many forms of local and professional discretion, imposed stricter regulations, focused support and incentive schemes and sanctions more precisely on achieving stated goals (Goertz 1986). By the end of the 1980s, their interventions were exhibiting three recurrent features (Malen and Fuhrman 1991, pp. 6–9):

- they had plural purposes, multiple effects, marginal gains and mixed reviews,
- the expectations raised by new curriculum and accountability policies had outstripped their capacities to deliver improvement, and
- the theory base of policy making and implementation in the areas of curriculum and accountability was thoroughly inadequate.

Other conclusions concern the closed systems theory in general use. It lacked an adequate account of effective policy making and implementation in professional conditions. It lacked communicative prerequisites. It could not offer threshold conditions of professionalism that boosted the external legitimacy of schools. It applied premature pressure to use education indicator results to hold systems accountable. The focus on process and outputs meant that public investment was not commensurate with demands for measurable quality. Its intrinsic plausibility was falling; stakeholders were no longer trusting indicators to represent the crucial relationships between inputs, processes and outcomes when policy making. It was encouraging accountability processes that were based on a belief that data proved the dominant theory of accountability.

Similarly, systems faith persisted in providing unsatisfactory answers to problems that had outreached its epistemic capacities. Accountability systems continued to discourage risk taking and entrepreneurial problem solving, distorting the impulse in educators to make improvements. Indicators continued to direct attention away from the underlying purposes, goals, and mission of schools. The standardization of pedagogy and curriculum encouraged teachers not to research children's learning. In the next chapter I will show that these 'usurpations' and their 'Despotic' effects continued in the 1990s, with little evidence of policy actors exercising 'their duty, to throw off such Government, and to provide new Guards for their future security'.

7

Testing, Testing, 1, 2, 3 ... Confusion in the Early 1990s

Introduction

There were two general trends to the focus of accountability and locus of obligation in the early 1990s. As the states became more assertive about decentralizing educational services or school-level reform, attention moved away from district results to the quality of school performances. These trends were accompanied by technical improvements as well as attempts to devise more fundamentally educative methodologies. These trends were constantly moderated and confused by the politics of accountability.

The trials and advances to practice were very uneven. New tests were devised in some jurisdictions to measure the extent to which the taught curriculum was being learned; mastery tests. Some states developed forms of 'authentic assessment' which added demonstrable understanding and appropriate application to the burden of proof of learning that students had to provide. Some states began trialing forms of inspection and peer review.

Simultaneously, public accountability for school performance began moving from the districts to the states and to the schools, although districts retained a significant role in evaluation. Monitoring also changed from ensuring that schools complied with regulations determined at federal, state and district levels of governance to comparing actual learning outcomes against expected or proscribed outcomes. The consequences of accountability also broadened; where once students felt the impact of 'high stakes' testing, there was an increasing propensity for states to intervene into poorly performing schools with the intention of improving the performances of professionals.

There were even some rare trials of school inspection models, most notably in New York and California. To illustrate, the New York School Quality Review process was modelled on the British HMI approach; the approach set aside by the OFSTED model in 1993. In brief, a team of expert educators in New York trialed a year-long formative evaluation process to review policies and practices, spent a week in the school to watch operations, inspected students' work and met stakeholders. The formative emphasis in this accountability scheme was seen in the team serving as 'critical friends', their using qualitative and co-operative methods to identify strengths and to build on them, and in how reports on practices (including stern questions for teachers to consider) remained confidential to the school. The aim of the model was to 'stimulate a culture of permanent reflective self-appraisal; after the year of external review, each school spends four years in a self-review process, and then the process begins again' (OECD 1995, p. 141). It is shown in Chapter 15 that this model relates to that proposed for 1997 in Tasmania. The scheme was not linked to district accreditation processes but it drew its aims and review framework from the New York education reform plan—the *New Compact for Learning*—and each school's mission. The extent of voluntary involvement grew steadily, encouraged with privileged access to professional development budgets after the first year, although it is not clear if all 4000 state schools in New York will decide to participate on a voluntary basis.

In this chapter I suggest that policy research in the United States in the early 1990s was learning the boundaries of test-based outcomes accountability, discovering the limits of reporting as a method of discharging obligations, and yet was still largely unaware of the problems that came with the systems theory that underpinned widespread indicator systems. I highlight the emergence of a desire to make US public schooling more accountable for its degree of fairness. I conclude that the disjointed nature of changes were due in considerable part to structural and political impediments to reform. I start by reviewing the nature and effects of the major mechanisms used across the United States to gather and report accountability data in the early 1990s; test-based accountability, and the role played by ubiquitous indicator systems and school 'report cards'.

Hitting the Limits of Test-Based Accountability

While the rhetoric of accountability broadened in the early 1990s, the evidence is that practices remained much as House (1975, p. 52) described them 20 years earlier; where 'tests are almost invariably used as a sole criterion measures in most accountability schemes'. The Director of the State Education Assessment Center of the CCSSO (Seldon 1994, p. 54) confirmed that while the United States was still 'probably the most data-based education system in the world, we are certainly out front in our confidence in the value of student testing data and other measures for planning and accountability'. On the other hand, this confidence was coming to be seen as misplaced.

Some of the major limits of test-based accountability were identified in research commissioned by the US Congress Office for Technology Assessment. Haney and Raczek (1994) were asked to evaluate the role of testing and assessment in educational programme accountability. They reported serious technical and conceptual problems. They defined the limits of 'outcomes accountability' largely by identifying the reliability and fairness of standardized tests and the validity of their results. They took the view that test validity needs to be determined both in terms of the inferences that may be drawn from the results, and the wider consequences of using such tests.

A major and unresolved problem concerned the aggregation or cross-level analysis of data. Schools and districts were typically held accountable by comparing average scores obtained by standardized tests. Significant differences in school averages were attributed to, and therefore, considered to explained by, the effects of schooling. Logically, significant differences between districts would need to be attributed to, and similarly explained by, the effects of 'districting', whatever that might be. This intriguing line of inquiry can be dismissed, however, since:

> considerable research has shown that analyzing group average test scores (or indeed group averages on any variable of interest) holds little potential to explain the processes which may have produced those outcomes ... [and] ... attempting to hold schools or educational programs accountable in terms of aggregate performance of students almost guarantees that accountability in terms of *explaining* patterns of individual students' learning will be largely impossible (Haney and Raczek 1994, pp. 16–17, original emphasis).

To go back in time to illustrate, the 1984 *Wall Chart* was based on 1982 data and had 36 columns, ten of them providing state average Scholastic Aptitude Test (SAT) and other college admissions test scores, along with ranks based on these means. The SAT was renamed the Scholastic Assessment Test in the early 1990s (and is not to be confused with the Standard Assessment Tasks (SATs) developed in England and Wales). The 1984 *Wall Chart* accelerated the trend towards outcomes assessment and away from inputs started by the Coleman Report. The annual *Wall Chart* series was eventually discontinued in 1991 by US Secretary of Education, Lamar Alexander, when the Trial State Assessment information from the National Assessment of Educational Progress came on stream. They were so controversial that, whatever their other effects, they encouraged all states to co-operate in finding better means of measuring, comparing and reporting student learning.

There were major and important technical objections to the *Wall Chart*. Nearly three-quarters of the variance in the 1982 SAT scores was demonstrated to be due to the percentage of the cohort that had taken the SAT (Powell and Steelman 1984). School average SAT scores were shown to be very poor indicators of 'school quality'. Indeed, the sponsors of college admission tests campaigned against the use of SATs in *Wall Charts* on the grounds that the methods violated the 1985 *Standards for Educational and Psychological Tests* (Haney *et al. 1993*). The practices were seen as unethical.

Apparently US Departmental officials had tried to take account of intervening variables, such as participation and graduation rates, but when they found that adjusted scores dramatically affected state ranks, they quickly reverted to the use of raw means (Haney and Raczek 1994, p. 19). It was hardly surprising that, a decade later, the methods were still exhibiting the 'ecological fallacy'; specifically, the belief that statistical relationships found in aggregated data would always hold true when the unit of analysis is changed (Robinson 1950, Galtung 1967). Estimates of the effects of 'schooling' and 'districting' were still being confused with the effects of grouping respondents and selecting units of analysis. The analysis of aggregated test results could still not provide a reasonable explanation of the educational outcomes for individuals. Similarly, in the absence of longitudinal data on variables known to make a difference, and ways of discounting the contextual effects, such as SES, it was concluded quite reasonably (Cooley and Bernauer 1991, p. 169) that 'assessment systems should not be reporting school comparisons. That may do more harm than good'. Again, the practices were seen as unethical.

The second major limit to test-based outcomes accountability was a case in point. It concerned the construct validity of the tests, specifically 'instructional sensitivity' (Haney and Raczek 1994, pp. 23–35). The SAT was constructed to be reliable and fair across all regions, to discriminate efficiently among test takers and to remain reasonably immune to variations in curriculum. Items had to be of moderate difficulty and to discriminate consistently (Donlan 1984, pp. 27–51). They therefore had to have 'minimal dependence on school curriculum ... [and were] ... limited to topics that are not normally covered in the scope of secondary school curriculums' (Angoff 1971, p. 22, Donlan 1984, p. 47). However, since the tests were deliberately disconnected from the formal curriculum, it was unethical for SAT results then to be used as an indicator of the quality of state, district or school provisions, or the quality of teaching or the degree of student learning.

The unethical practices did not stop there. In a situation where such tests became the basis for 'high stakes' outcome accountability, and where test results could not be shown in a clear and reliable manner to change as a direct consequence of teaching, schools were encouraged to adopt methods that boosted scores, whatever their educational merit. Methods developed that ranged from coaching (Haney *et al.* 1993) to cheating (Haney 1993). Such practices were shown to pollute the inferences drawn from results (Haladyna *et al.* 1991). Surveys also showed that students were poorly motivated, using counter-productive test-taking strategies and deeply sceptical about the validity of standardized tests (Paris *et al.* 1991).

The third limit to test-based outcomes accountability has been described as 'technological myopia' (Haney and Raczek 1994, pp. 36–44). It is inevitable, in the Information Age, that tests cannot represent the scope of valuable learning. Trying to measure rapidly changing vocational education using performance standards opened a veritable 'Pandora's box' of problems (Hill *et al.* 1993). The traditional boundaries between fields of knowledge had dissolved.

Tests have also become obsolete more quickly than ever. Public values concerning valuable aspects of schooling have also moved. In the 1990 Gallup/*Phi Delta Kappan* polls (Elam 1990, p. 42), the US public gave top rating to the sixth and non-measurable national goal: to free every school from drugs and violence and offer a disciplined learning environment. In the event of school choice, most respondents rated 'quality teaching staff', 'maintenance of school discipline', 'curriculum and size of classes' as more important than 'grades or test scores of the student body' (p. 44). In 1992, 26% of respondents took the view that curriculum should be improved by placing 'more emphasis on the basics' (Burgher and Duckett 1992, p. 26). By 1993, 'inadequate funding' was seen as the biggest problem facing local public schools (Elam *et al.* 1993, p. 138).

It can be concluded that while policy discourse in the early 1990s about accountability tended to focus on testing students' cognitive and transferable vocational skills, public attitudes and the perceived longer term effects of education more often reflected transient social attitudes and values than proven consequences. The persistence of major ethical problems with the aggregation of data, instructional insensitivity and technological obsolescence, however, comprised massive political and educational challenges to current accountability policies. This applied particularly to the use of school report cards.

Turning over School Report Cards

The publication and dissemination of 'school report cards', 'school reports' and 'school profiles' were commonly regarded as appropriate methods of publicly discharging obligations in many states well into the 1990s. Their form and content, however, varied so much from state-to-state that they may have had limited relevance to public accountability, with the real possibility that they were actually serving other purposes. The degree of relevance can be derived from the surveys conducted and reported by French and Bobbett (1995). They used five categories to identify differences in eight Eastern states; instruments used to measure school performance, student-outcomes reported and the procedure for reporting them, student-characteristics reported, school and community factors reported, and statistical procedures used for evaluation data. They then compared the findings to the results of a survey of 11 Southern states.

There was little consistency in instrumentation found in the eight Eastern states. Two states did not report student-outcome data at all, five used state-developed tests and one used a national achievement test. Five states used indicators in addition to test scores. No two states appeared to have the same view of 'student outcomes' or similar procedures for reporting data. Performance data were being aggregated to school and or district level. Reports were even being delivered without interpretation or sophisticated explanation. Pennsylvania provided a *Manual of Strategies for Release to Press and Public* for educators. No two states reported using the same student, school and district, or community and district financial factors. Statistical analysis was limited to frequency distributions, means, percentiles, percentages (achieving locally-defined achievement and process standards), ratios and trends (resource and instructional), and there were no analyses reported concerning the impact of student, school or district variables. Since the same inconsistencies were found in the 11 Southern states surveyed, three tentative conclusions may be drawn.

First, since most states provided distributional data on learning obtained by state-designed instruments, and accompanied them with socio-economic and expenditure data, it can be not assumed that interstate comparisons, school improvement, curriculum or teacher development purposes were being served. Second, idiosyncratic forms, contextual variables and varying procedures suggested that incremental design processes reflected local contingencies rather than a general and shared view of the obligations to be discharged through accountability processes and criteria. Third, as 'there is no information provided that offers educators, parents and others insights into the factors that have actually contributed to higher or lower student performance levels in their own and other schools' (French and Bobbett 1995, p. 22), there seemed little likelihood that remedial or educative purposes were being served by school report cards. Given their survival in the face of such criticism, there is reason to suppose that they were relevant for other purposes, and that the policy challenge is really of differentiated relevance.

It was relatively common in the early 1990s for the efforts to improve public schools in the United States to rely on holding school, district and state officials accountable. Student performance was typically used to indicate the quality of policy making and implementation at all levels. This had proved implausible for a number of reasons, some indicated above. One is that the aggregated nature of student outcomes made them a very poor indicator of governance and leadership services. Another is that teachers alone had direct contact with students. Third, the duties of district boards were usually limited to policy making and the design of policy implementation. Fourth, officials' duties were often about enforcing regulations, providing legislated access for elected and appointed representatives and implementing fiscal policies; policies proscribed in prosaic and bureaucratic terms. Fifth, politicians have to articulate interests and public sentiments, and determine relative funding priorities.

Accepting this context of differentiated relevance led Wong and Moulton (1996a) to suggest that institutional indicators were required that could measure the degree to which officials' actions and decisions create conditions which actually facilitated teaching and learning in the classroom, and hold political leaders (as well as educators and educational administrators) accountable for the success for Chicago's public schools. Chicago had introduced Local School Councils (LSCs) and district-based governance from late 1988 in an attempt to reform schools. A depth case study in a Chicago Latino school community (Yanguas and Rollow 1994, p. 50) found that the reforms had

> ushered in a long and difficult period; a war over power and control. But this battle also energized the broader parent community, catalysed their social resources, and focussed their attention on issues of school improvement. Five years after the passage of reform Sprague had an activist principal, a supportive council and broader parent community, and a newly roused faculty. There was cause for hope.

Wong and Sunderman (1994) evaluated the extent to which Chicago's LSCs and district-based governance had provided accountability. The context featured court rulings, state laws and federal mandates, schools providing a wide range of social services using intergovernmental revenues, and increasingly influential corporate actors and organized interest groups. They reported the fragmentation of each districts' central office as lateral institutions and groups competed for influence and as higher levels of government expanded their programmatic demands. Fragmentation was traced to turf and boundary disputes, managers of state and federal programmes operating independently and policy decisions being disconnected from classroom teaching and curriculum. Local influence was achieved mostly by (a) the LSCs using the 'Nominating Commission' to control membership of the Illinios State Board of Education, (b) community organizations training people to run for the LSCs, and (c) LSCs having the power to contract and fire school principals. Three reforms were recommended; reconnecting central office to the classroom (by integrating categorical programme resources with general support for teaching activity), developing broader and more collective performance indicators (rather than using student achievement as a proxy for school success), and additional funding justified as human capital investment.

The conceptual and methodological problems associated with such reform were found to be immense (Wong and Moulton 1996a). The first problem was that accountability had long been practised in Chicago in three principle ways; as a narrowly-defined set of student learning outcome measures, fiscal and administrative compliance with external mandates, and leadership selection and appointment procedures in various committees. All were shown to be poor indicators of the improvement of learning and teaching. It was decided that 'current calls for institutional accountability which are based on student achievement suffer from low construct relevance. Governance performance has not been convincingly linked to student performance' (Wong and Moulton 1996b, p. 3). Instead, it was argued, policy actors should be held accountable for discharging the parts of their formal duties that actually relate to the improvement of teaching and learning. This meant holding 'educational policy actors accountable for only those tasks and duties which properly fall within their purview' (p. 1). This also implied devising institutional indicators of governance performance that effect teaching and learning for *each* set of policy actors, indicators that could be aggregated into a 'governance report card'. The first task was to define criteria, the second to develop methods of measuring the performance of 'policy actors'.

Wong and Moulton (1996b) identified 74 duties that were simultaneously expected of particular actors, applied to more than one actor, and, all shown to be relevant to a common construct; teaching and learning. One hundred experts were then asked to rate the performance of six governance actors (the Governor, Democrat lawmakers, Republican lawmakers, Mayor, Board, Central Office, and Teachers Union) at two points in time; June 1995 and January 1996. Three-facet Rasch analysis methods were used so the sample only had to be diverse, well informed and sincere, and could legitimately leave out parents. While the effects of the research process on the policy actors was not clarified, the interim findings were interesting.

All policy actors showed at least modest gains in performance from 'Poor' to 'Satisfactory' except the Chicago Teachers Union (CTU), with statistically significant gains achieved by the Board, Central Office and the Mayor. This outcome was attributed to the CTU losing some influence with the passing of the 1995 Chicago School Reform legislation and public euphoria over the appearance of change. Item analysis showed that duties that required routine compliance were attended to by all actors in a satisfactory manner. The second tier of duties required actors to take initiatives in support of school improvement. Analysis showed that policy actors saw these obligations as their primary work. The third tier comprised funding and political duties. Policy actors dreaded these obligations, especially the handling of teacher incompetence and obtaining more funding from the legislature for public education.

Interim 'Report Cards' were provided for each policy actor. They show that the Governor barely provided adequate funding with little improvement over the period. The Republicans were consistently rated as abysmal. The Democrats were rated only marginally better and advised to review their relationship with the CTU. The CTU was asked to review its approach to teacher incompetence and to consider policy issues broader than the economic needs of members. The Mayor's gains were traced to four forms of performance; negotiating industrial agreements, using expertise by setting up an expert Board of Trustees, controlling expectations by articulating clear and realistic policy goals, and building systemic capacities by appointing his former staff into senior school system positions. The improvements in the performance in the Board and Central Office were traced to better leadership, their sharper focus on service to schools and cutting costs and waste.

This study was remarkable for a number of reasons. It teased out the functional relationships believed by educational experts in Chicago to link governance, leadership, teaching and learning. It identified service and support shortfalls that apparently retard educational enterprise. It was a extremely rare case of political and governance accountability being driven by educational expertise, instead of the far more common converse case. The indicators developed proved to have high construct relevance to educational activity. And by comparison, as the next section illustrates, it showed that most other indicator systems in use were conceptually impoverished and educationally irrelevant.

The Poverty of National and State Performance Indicator Systems

In 1991 the NCES published an influential report by the Special Study Panel on Education Indicators; *Education Counts: An Indicator System to Monitor the Nation's Educational Health*. As foreshadowed above, it recommended that standardized multiple-choice questions be set aside in favour of more 'authentic assessment' which would expect students to apply their knowledge and reasoning to new problems. It helped accelerate the move away from monitoring compliance to regulations towards evaluating the output of schools using 'performance indicators', while admitting that there were formidable outstanding technical problems. The report concluded that a credible indicator system would have to be based on student learning and the quality of schools, and require more complex information than inputs and resources; information such as students' learning opportunities, teachers' qualifications and competence, and the character of the institution.

The report also listed fundamental limits to these proposals; no agreed model of an effective education system existed, there were doubts over the validity and reliability of data, and that it was unfair to compare crude quantitative data on learning when students, schools and districts had very different access to resources and faced different problems. The need to rebalance attention given to three competing policy values, specifically excellence, choice and equity, was identified but left undeveloped.

In time it also became clear that school performance evaluation by states was not being transformed by *Education Counts*. The methods and data used by all 50 states to report on public education were gathered by a telephone survey of state representatives in spring 1994 (Glascock *et al.* 1995). Thirty-nine (78%) of the 50 states were found to produce a report as specified by their state legislature. The use of indicators varied widely. The most common educational indicator was still standardized test results, followed by empirical demographics data (enrolment, number of faculty, name and location of school, brief summary of the system or schools), financial information (current expenditure per pupil, school-district status, teacher average salary, federal, state and local sources of revenue, expenditure by category) and social indicators (dropouts, suspensions, expulsions, school climate, parental involvement).

With regard to levels of reporting, Glascock *et al.* (1995) found that 17 states had retained school reports, 30 issued school district reports, 29 produced a state report, with only 12 states providing school, district and state reports. The data for these reports were collected by electronic disk or transfer, paper templates or abstracted from previously published reports. Verification was done manually or by electronic variance checks. Thirteen states were unclear about collection and verification procedures. The primary target audiences were found to be department administrators in 30 states, the media in 23, legislators in 20 and parents in ten states.

Given the implications of Wong and Moulton's (1996b) study above, the next issue was the likelihood of accountability policy reform and how data collected were actually being used. Glascock *et al.* (1995) found that only about one-quarter of all states evaluated their indicator programme, one-quarter were yet to do so and the remaining half were unsure if they did or not. The methods used to evaluate indicator policies were instructive. Louisiana was conducting stakeholder focus groups and external evaluators. Michigan was using telephone polling to review their indicator programme. Connecticut had a stakeholder advisory council that reviewed and developed the format and content of their report.

With regard to the use of data, 16 states disaggregated their data, ten by gender and ethnicity, and nine disaggregated data so that similar school comparisons could be made. Fourteen states did not disaggregate data, nine were unsure and a number of states were attempting to disaggregate the data using other categories, such as SES status, urban/ rural, etc. Like California, Oklahoma reported identifying schools 'at risk' in its report, provided developmental grants, intervened after three years of 'at risk' status, and had actually moved students out of one school prior to closing it.

A number of tentative conclusions follow. Based on self-reported actual practices in 1994, which would have tended to be more generous than critical, there was very limited agreement between states on what should count as an educational indicator system. There is surprisingly little evidence that recommendations from the Rand Report or *Education Counts* had been implemented. Accountability processes that did exist by the mid-1990s principally used standardized test results and left the reader to interpret them using demographic, financial and social information. Test results were still serving as a near-universal proxy for governance, school performance, the quality of teaching and learning outcomes. However popular this remained a technically implausible public policy.

Second, given the variances in reporting by level, audience, data, verification, disaggregation, stakeholder involvement and programme legitimation, there was very low coherence between states' accountability policies. There were many different theories of accountability in use. Multiple and contradictory theories continued to coexist. The nature of and the relationships between what were identified as 'educational indicators' were not being contested, except in the relatively rare cases where states had actively promoted debate, as in Louisiana and Connecticut, or where the public had been polled and the results published, as in Michigan. Hence, not only were the states harbouring a confusion of accountability policies, there was little prospect of active competition between the policies that might stimulate concerted analysis and policy development.

Third, many states employed strategies for reporting that were unconnected with other accountability processes. In most of these cases reporting appeared to satisfy mandates; a reform strategy shown in Chapter 6 to be educationally counterproductive. Few states were trying to apply the findings, as in Oklahoma. Very few states had in place a reporting strategy that both accounted for and contributed meaningfully to the improvement of learning, teaching, leadership or governance. With some exceptions, the state-based educational indicator systems surveyed by Glascock *et al.* failed to qualify as educative forms of accountability. This, once again, begged the question of the purposes actually being served by state accountability policies and practices.

Learning from Indicator Systems

A State Accountability Study Group (SASG) was established by the Office of Educational Research and Information (OERI) in the US Department of Education to explore means of creating more responsible and responsive accountability systems in big city education systems. The SASG investigated the need for structures (for teachers, parents, school and districts) in order to explicate responsibilities, purposes and audiences (OERI SASG 1988). This research was originally intended to replicate the 'natural accountability' found in small schools in large city schools that were serving many and unique constituents while relating to many layers of bureaucracy. It found that the methods being used to develop more responsible practice and greater responsiveness to clients were often impeding the achievement of these purposes. The SASG came to two blunt conclusions:

- policy makers were tending to use the easiest strategies available, such as monitoring students' test scores, promotions, graduation, teacher incentives, and school rewards or sanctions, and
- they were confusing the presence of such indicators with the need for accountability systems that could interpret and act on the implications of such information.

In case these conclusions are regarded as unfair, it can be noted that Darling-Hammond and Ascher came to much the same finding, and further, pin-pointed the basic problem (1992, p. 11):

Massive testing or any other data collection effort does not create an accountability system, nor does it guarantee improvement in urban or non-urban schools. A school or school district creates various policies and practices that make it more accountable by using many different tools, including methods for teacher and parent participation in decision making, bureaucratic regulations, legal recourse, safeguards and support for the competence of staff, and options for choice. Data about the student and school progress should inform the system so that responsible decisions are made and problems are corrected when they arise.

The interesting policy research issue that follows is why the OERI persisted with an investigation of structural indicators rather than reviewing the deeper need for policy governance and decision processes that might give carriage to reform, and to reform in classrooms. The OERI simply pressed on with a systems model of inputs, processes and outputs to identify indicators. It wanted to investigate how the relationships between them might result in greater 'school effectiveness'. While this systems and production function approach failed to produce definitive casual stories, it proved very influential with some interesting results. For example, Archbald (1996) proposed that the OERI indicators be modified to boost 'school choice', that is, parents capacity to make choices between schools. He began with the indicator system model shown in Table 7.1.

Archbald defined input variables as 'givens' (p. 92), which, at a stroke, gave procedural forms of equity absolute dominance over distributional equity. He also divided process variables into supply-side (producer) and demand-side (consumer) variables, which, at another conceptual stroke, reduced the complexity of schooling processes down to a simplistic and economic dualism. Producer indicators such as 'educational options', 'accessing options', 'information dissemination' and 'transportation' were then proposed to show how district and school policies and practices support and regulate parent choice. Demand-side indicators such as 'parent/student knowledge of options and preferences' and 'parent choices' were also proposed to show how information is used, how choices are made and how parental preferences are expressed. When it came to output indicators, however, the partial values involved, the doubtful construct validity of indicators and the limited relevance between this approach and educative consequences of accountability became particularly apparent. On the other hand, it must be conceded that Archbald (p. 100) provided a penetrating evaluation of the methodology involved when he concluded that:

Table 7.1: Archbald's model of an indicator system for school choice

Inputs	Process	Output
Fiscal resources	Organizational characteristics.	Achievement.
Teacher quality	Curriculum characteristics.	Participation.
Student background	Instruction characteristics.	Attitudes.
Other exogenous variables	Other school and district characteristics.	

the school choice movement reflects the goals of individual choice and liberty ... [and these goals] ... are difficult to measure directly. Proxy indicators must suffice. Indicators should reflect the extent to which parents are successful in getting schools they have chosen, parents and students are satisfied with the school choice process, and students are learning at appropriate rates. The question of whether school choice can produce greater accountability among educators cannot be answered directly by indicators.

Nevertheless, a good deal of indicator research based on systems models tried to do just that. Indeed, attempts to generate key indicators of school processes sometimes tried to serve purposes wider than just 'school choice'. Willms (1992), for example, started with four criteria for selecting a 'key set' of indicators:

- Which indicators provide a *balanced picture* of schooling processes across levels of the system, and across types of constructs?
- Which indicators facilitate self-examination and the process of school renewal?
- Which indicators are seen as tractable variables by school staff and administrators?
- Which indicators are easy and inexpensive to measure?

While the nature and justification of a 'balanced picture' were not clarified, he then used a literature review on school processes to derive the 'key indicators' summarized as Table 7.2.

The crucial point here is that a great deal depended on Willms's values, his interpretation of selected works, and his adoption of an analytical model of organizational climate. For example, if quality, choice and equity are considered to be three major policy values in US public education, Willms rather favoured the first two over the third. This is not to criticize his 'balance' of values but to draw attention to the positivist epistemology involved. His methodology demonstrated that what emerged as 'key indicators' depended on what counted to him as 'key' purposes, schooling, valuable knowledge and how it might be demonstrated. Hence, if 'reform' in 'a school' was defined as the objective, the methodology he used implied the value of such an approach being embedded in a school community's policy action research process. It is also possible that Willms might have overlooked a strand of accountability research developing in the early 1990s that gave a much higher profile to equity, albeit defined as 'equal opportunity to learn'.

Opportunity to Learn (OTL)

It was noted above that a special feature of policy settlements in the early 1990s was the concern over the unfairness of comparing schools which had different resources and intakes. Instead of taking the British policy path into the 'value-adding' capacities of schools, most US policy analysis focused on what this might mean for the individual. This is consistent with the hyper individualism discussed in Chapter 2. A broad expectation evolved in many states that accountability processes should demonstrate the extent to which public schools were providing each student with a 'fair opportunity

Table 7.2: Willm's key indicators of school process

Constructs	Indicators
Ecology and milieu	The physical and material environment, characteristics of teachers and principals, and intake composition of the school.
Segregation	The extent to which pupils are separated according to social class, ability, academic achievement, gender, race, and ethnicity.
Disciplinary climate	The extent to which there is a clear set of rules, how rules are perceived and enforced, and the extent to which students comply with them.
Academic press	The extent to which pupils, parents, and school staff value academic achievement, and hold high expectations for academic success.
Intended *versus* enacted curriculum	The strength and intent of the curriculum policy instruments, coverage of instructional topics.
Pupils attitudes:	
Sense of efficacy *versus* futility	The extent to which pupils feel they have control over their school success and failures, whether teachers care about their progress, and whether other pupils punish them if they do succeed.
Toward school	Pupils' satisfaction with their schooling.
Quality of school life	Pupils' general sense of well-being, quality of pupil–pupil and pupil–teacher relationships, academic behaviour and life plans.
Teacher attitudes:	
Sense of efficacy *versus* futility	Teachers' confidence that they can affect pupil learning and manage pupil behaviour.
Commitment and morale	The extent to which teachers believe their work is meaningful, and the extent to which they accept organizational goals and values.
Working conditions	Autonomy in accomplishing school goals, opportunities for professional growth, type and frequency of evaluation, salaries, class size, time available for non-instructional activities, extent of collegial decision making.
Instructional leadership of principals	The extent to which they: • set high standards and create incentives for pupil learning, and transmit the belief that all pupils can achieve at a high level, • set clearly-defined goals and priorities for pupil learning, and prescribe means to achieve them, • enhance teacher commitment by involving teachers in making decisions, by providing opportunities for teachers to improve their skills, and by evaluating teachers consistently and fairly, and • minimize extraneous and disruptive influences that keep teachers from teaching and pupils from learning.

to learn'. How this translated into practice can best be explained by reference, again, to California.

As noted in Chapter 6, California was one of the first states to make its schools directly accountable. Concerns over poor achievement standards in 1983 saw the state set performance indicators, rate each school annually, put the results in the public domain and reward 'merit schools'. With such clear targets and incentives, the measured levels of achievement rose, so much so, that in 1989, the statistical norms had to be raised. In the early 1990s, much as elsewhere in the United States, the focus of reform began to broaden to include the quality of teaching, school planning and school governance. The view was gradually taken that school accountability for demonstrable learning must take into account, and make more fair, the degree of appropriate resources and support, or else students would not have the same real opportunity to learn. The twin ideas that each student had an equal right to (a) a high quality education and (b) to compensatory public investment that moderated the effects of personal background, were crystallized by the 'opportunity to learn' (OTL) movement.

In 1993, California's state OTL taskforce specified the principles of OTL (OECD 1995, p. 150) (see Table 7.3).

These OTL principles strongly influenced the way that California's accountability mechanisms developed in the early 1990s. The first example was the California Learning Assessment System (CLAS) first piloted in 1992. The CLAS replaced the California Assessment Program, noted in Chapter 6, that was developed in the 1980s to collect quantitative achievement data. The CLAS collected a broader range of data, including portfolios of student work, and used the OTL principles in Table 7.3 to evaluate services. Problems with sampling and reliability, and gaps in the data (when financial and political turbulence in 1992 and 1995 suspended the scheme), limited the capacity of CLAS to provide public accountability and effective school review processes based on student achievement.

A second accountability mechanism is an annual publication first developed in 1989; the *Performance Report Summary*. It became a collation of test results and CLAS findings

Table 7.3: California's OTL policy

Principle	Specification
Educational	Access to a rich curriculum and high quality instruction.
Compensatory	Extra support for students with special needs.
Technological	Access to effective technologies for teaching and learning.
Secure	A safe environment to learn in.
Co-ordinated	Well co-ordinated support services.
Equitable	Fairly distributed resources.
Coherent	Coherent policies at state and district level.
Standards-based	State-wide and district standards, with school standards being developed when intervention was justified by poor results.

by school and by district. Schools were compared against themselves over time, against other schools in the district, and against other schools within bands of schools with similar SES contexts. Schools received a full version of their performance in addition to the summary that appeared in the *Performance Report Summary*.

The third mechanism is a School Accountability Report Card. Since 1988, each school has been required by law to report to the community. While most schools report student achievements, dropout rates, class sizes and professional development, the information varies to the extent that inter-school comparisons of performance cannot be made.

As noted in Chapter 6, the fourth mechanism, the Program Quality Review (PQR), dates from the early 1980s. Every three years schools were expected to conduct self-evaluation using state standards. These evaluations were validated by an external review team. These visits have been used increasingly in the 1990s to indicate the need for new teaching methods implicit in new curriculum frameworks, to encourage schools to relate school outcomes to state standards, and to extend the use of external 'critical friends' and ideas about continuous development.

The fifth mechanism, the 'Focus on Learning Program', is a voluntary accreditation system for high schools jointly managed by the Western Association of Schools and Colleges and state authorities. Every three or six years, a team of educators from outside of the district visit and review systematic documentation of purposes and operations, performance indicators, and then with school personnel, address five criteria (vision, leadership and culture, curricular paths, powerful teaching and learning, support, assessment and accountability). As with the PQR, the process begins with a year-long internal and formative evaluation process. Committees, focus groups and leadership teams consult and identify expected learning outcomes, collect and interpret evidence about performances, and build a school action plan. A major concern is the quality of evidence concerning teaching and learning. It is gathered by reviewing students' school work, observing lessons, interviewing students, analysing test results, and collecting the views of parents and former pupils. The accreditation team's visit then helps refine the action plan, and awards accreditation for up to six years.

The sixth approach was to maintain incentive and intervention regimes. The 'Californian School Recognition Programme' was set up in the mid-1980s to celebrate publicly examples of outstanding school achievement. From 1991, unsuccessful schools (determined by CLAS results, attendance rates and dropout rates) were taken through a three-phase process of increasingly external forms of intervention; self-selection of an in-house 'instructional improvement team', appointment of an external review and advisory team with access to a special improvement grant of up to $50,000, or, appointment of an external consultant manager who evaluated, recommended the reorganization of school and district resources and personnel, and then managed the school until its performance rose above the 'at-risk' threshold.

This array of strategies in California, clearly influenced by the OTL principles listed, had three common themes; 'to improve accountability, establish a climate of self-review in the schools, and to involve students, parents and the local community in the whole school-improvement project' (OECD 1995, p. 152). Whether this combination of strategies will survive and cope with the partisan politics and burgeoning problems that California faces is, however, another matter. It has a rapidly expanding school population, growing poverty, declining literacy and escalating student mobility. The economic and political context is marked by crisis, high unemployment, growing budget deficits, fiscal restraint, and ideologically driven political turmoil. Attempts to decentralize support to 7500 schools in more than 1000 school districts have, understandably, encountered major problems of structural and regulatory complexity. In 1995–1996, all California's high schools began the accreditation cycle noted previously, and therefore, all will be affected within the next six years. The OECD's conclusion (1995, p. 155) appears warranted:

> In spite of some clear improvements in scores in the mid-1980s, there is still widespread public concerns about standards in California schools; but it is safe to say that without the reforms described above, the quality of schooling in the state would be a great deal worse.

The efforts to reform accountability policies and practices in California also suggests that there are limits to the capacity of governance in education. One is the degree to which educators will tolerate comparisons being made between students and schools for what are construed to be political purposes. These and other limits are discussed in the next section.

Political Limits to Accountability Policy Making

Corbett and Wilson (1993) examined the relationship between testing and reform across two states and found that a test-based approach to accountability became antagonistic to reform whenever assessment data were used to make comparisons. They found that such assumptions presumed that a high level of uniformity actually existed between school districts, and moreover, that making such comparisons would provide stimuli for change. But, having weighed the evidence from two state's minimum competency programmes, they came to the contrary view that 'the wedding of statewide standardized testing and reform intentions has serious, deleterious repercussions for local school improvement' (p. 4). The programmes were found to combine high stakes consequences and high pressure on teachers and to side-track teachers into practices that raised test scores quickly. While there was 'no question that policymakers can control education through such policies; there is serious doubt that they can reform education through such policies' (p. 130). They concluded that it was more that a question of coping with rebellious colleagues:

> when the modal response to statewide testing by professional educators is typified by practices that even the educators acknowledge are counterproductive to improving learning over the long term, then ... to maintain the policy ... and label the response as educational 'misuse' is irresponsible (p. 147).

The actual policy outcomes, specifically a narrower curriculum focus and crisis-orientated decision making, were found to be antithetical to systematic, sustained examinations of educational purposes, processes and structures. If this policy was so clearly anti-educational, why was the policy retained? The answer was found in the political culture of education policy making. Corbett and Wilson (1993) demonstrated that, although combining accountability, uniformity and political motivations decreased the probability of improvement, state-wide testing programmes cohered well with the state political culture, and thus were retained and defended vigorously. Five major reasons were cited;

- such programmes reinforce the popular idea that learning is knowledge acquisition and easily measured,
- the results are publicly consumable,
- however invalid, the comparisons drawn are of wide public interest,
- the testing cycle has a closer fit to political cycles of values advocacy that to the longer term process of planning, implementation and evaluation integral to successful improvement, and
- action stimulated by testing results can satisfy the political impulse to create the impression of change, although this activity may not lead to improvement.

This finding also suggested that when policy makers insist on a policy in the face of covert or overt professional rebellion, it could be signalling the influence of the culture and priorities of broader polities. It can also be flagging the existence of conceptions of accountability favoured by interest groups who have the capacity to express their view in an influential manner. Policy settlements have been found (Cibulka 1991) to be influenced by three competing conceptions of accountability. Each of the three models has its own suite of assumptions and rationale, which, in turn, affects the nature of policy design and policy settlement processes in different ways. While Cibulka's analysis relied on a now dated evaluation of the policy settlement process in England and Wales, in order to provide an exemplar of 'consumer initiative', his still helpful analysis is summarized in Table 7.4.

Another limit to this modelling was that accountability processes were defined as assessing and reporting student achievement. The United States was slow to develop school inspections as part of its national accountability infrastructure, despite it being relatively common in other countries. This meant that public-school management and governance proceeded without data typically created by whole-school evaluation, without the information offered by expert and peer outsiders and without the knowledge typically generated during interactive and formative school evaluation. And when school accreditation and inspections were conducted by the states in the early 1990s, they tended to report on the practices of school districts and to discuss curriculum compliance, rather pedagogy, curriculum and school development. Finally, although districts, states, federal government and commissioned accreditation agencies each visited and evaluated aspects of schooling for their own purposes, there was little co-ordination between them (Blank and Schilder 1991). Such structural impediments only go so far in explaining the sources of political demand for accountability in the early 1990s.

Table 7.4: Cibulka's three models of accountability policy settlements

	State oversight	Local citizen	Consumer initiative
Source of legitimacy	Bureaucratic.	Responsiveness.	Market.
View of accountability reporting	Tool for superordinate government to improve the efficiency in the system.	Source of political power and a stimulus for political action by individuals and interest groups.	A means of maximizing choice.
Exemplar state	South Carolina.	Illinois.	England and Wales.
Policy design methodology	Strong. Consensual and participative processes offered to stakeholders.	Weak. School report cards relied on the power of public disclosure.	Moderate. Involved professionals in early stages, then centralized.
Policy impact	Strong. Accountability system integrated with other policies.	Modest. The initiative for action relied on local patrons or school officials.	Compromised. A mix of regulations with political and market mechanisms.
Advantages to policy settlement	Coherence and popularity during implementation.	Builds community and customizes policies during implementation.	Centralist determination of conditions promises policy fidelity.
Limits during policy settlement	Norms tend to discourage evaluation of the policy and practices.	Slow and moderated by local expertise and values conflicts.	Creates a politics of local and professional resistance.

Despite the confused use of educational indicators in practice examined above, there is a good deal of evidence that performance reporting has had a high profile in state education policy discourse. This profile has been found to be due to increasing political pressure for greater accountability. Cibulka and Derlin (1995) reviewed performance reporting policy documents from all 50 states and interviewed officials by telephone in 48. They asked about accountability policy sources, sources of controversy and links between performance reporting and other policies. They identified sources of demand for accountability and impediments to accountability as a state policy tool.

Three major sources of demands for accountability were reported. First, political and economic élites that influence government policy had used accountability as a theme when they had promoted efficiency, rational management and entrepreneurialism in the public sector (e.g. Osborne and Gaebler 1992). Second, populist support for greater accountability rose in the 1960 and 1970s never to abate. Trust in 'big government' fell after the Vietnam War, taxpayer resistance rose and the demand for minimum competency testing for advancement or graduation persisted. Third, the education profession itself regularly reiterated a technical rationality that offered greater accountability through bureaucratization, rather that creating politically critical strategies. The systemic reform movement (Smith and O'Day 1991, Fuhrman 1993), for example, argued that state and federal policies suffer from fragmentation and contradiction, and that all aspects of accountability, such as goals, curricular frameworks, standards, assessments and professional development, should be better co-ordinated.

The reasons why accountability continued to have a high profile in policy discourse, yet played a relatively modest role in state policy making, were traced by Cibulka and Derlin (1995) to political, populist and professional impediments. First, while partisan politics played a minor role, they found that accountability policies had been typically promoted by an appeal to productivity, which, in turn, was contested by three other political values; equity, economy and 'local control'. The reporting of test results, they said, continued to be controversial, no matter how data were disaggregated or contextualized. The media had often undermined their own campaign for full disclosure with unsophisticated interpretations of performance data. Hence, despite its high profile as rhetoric, performance reporting in the early 1990s had often been construed as unfair, non-remedial, too costly or authoritarian, and had become politically vulnerable and susceptible to competing policy commitments.

'Local control' re-emerged as a major campaign theme in more recent times. In Idaho, for example, the newly elected Republican State Superintendent attempted to return all *Goals 2000* funds to the federal government on the grounds that accepting the conditions concerning standards implied an endorsement of outcomes-based education (OBE). These political impediments were also accentuated by structural ambiguities, in addition to those listed above; the fragmented patterns of K-12 educational governance, some state education officials not being answerable to their governor and different parties controlling bicameral houses of legislature.

Populist impediments to a greater role for accountability in state policy making were also reported. Cibulka and Derlin (1995) also found that public concerns over rising costs and perceptions of falling standards were being held in check by other factors. While élites claimed popular mandates, there was little organized public pressure or popular movement apparent demanding accountability. Instead, for example, the campaigns against OBE in Pennsylvania and Colorado (Boyd *et al.* 1996, Rothman 1995) had valued particular obligations favoured by the fundamentalist Right; teaching traditional curricula, localist self-sufficiency, mastery learning and competitive excellence. With regard to these campaigns, Cibulka and Derlin suggested (1995, p. 15) that:

Their greatest success may well be to have transformed OBE, performance testing, and more broadly, systemic reform from a technical-rational frame to a political one. By broadening the political debate and widening the arena of relevant actors, they may alter the political settlements.

The third form of impediment to accountability affecting state policy making identified by Cibulka and Derlin (1995) originated in professional realms. Proposed accountability reforms, especially those expressed in technical-rational terms, still triggered automatic resistance by educators. Despite the ubiquity of testing, the role and value of tests remained controversial issues among teachers. In addition to the details of accountability policy being contested, many teachers reportedly doubted that reform should be seen in systemic terms or managed from state level. There was considerable support encountered by Cibulka and Derlin for Clune's (1993) view that reform should be differentiated by jurisdiction and political history and decentralized to better respect the pluralism of the United States. On the other hand, there was little evidence of professional support found for the levels of coherence and integration in systemic policies that Smith and O'Day (1991) saw as essential to school improvement.

Cibulka and Derlin's (1995) follow-up research confirmed the situation described above concerning policies and practices in the early 1990s. The major source of the confusion in practices is in the nature of each state's politics of public education. Improved access for accountability information had occurred simultaneously yet independently from other policies designed to influence improvements in state education systems. Despite the rhetoric, accountability policy development remained largely marginalized in the 'push' and 'pull' of interests between government, élites, populist and professional groups. There were minor trends towards integration and coherence in state educational policies, collecting performance information and involving stakeholders and the media in methods of improving schools. Performance reporting endured as a potent symbol of attention in state politics. This led to the fifth and final conclusion; that accountability had acquired a major role as a policy tool in the symbolic politics of state's public education.

Summary

Accountability policy making in US public education in the early 1990s has been shown in this chapter to have been increasingly targeted at three issues (OECD 1995, p. 138):

- a widening anxiety over the unfairness of comparing schools with different resources, intakes and capacities,
- devolution of greater accountability to schools, usually by scrapping compliance regulations, although sometimes at cost to equity and probity, and
- the fundamentalist religious Rights' resistance to the idea of testing students' performance against defined developmental criteria.

The states became more assertive. The attention of policy actors focused more on the quality of school performances. New forms of 'authentic assessment' emerged. Some states trialed forms of inspection and peer review. Monitoring policies shifted from compliance to comparing learning outcomes against expected or proscribed outcomes. Public accountability for school performance moved from the districts to the states and to the schools. 'High stakes' accountability broadened to include intervention and the forced development of schools. Technical and methodological advances were, however, constantly moderated by the politics of accountability.

Policy research in the early 1990s tended to rediscover the limits of test-based outcomes accountability while remaining largely unaware of the problems caused by systems theory. The largely fruitless search for comprehensive indicator systems continued although proposed reforms were also impaired by structural and political impediments. Political confidence in the role played by ubiquitous tests, indicator systems and school report cards came very slowly, if at all, to be seen as misplaced.

Research into the limits of test-based accountability remorselessly reiterated serious technical, conceptual and ethical problems. The inferences drawn from the results and the wider consequences of using such tests remained severely limited by three unresolved problems:

- aggregated data not being able to provide a reasonable explanation of the educational outcomes for individuals,
- the low construct validity of the tests concerning instruction and curriculum, and
- the inevitable obsolescence of tests when societal views of valuable knowledge constantly change.

Some policy work in the states appeared to focus on improving techniques. Research confirmed, however, that school report cards were not likely to serve remedial or educative purposes, and indeed, that reforms should try instead to reconnect central office to the classroom, develop broader and more collective performance indicators, and seek additional funding as human capital investment. It seems as if many state accountability practices were little more than a narrowly-defined set of student learning outcome measures, fiscal and administrative compliance with external mandates, and leadership selection and appointment procedures in various committees. Institutional accountability based on student achievement continued to suffer from low construct relevance. There was no convincing link between the performances of governors, leaders or teachers and students' learning.

The possibility of educative indicators was demonstrated. Wong and Moulton devised indicators of performance for sets of policy actors that related their services to teaching and learning. Their research identified the relationships believed by educational experts to link governance, leadership, teaching and learning, and as important, highlighted service and support shortfalls that retard educational activity. It was a rare example of political and governance accountability being driven by educational expertise. The indicators proved to have high construct relevance. Other indicator systems were shown to be conceptually impoverished and educationally irrelevant.

The research in the early 1990s tended to reiterate many earlier findings and to report that the evaluation of state indicator policies was still generally rudimentary and cursory. The disaggregation of data was idiosyncratic to each state and the interpretation of tests by clients tended to be a voluntary and ill-informed exercise. Test results were still being used as a near-universal proxy for education. There was continuing low coherence between state's accountability policies. Multiple and contradictory theories continued to coexist in policies with little prospect them being refined. Many states were employing strategies to satisfy mandates that were unconnected with other accountability processes. Why did these accountability policies not change?

The SASG concluded that policy actors used the easiest strategies available to satisfy mandates and that they confused the presence of indicators with the need for accountability systems that could make use of such information. Darling-Hammond and Ascher confirmed this judgement and showed that the collection of data was essential to but an insufficient condition for actually providing accountability or guaranteeing improvement. Despite this increasingly repetitious advice from policy research, many government agencies like the OERI pressed on with a systems model of inputs, processes and outputs to identify indicators, usually intending to investigate how the relationships between them might result in greater 'school effectiveness'. Their systems and production function approach, again, failed to produce definitive production casual stories, and yet was repeatedly pressed into service to design improvements. The methodology used, however, guaranteed elaborate personal theories rather than 'key indicators' in school communities or systems of schools on what should count as purposes, schooling, valuable knowledge and how it might be demonstrated. The potential value of school community policy action research was highlighted but rarely acted on.

Policy settlements in the early 1990s also featured a fresh concern over the unfairness of comparing schools with different intakes and resources. Instead of considering the value-adding capacities of schools, as in Britain, most US policy analysis struggled to provide each student with a 'fair opportunity to learn'. In California, OTL policy was elaborated as eight principles with specific practical implications and attempted to improve accountability, to establish a climate of self-review in each school, and to involve students, parents and the local community in whole school-improvement projects.

There was also a good deal of evidence in the early 1990s that suggested that educators resisted state efforts to reform accountability policies whenever comparisons were proposed for what were construed to be political purposes. Educators questioned the assumption that a high level of uniformity actually prevailed between school districts and that making such comparisons would provide stimuli for change. They had responsible reasons for doing so. 'High stakes' consequences for schools and students and high pressure on teachers were still encouraging teachers to raise test scores quickly by whatever means possible. Actual policy outcomes included a narrowing of curriculum and crisis-orientated decision making and excluded systematic, sustained reviews of educational purposes, processes and structures.

Why were such accountability policies retained? State-wide testing programmes cohered with states' political cultures. Such testing programmes reflected popular ideas about learning and its measurement. Results and comparisons created wide public interest. The testing cycle fitted the political cycles of issue management and values advocacy better than the longer term process of successful improvement of education. Action created by testing results satisfied the political imperative for the symbolism of change without inevitably leading to improvement.

On the other hand, it is also the case that politicians responded to more plural conceptions of accountability in the early 1990s than did educators, even in the face of covert or overt professional rebellion. Policy settlements were found to be influenced by three competing models of accountability; a bureaucratic state oversight, responsiveness to local citizens, and a market response to consumer initiatives. Demands for accountability were traced to three major sources; (a) political and economic élites using accountability to promote efficiency, rational management and entrepreneurialism in the public sector, (b) populist support giving expression to taxpayer resistance and 'local control', and (c) educators regularly using a technical rationality that offered greater accountability through bureaucratization.

The startling differences between the high political profile and the minimal influence of accountability policies was also attributed to political, populist and professional impediments. First, when accountability was promoted by an appeal to productivity, it was quickly contested by three other political values; equity, economy and 'local control'. It was branded as unfair, non-remedial, too costly or authoritarian, and thus was easily displaced by competing policy commitments. Populist impediments to a greater role for accountability in state policy making included public concern over rising costs, 'big government' and OBE. Third, proposed reforms expressed in technical-rational terms, and the role and value of tests, remained controversial issues among teachers.

Cibulka and Derlin's conclusions merit repetition. During the early 1990s, improved access for accountability information remained disconnected from other reform policies. Despite the high rhetoric, accountability policy making was marginal to the reconciliation of interests between government, élites, populist and professional groups. Accountability reporting seemed to have acquired status as a major policy tool in the symbolic politics of public education.

The implications for practice in the mid-1990s were clear and may be summarized. State education systems would have to attend to particular policy issues if accountability processes and criteria were to become a more potent policy reform lever and serve more educative purposes. Performance reporting would need to be used as an integrated policy tool. Feedback would need to serve internal improvement intentions as much if not more than external accountability purposes. The collection of contextually-driven data would have become greater than prescribed data, the latter to be derived from the former. Performance reporting would have to become part of a comprehensive indicator system and be used in furtherance of greater equity rather than to merely demonstrate greater school productivity. How far policies and practices were actually advanced by the mid-1990s is the subject of the next chapter.

8
The Adequacy of US Policies and Policy Making

Introduction

The most recent trends in the United States do not constitute reforms to accountability policies. They seem to comprise an intensification of past practices with marginal improvements. On the other hand, accountability policy researchers appear to be considering and developing substantial policy and methodological alternatives. With regard to practical trends, an OECD evaluation noted (1995, pp. 139–140) that:

- states are becoming the most important funders, regulators and evaluators of schooling,
- districts are increasingly inspecting schools and strengthening reporting requirements,
- federal authorities are using programme evaluation to hold schools, districts and states accountable, and
- prestigious private agencies are increasingly being commissioned to accredit, high schools, colleges and universities.

In this chapter I demonstrate that this general intensification of practice continues to be driven by a presidential politics of symbolic reform. A widening concern for more equitable outcomes is encouraging policy research into more authentically educative forms of accountability. It is concluded that this emergent phenomenon is signalling the need for more appropriate policy research methods.

Presidential Accountability Politics

As indicated above in Chapter 7, the most recent phase in the evolution of accountability policies in the United States has been driven by President Bush's *America 2000* education strategy, announced in April 1991, and President Clinton's *Goals 2000: Educate America Act*, which passed in March 1994. There were a few interesting changes in the interim. Section 306(e) of the electronically distributed version of *Goals 2000*, for example, celebrated outcome accountability and gave it an added role in the development of governance and management:

in the area of governance and management, each State plan must establish strategies for improved governance and management of its education system, such as: (1) aligning responsibility, authority, and accountability throughout the education system ... [thus] ... focusing accountability on educational outcomes rather than monitoring compliance with input requirements.

Another change in *Goals 2000* was the proposal that OTL standards, conceived much as reported in Chapter 7 (i.e. as school delivery standards), be added to the repertoire of curriculum content and student performance standards listed in *America 2000*. This also meant that, despite the American tradition of local control of education, new national goals and state curriculum frameworks were given presidential legitimacy and, where backed by state governors, had to be implemented in public schools (Kirst 1994). This reaffirms another point made in Chapter 7; that education policy making has been increasingly centralized and that the power and requirement to implement has been devolved to school governors. The 'high stakes' testing of learning outcomes has been further intensified although the nature of the problems being encountered suggest the persistence and limits of the systems theory in general use. Boyd and Halpin (1994) provided a penetrating analysis of the problems associated with these presidential politics of accountability.

First, there is still no convincing causal story that links the now standard international justification for changes (that a better educated work-force is needed to enhance economic competitiveness) to the strategies being used: the simultaneous centralization of curriculum policy powers and the decentralization of school governance powers. Indeed, these structural and market strategies appear to increase the likelihood of inexpert and ideological forms of knowledge being taught by teachers who will be encouraged to instruct 'by numbers' to 'numbers' for 'numbers', and for poorer forms of science to gain credence in the production of knowledge.

Second, the continuing crudity of the technical mechanisms used signalled the continuing presence of a simplistic production function model that sought to control schooling through the manipulation of inputs, processes and outputs. Tighter curriculum controls were further de-skilling teachers. Testing even narrower wedges of knowledge, and publishing aggregated results, were not able to change the effects of the social intake of schools or to alter the value that specific schools added. And while the publication of results was meant to add to the quality of data that parents might use to select schools, and thus force schools to improve when parents exercised such choices, the public ranks could not alter parents' capacity to exercise choice. Indeed, pending further research, it was speculated by Boyd and Halpin that the ranks are just as likely to help advantaged schools to select students, further improve their market position, and thus compound the problems faced by people in 'poorly performing' schools.

One general upshot of presidential politics and earlier trends is that the term 'accountability' has fallen into disrepute among educators. Wohlstetter and Odden (1992) also implied that it is a 'politically incorrect' concern in school-based management. It tends to be accorded meanings in mercantile, industrial or political metaphors rather than be aligned with notions of morally responsible schooling. This unfortunate situation seems to be made worse whenever accountability policies put undue trust in market forces and forego professional, technical and equity considerations. More, the sheer persistence of the dilemmas reviewed above suggest that the problem could be 'locked in' by the limits of a 'systems' theories of institutions (Crowson et al. 1996). In the next few sections more diverse perspectives are considered to help develop an alternative approach to researching accountability policies.

Accountability and the Coherence Movement in Systemic Reform

A number of arguments in Chapters 6 and 7 concluded that reforms to accountability policies and practices need to be reconnected to reform in other policy domains in education systems. Two conditions that seem to prevent this happening more commonly are overload and fragmentation; the former a vicious mix of intervening contextual conditions and rising expectations, the latter caused by conflicted demands and disjointed policies and practices. A third more hidden condition is the rationalist assumption that reform will follow when systems have greater coherence between inspirational learning goals, focused instruction, and supportive governance and accountability procedures. Many 'reform' programmes are built on such an assumption. But when is enough systemic coherence enough? Or, at what point, and why, does systemic coherence suddenly create reform? It appears that degrees of systemic coherence are a necessary but not sufficient condition of reform. Instead, instances of reform happen when the hearts, heads and practices of educators change, and the relationships between learners and educators improve.

Fullan (1996) warned against the 'rational trap' of making objective and systemic coherence an aim of reform. Noting the fundamental non-linearity of evolving systems, he argued that working on alignment, clarity and consistency would help but that a preoccupation with policy orchestration would not achieve reform. Reform will only happen, he argued, when a majority of educators achieve greater clarity and coherence over purposes, policies, theories and practice. Two sets of strategies were recommended to achieve the broad scale of change required; networking, and reculturing and restructuring. Honig (1994) located the first strategy, networking, in four principles derived from a systemic action research project in California:

1. Any effort to improve schools must be designed to meet the goal of creating an active, thinking curriculum in specific disciplines, and success should be judged by whether increasing numbers of students reach agreed-upon performance standards (p. 791).
2. School efforts must be collective, not just individual; they must be comprehensive, not piecemeal; and a high percentage of the staff must commit to improving student performance and to making the changes necessary (p. 792).
3. State- and district-level strategies must be comprehensive and make use of the major leverage points in the education system to create a context in which effective learning communities are developed at each school (p. 793).

4. Large-scale implementation of school improvement efforts depends on networking strategies that can organize large numbers of schools around a powerful idea for institutional improvement, provide assistance, and allow conversations with sister schools (p. 793).

The key strategic and tactical features of purposeful and effective networks were proposed by Fullan (1996, p. 422) as:

- ongoing, systematic, multilevel staff development (usually involving identified teacher leaders within each school and external staff developers),
- multiple ways to share ideas, including telecommunication, cross-visitation, and workshops,
- integration with school-wide and district-wide priorities and mechanisms, including leadership of school principals, collective actions by the majority of teachers, community development, school improvement plans under district auspices, growth-orientated performance appraisal schemes, and teacher union interest in professional development, and
- a commitment to and a preoccupation with inquiry, assessment of progress, and continuous improvement.

Networking has at least two limits. Fragmentation can follow the availability of too many networks. Networks generate ideas and initiatives that do not replace but are typically grafted onto existing assumptions, theories and practices. Hence the need for reculturing and restructuring. Reculturing, Fullan argued, involves developing new professional and administrative values, new theories about learning, teaching, leadership and governance, and new norms of practice. Restructuring would follow; new relationships between people at work and new assumptions about relationships to facilitate the growth of learning, problem solving and organization. Logically, the changes in relationships, restructures, occur as people in organizations reculture.

Following Fullan's advice, there are five major implications for systemic accountability reform. First is the need to review the level of objective alignment between accountability purposes, policies, theories and mechanisms. The purpose is not to achieve absolute objective coherence but to eliminate obvious cases of contradiction and to build links with other policy realms such as evaluation and planning. Second, learning progress needs to be monitored using data aggregated *from* the classroom, not *to* the classroom. Third, proxy indicators of systemic reforms will need to be developed to measure the contribution that accountability networking is making to classroom, school and system improvement. Fourth, proxy indicators of reculturing and restructuring will have to be refined to measure the contribution of accountability processes and criteria. Fifth, systemic accountability processes should regularly map teachers' understandings of expectations, their purposes, the level of coherence they perceive between policy domains and with practice, and the extent to which they are critical consumers of new ideas. An example now offered involves reconstructing an 'accountability for productivity' policy so that it features 'educative resource accountability'.

Productivity and Educative Resource Accountability

It is interesting how little accountability theories and practices have actually changed in public education when compared to other professions. The development of very different views on accountability in the 1990s helps illustrate the conceptual limits of current assumptions in education (Haney and Raczek 1994). In financial accounting, for example, accountability is now regarded as much an art as a science. The results of 'creative accounting' in any situation can depend on who is the client or the purposes to be served. Contrary to public perceptions, the bottom line of financial accounting is not simply profit but depends on the time line adopted and assumptions about the future value of current assets. This is a useful reminder that accountability in education can serve any combination of education's plural purposes.

There is often an unspoken assumption among educators that education is a valued end in its own right; a universal public good of intrinsic value. Setting aside the possibility of ego extension, it is interesting to note that economists have largely abandoned efforts to ascribe intrinsic value to goods and services in favour of comparative outcomes. Welfare economists such as Schumpeter (1954) have raised the theoretical possibility that the 'value in use' of education, its utility, could achieve more through government intervention than the maximum attainable under a *laissez-faire* market system in a perfect state of equilibrium. This, presumably, is a reason why most civilized societies have established public education services. This possibility of attaining greater gains through public governance depended on a method of measuring, or having a proxy for, the utility of education. US public education has long used student achievement as a measure and proxy of valuable educational outcomes. Unfortunately for education, two theoretical advances in the economics of social welfare have shown that such an approach is most unlikely to be trustworthy.

First, in 1906, Pareto demonstrated that utility was immeasurable in social welfare and would have to be replaced by a scale of preferences (Hicks 1981). In 1950, Arrow (1950) showed mathematically that if there are at least three preferences which members of a community are free to order in any way they choose, the selecting of any social welfare function that creates a social ordering based on those individual preferences is too circular to be rational, even when trivial and dictatorial methods of aggregation are ruled out. The practical implication is that even if US public education were to give away its complex armoury of measures of student achievements in favour of (say) greater trust in professional judgements, in order to acknowledge Pareto's Immeasurability of Utility thesis, Arrow's Impossibility Theorem would then come into play and insist that education will never have an analytic method of determining a policy of accountability that is to be generally preferred. The best that will be possible is an array of competing policy preferences that might be reconciled or prioritized. Policy choices remain moral choices. And if the possibility of ever being able to evaluate an educational policy in any fundamental economic sense is very low, policy actors will be well advised to abandon the analytic search for an education production function, for a medium of exchange in education and for the indicators of the theoretical line between public and private goods.

While the possibilities of 'accountability for productivity' policies have been closed off by Pareto and Arrow, a path appears to have been discovered in the state of New York that might yet lead to an educative form of 'resource accountability'. Monk and Roellke (1995) argued that the origin, disposition and uses of resources should be traced for accountability purposes through to the quality of the delivery of educational services and to the production of educational outcomes. They proposed a resource-based approach to accountability using indicators of actual educational opportunities to measure the productivity of districts. Some detail is warranted at this point to illustrate the method.

In Phase I, the New York State Department of Education selected five indicators with organizational origins that were believed to account for district productivity; lack of K-12 continuity, reduced enrolments, high expenditures, high cost to the state, and high tax yet low wealth. One hundred and thirty-nine districts fell below the normative threshold established for at least one indicator. Consultations in Phase 2 eliminated 'false positives' and verified that each district would benefit from inclusion in the study. No 'false negatives' volunteered to be involved. During Phase 3, the state and district collaboratively evaluated alternative solutions. In Phase 4, the state and district resolved any differences concerning local strategic plans.

The features of all four phases that proved critical in methodological terms were the absence of penalties and rewards, the presence of co-operation and communication, and the ongoing development of 'thresholds' and districts' indicators of productivity. Apart from its growing technical merits, this Organizational Change Study also revealed four issues important to resource-based accountability policies and policy research. It showed that the people organized as 'education systems' can learn to relate purposes, resources and services delivery to the production of educational outcomes. They can develop co-operative accountability policy making and implementation processes with an emphasis on open inquiry and collaboration between 'the state' and its constituent units. And whatever indicators are selected at the outset, accountability systems in education need to remain flexible and respect changing plural interests and values that are in competition. In this regard, Monk and Roellke also recommended (1995, p. 502):

> resource allocation indicators that capture important aspects of the kinds of educational opportunities that are actually being delivered to (or produced for) students. We also recommend designing an accountability mechanism that is steeped in partnership and mutual respect. We believe that this can be accomplished by building an accountability dimension into a broader study process.

The study found that districts handled enrolment declines in very different ways but that they were very interested in comparative data and strategies. Some indices were then substantially refined. Several hostile or suspicious state–district relationships were gradually transformed. Monk *et al.* (1996, pp. 31–32) observed that 'this kind of study process works best when there are opportunities to move back and forth between the cross sectional (and perhaps longitudinal) indicator data and the site-based qualitative data'. To clarify, the study used (p. 3):

> statistical analysis of indicator data to identify predictors of good schooling outcomes. The study also includes ... qualitative research methodology where the goal is to understand what lies behind the indicator numbers and where the emphasis is on understanding the unique features of individual sites that may have a bearing on the success of an organization at realizing educational results. Second, the study places a significant emphasis on collaboration across levels of school governance ... to engage the State and local school districts in a constructive dialogue about how best to improve the performance of students ... to provide feedback to districts ... to conduct 'action research' in the best sense of that term.

Given the extent to which 'policy' was regarded as a social construction, I will come back to action research as a potentially useful method in accountability policy research below when I have examined equity-driven advances to accountability policies.

Equity and Accountability

Many reforms in the 1990s have implicitly or explicitly linked poor test scores to the lack of student discipline, vague curricula, lack of focus on 'the basics' and poor teaching. Some reformers have accepted and promulgated a public myth by placing most of the blame on teachers, seeing them as recalcitrant or as lacking the capacities to carry through reforms (Clark and Astuto 1994). This myth became progressively more difficult to sustain as studies showed that poorly performing schools and districts remained largely impervious to a battery of increasingly hard-edged threats, incentives, sanctions and interventions. Similarly, increased investment in staff development and better alignment of the curriculum and testing did not make a difference (e.g. Dornun *et al.* 1995). A persistent feature of these apparently recalcitrant districts is that they 'tended to primarily serve poor and/or minority students' (Berry and Ginsberg 1996, p. 2).

Since the late 1980s there has been a growing alternative accountability literature in the United States arguing that a more ambitious framework for reform is required, a framework that assumes that accountability policies need to do more than impact on instruction and test scores. A number of US researchers have called for a better balance between excellence and choice and equity in accountability policy making with educative effects in mind (Oakes 1989, Shepard 1991, Darling-Hammond and Ascher 1992, Darling-Hammond and Snyder 1992, Elmore 1994). The recurrent key arguments in this equity-driven literature, which reiterate the findings in Chapters 6 and 7 above, can be summarized:

- be it basic skill, achievement or performance-based, tests continue to dominate evaluation and accountability processes and criteria,

- the scores from aggregated standardized minimum competency tests are drawn from overly narrow conceptions of cognition,
- mean test scores cannot be applied at the level where learning takes place and remain peripheral to educative processes,
- tests tend to be obsolete and to be contaminated by teacher and student mobility,
- high-stakes testing narrows the curriculum, misdirects instruction, deprofessionalizes teaching, and puts hard-to-teach students at risk,
- uni-dimensional measures of individuals and institutions are not found in other professions and are inappropriate in education,
- accountability systems should increase the chances that best practices will occur, decrease the chances that 'harmful' practices will occur, and encourage critical reflection that will identify and change policies and the practices that are 'harmful' or ineffective, and
- racial and ethnic differences in achievement cease being significant when minority students have access to comparable curricular opportunities and experienced, qualified teachers.

Oakes (1989), for example, had taken this to imply that 'context indicators' were needed to help contextualize and interpret 'performance indicators'. She called for a multiple-indicator accountability system based on the three domains summarized in Table 8.1.

A vital clue here is the need for multiple contextual indicators. They were initially thought to be technically necessary to explain away variances due to the assumed-to-be-unchangeable characteristics of a population, such as poverty, race and ethnicity. It was later realized, from case study research in 'poorly performing' districts, that, 'despite considerable efforts to redesign the ways practitioners collect and examine a range of data, innumerable barriers [to improvement] exist—most notably when the lack of time and professional development render the new information virtually useless in reforming practice' (Berry and Ginsberg 1996, p. 5).

Berry and Ginsberg's research also found that many other complex political, economic and cultural factors impeded 'improvement'. These factors were expressed in five ways; a politics of race, an economics of poverty, vacuums in communications and leadership, an uncritical mass of human resources, and a culture of unchanging

Table 8.1: Oakes' multiple-indicator system of contextual indicators

Contextual domains	Criteria
Access to knowledge	The quality of teachers and curricular resources provided.
Press for achievement	The quality and type of assignments and recognition for academic accomplishments.
Professional conditions for teaching	The extent to which teachers work well together and they are prepared to participate meaningfully in decisions affecting students and the curriculum.

practice. The implications for school reform provided strong endorsement of Darling-Hammond's (1993, p. 754) earlier advice, specifically that:

> The changed mission for education requires a new model for school reform, one in which policy makers shift their efforts from *designing controls* intended to direct the system to *developing the capacity* of schools and teachers to be responsible for student learning and responsive to student and community needs, interests, and concerns. Capacity building requires different policy tools and different approaches to producing, sharing, and using knowledge that those traditionally used throughout this century.

This view recognized the differences between contractual and responsive forms of school accountability and the vulnerability of legitimacy based on one indicator, test results, compared to the more compelling intersecting zones of legitimacy created by multiple indicators. It also foreshadowed the need for the political, economic and social development of school communities to the stage where they could each develop locally shared goals and understandings over the education they want for their children, and then translate those expectations into accountability policies, criteria and practical processes (p. 760). To illustrate, again using Berry and Ginsberg's (1996) case study and the dismaying extent to which a politics of race was found to be impeding school reform, it was reasonable to conclude that (p. 28, original emphasis):

> what is called for here is a form of community *therapy* as a cornerstone of accountability and school reform. Equity—in terms of creating the kinds of learning opportunities necessary for the children of Sylvan—will be a moot point and a hopeless goal—without such a transformative political development.

Were this case study to be replicated, and Darling-Hammond's generalizations found to be warranted, four conclusions might be drawn. A testing programme in such settings would be hostile to building a co-operative politics of inquiry between stakeholders. A mandated school exchange system of rewards and punishments would be inimical to an economics of mutuality that featured trust and collaboration. A bureaucratic set of rules and regulations would be particularly antagonistic to an educative discourse between teachers, administrators and the community that was intended to clarify and serve children's needs. This leads to a fifth and more general possibility; that the current methodology of accountability policy making could be a central part of the problem in US public education.

This possibility was examined by considering research that compared performance indicators between states with similar demographic characteristics (Salganik 1994). Many states provided three types of data along with each school's and district's test scores; demographic data, means from similar states, and national data, and then left the reader to interpret the comparisons. The other method used to predict graduation rates was regression analysis which employed various background factors, most commonly poverty status. The adjusted r^2 of explained variance in graduation rates was found to be 0.34 controlling for youth poverty and 0.50 controlling for youth poverty and minority enrolment. Each method, however, was found to have advantages and disadvantages. While the performance of comparable groups was easily reported in clear and understandable terms, usually as a difference in mean or percentile, the particular characteristics of some states made some groups of states invalid. While regression methods used the relationships between background characteristics and outcomes over the entire range of the data, and were better at achieving comparability on more than one background characteristic, the *choice* of demographic characteristics used to specify similarity was found to have a larger effect on indicators that did the methodology used.

Salganik (1994, p. 137) concluded, quite reasonably, that 'the decision about which characteristic to include is largely a policy matter, determined by which aspects of similarity are desired'. In sum, be it based on group comparisons or regression analysis, the dominant methodology of accountability preferred by most states in the mid-1990s shared high degrees of vicious circularity. It is time for a fundamental change to theories in use. A promising candidate, authentic assessment, is now examined.

Authentic Assessment

The conceptual impasse noted above was realized and traced to a limited understanding of the sources of legitimacy, the strengths and weaknesses and the purposes, processes and focus of different types of accountability. An early useful step was to characterize and contrast the models evident in practices, and thus, identify the competing theories in use. Five types of accountability current in big city systems were identified (Darling-Hammond 1989, pp. 3–9) and are summarized in Table 8.2.

This clarification of theories-in-use was not a proposal for selecting one best universal style by system. Instead it helped develop Wise and Gendler's (1989) earlier advice that the most appropriate mix of accountability types was to be determined by level and by policy issue. Wise and Gendler used equity and productivity policies as examples to show that the purpose of accountability issue by accountability issue varied, and each required a customized mix of mechanisms. Equity policies should be determined, they argued, at higher levels of governance to avoid capture by local and discriminatory majority politics. Conversely, productivity policies were more likely to be effective if determined by local and participative governance, and by a greater emphasis on parental choice and school-based management. They also noted that by guaranteeing equitable access to resources, states encouraged local initiatives, fairer distributions of teaching staff and more customized curriculum and support options. However, when states simply mandated outcomes and defined them in terms of standardized test scores, the aim of 'equal education ends up degrading learning for all. Individuality, creativity, and depth are lost; all that is retained is uniformity, conventionality, and trivial skills' (p. 36).

This multiple purposes, means and legitimacies approach then had to be reconciled with the continuing existence of the apparatus of state. Darling-Hammond and Ascher (1992, p. 11) recommended 'the right mix of tools to provide support for school

Table 8.2: Darling-Hammond's five types of accountability in US big-city systems

Dimension	Political	Legal	Bureaucratic	Professional	Market
Basis of legitimacy	Mandate.	Legislation.	Hierarchy.	Expertise.	Consumerism.
Approach to accountability	Legislators and school board members regularly face their electorates.	Courts determine which practices best cohere with laws concerned with schooling.	District and state offices regulate schooling with set standards and procedures. to professional standards.	Specialist knowledge, qualifications and adherence	Choice by consumers is exercised over courses, schools and school policies.
Strengths	Establishes general policy directions.	Establishes and defends individual and group rights.	Standardizes procedures to produce desired outcomes such as equity.	Applies complex knowledge and subtle decision making to individual needs.	Articulates purposes when state has no interest in controlling choice.
Limits	Decisions of elected officials can be shrouded. Politics do not guarantee minority rights.	Not all educational decisions are subject to the law. Access to law is uneven.	Not sensitive to unique and plural needs. Does not encourage learning.	Not sensitive to competing public goals such as cost containment and productivity. capacities.	Cannot guarantee equity of access to services or services of a given quality.
Purpose of accountability	Coherence between public policy and services.	Equitable access to public resources, such as rights, under the law.	Equal and standardized education.	Improving learning and the growth of knowledge about practices.	Responsiveness and improvement of teachers and schools.
Processes of accountability	Stakeholder sensitive evaluation.	Evidence, argument and interpretation of the law.	Regulated practice. Inspection and testing systems.	Professional reflection of the quality of practice.	Parents and students chose programs, policies and schools.
Focus of accountability	Reconciliation of stakeholder priorities and outcomes.	Legislated rights, access to, and quality of, public services.	Adherence to standardized operating procedures.	Professionalism in response to client's needs.	Responsiveness to consumer demands.

Table 8.3: Darling-Hammond and Ascher's recommended shares of responsibility and accountability

Level	Scope of responsibility and accountability
State	Equal and adequate resources to schools and for ensuring the enforcement of equity standards and standards of professional certification.
School district	All adopted policies, for equity in the distribution of school resources, and for creating processes that make them responsive to the needs and concerns of parents, students and school-level staff.
School	Equity in the internal distribution of resources, for adopting policies that reflect professional knowledge, for establishing means for continual staff learning, for creating problem-identification and problem-solving processes that drive continual improvement, and for responding to parent, student and staff ideas.
Teacher	Identifying and meeting the needs of individual students based on professional knowledge and standards of practice, for continually evaluating their own and their colleagues practices, for seeking new knowledge, and continually revising their strategies to better meet the needs of students.

improvements that will encourage responsible and responsive education' at each level in systems in order for them to assume appropriate shares of responsibility, as summarized in Table 8.3.

With regard to school accountability, Darling-Hammond and Ascher (1992, p. 28) recommended particular processes and indices. As noted above in Table 8.3, they emphasized professional accountability processes that were learner-centred and knowledge-based. By this they meant strategies that would raise the sense of responsibility of competent professionals, use expert knowledge in making decisions, and where certainty did not exist, action research would be used to continually discover more responsible professional responses. They also recommended attention to particular generative conditions; (a) sensitive personnel policies, (b) opportunities for professional development, action research and consultations about the problems of practice, (c) the evaluation of student learning, classrooms and school practices, and (d) the creation of incentives to sustain teachers' learning, self-evaluation, and their willingness to teach and lead in the classroom and staffroom. Regular and collective problem-solving or action research was repeatedly emphasized in order to build a climate of inquiry and a culture of critical reflection with regard to purposes. I will come back to an appropriate form of action at the end of this chapter.

A series of systematic case studies (Darling-Hammond *et al.* 1993) led to the view (Darling-Hammond and Snyder 1993, p. 17) that learner-centred schools should hold themselves accountable for the quality of the interaction between the six elements of accountability systems (i.e. policies, structures, processes, feedback and assessment mechanisms, safeguards, and incentives) involving significant educational players (students, staff, school, family and larger community):

- relationships and voice in school—for example, governance, decision making, and communication mechanisms developed to ensure that important needs and issues are raised and addressed,
- school organization that helps to personalize relationships, ensures attention to student needs and problems, and brings coherence to teaching and learning,
- vehicles for staff interaction, shared inquiry, and continual learning that strengthen practice and create opportunities for continual evaluation and improvement of teaching,
- forms of student assessment that reveal student strengths, talents, abilities, and performance capacities, including methods by which teachers observe and evaluate student growth along with formal exhibitions, performances, examinations, and portfolios, and
- strategies to promote ongoing evaluation of school functioning utilizing input from parents, students, staff, and external reviewers.

It was clear from Darling-Hammond and Ascher's analysis that the student-centred imperatives were being emphasized far more than the knowledge-base or the need for accountability to traverse levels in a coherent manner, and for equal attention to be paid to the quality of learning, teaching, leadership and governance. The need for such policy proposals also directed attention to the adequacy of current policy-making capacities.

The Adequacy of Policy Research

The four sections above illustrate how accountability policies are both the medium and outcome of policy articulation processes whereby historical commitments and knowledge are continually reconciled with changing priorities, local and national events and emerging perspectives. Various attempts have been made to theorize the evolution of accountability policies. As explained in Chapter 1, Elmore and Associates (1990) argued that they are a by-product of ongoing competition between three fundamentally different theories of accountability based on technical, client or professional assumptions.

Where Elmore's account rests on a typology of political interests and mutually exclusive epistemologies, others have started directly from metaphysical assumptions to provide very different notions of accountability. If, for example, the multiple definitions of the 'real world' in education are regarded as social constructions (Greenfield 1975, 1988), criteria become the standards, benchmarks or indicators invented by well-intentioned people engaged in the process of making judgements. The data they use cannot be treated solely using a logical empirical epistemology since they inevitably embody objective, subjective and normative dimensions. Radical humanists add yet another layer to these constructivist assumptions. They argue that administrative practices can either control or liberate the thinking of people in organizations (Foster 1986a). An important implication here is that accountability criteria and processes are a major potential means by which the 'moral economy' of an organization is established, reproduced and, if educative, transformed.

The possibility that the positivism in policy research could be a central part of the problem of accountability was examined in passing by a study in Dade County Public Schools, Florida. Herrington (1993) examined the extent to which schools' report cards help with the education of continually underachieving minority students. She selected Dade County for four main reasons; school report cards were mandated in 1976, it has a highly prescriptive legislative record, it has an even longer record of devolutionary reform strategies, and while it makes data available on attendance, dropouts, corporal punishment, suspensions and expulsions in terms of race and gender, it does not require student achievement data to be reported by race or gender. Miami also has the fourth highest poverty rate of all US cities and a politically volatile and diverse ethnic mix. Dade County is the fourth largest school system in the United States and has extremely high rates of student growth and mobility. In this setting, school report cards were designed to serve three purposes; inform parents about the performance of their children's schools, give parents a basis for seeking improvements, and provide school-specific diagnostic data. The policy-making process was held to be sound until the policy outcomes were evaluated.

Twenty-one school and community officials familiar with community relations, minority performance and public schooling were interviewed. The findings were clear; the race and ethnic data categories were very insensitive, parents were not reading the reports, school administrators were insulating reporting from school improvement strategies, and none of the three primary purposes above were being served. Herrington (1993) concluded that although a 'numbers-based strategy' intended to challenge race or ethnicity-based institutional biases was held by stakeholders to be fundamentally at odds with an environment conducive to school-level improvement, 'the development of a two-tiered strategy may be necessary to capture the benefits of school cohesiveness and community-building at the school level without losing the monitoring and oversight critical to identify and alert educators and others to continuing poor achievement among racial and ethnic subgroups of students' (p. 45). In the interim, report cards lacked educational authenticity among parents and educators and among the managers of improvement. Simultaneously the methodology of policy making was shown to be flawed.

By the mid-1990s this finding appears to have been generalized across all polities and, once again, a presidential politics of symbolic reform was evoked. In March 1996, the nation's state governors met in a second historic Education Summit. It was an open question as whether repeating the cycle would create reform. To some this meeting symbolized that education reform had, once again stalled, or worse, lost its way. Others interpreted the policy impasse in terms of a Democratic President confronting new Republican majorities in both federal houses. A third possibility was that reforms had been proceeding steadily but not at a pace that gave political satisfaction.

Cibulka and Derlin (1996b) used cases of performance accountability in Maryland and Colorado to explore a fourth possibility; that the public perception that education reform policies are incoherent and confused is due, in part, to the absence of agreed criteria by which to judge the adequacy of such policies. This hypothesis was consistent with the high rhetorical profile of accountability in symbolic politics, as discussed in

Chapter 7, since metaphors were used to accommodate multiple interpretations. Cibulka and Derlin focused on the extent to which performance accountability in the two states cohered with the central aim of systemic reform promoted by Smith and O'Day (1991): the improvement of student outcomes. The details of the two cases are important for how they illustrate the extent to which policy adequacy appears to be substantially moderated by the sophistication of the methodology.

Like other states, Maryland had adopted state-wide, norm-referenced tests and competency-based graduation tests in the 1970s, and demonstrated comparative success, only to find that, in the 1980s, accountability demands from the business sector escalated. The Maryland Commission on School Performance reported in 1989 and the State Board of Education moved immediately to 'fast-track' adoption without waiting for legislation or for the State Department of Education to plan implementation. The tactics they used were instructive. Learner outcomes (proficiency standards) were developed with the help of teachers in six months, for grades three, five and eight, in reading, writing, language usage, mathematics, science and social studies. Within a year, using external consultants, assessment instruments were developed that were independent of Maryland's curriculum frameworks. Districts soon adapted their curricula to fit the instruments and standards. School support was guaranteed by instructing the 24 superintendents each to contribute 20 days time each year for 15 participating teachers. Union criticism was avoided by paying the teachers extra to work in vacations and ensuring that instrumentation could not evaluate individual teachers. In sum, the policy-making process was iterative, professionally pragmatic, well resourced and stakeholder sensitive.

The instruments could also, report on learner outcomes and performance standards across the state, for each school district and for each school. After some expert debate, standards were classified in three bands; 'excellent', 'satisfactory' or 'not met'. Students had to demonstrate that they both understood and could apply knowledge. The State Board also set minimum targets, such as 70% were to achieve 'satisfactory' ratings, 94% attendance and 3% maximum dropout. However, when no district system met them by 1996, the achievement date was moved back to coincide with the national goals date of 2000. Between 30% and 35% typically achieve satisfactory standards and less than 20% are rated excellent. Put simply, the policy-making process created criteria and processes that were regarded by all stakeholders as appropriate.

The implementation process used in high schools since 1995 began with even more extensive consultations with school districts, teachers and other stakeholders. Concerns over possible inequitable effects were considered but did not slow the process. Each school was required to mobilize a 'school improvement team'. In 1996, 38 schools did not meet the set standards of achievement, attendance and dropouts. They were then directed to develop and submit a 'transition plan' and a 'long-term reconstitution plan' to the State Board for approval. Cibulka and Derlin (1996b, pp. 20–21) noted how 'the right combination of programme design decisions, implementation skills, institutional support (particularly the state board) and political support, along with continued public interest in reform, has made the Maryland reform a sustainable one'. In sum, accountability policy making in Maryland both used and proposed sophisticated accountability policy making and implementation processes, expecting school communities to make and enact customized policies.

Accountability reforms in Colorado have also been sustained by a state governor, a commissioner of education and a state board all providing consistent direction and support, albeit in a setting where local school districts have constitutional authority over curriculum development, policy and operations, and the assessment of student learning. Although a state-testing programme was devised in the mid-1980s, it ran up against the political value and tradition of local control. The attempt to compare districts was abandoned after two years.

An alternative political and structural strategy began in the early 1990s. School improvement legislation established a network of state, district and school accountability committees, each with a broad representation of stakeholders. This network was then used to give carriage to standards-based education policies from the State Assembly. There was opposition from the Right, which favoured local control, and from the Left, which preferred compensatory education for the disadvantaged. As the debate abated in 1993, the state level committee was reinforced and relaunched, using legislation, as the Standards and Assessment Development and Implementation Council (SADI).

SADI's standards in reading, writing, mathematics, science, history and geography for grades 4, 8 and 11 were approved by the State Board in March 1996. Mindful of local power traditions, all districts have been asked to adopt their own standards by January 1997, notably standards that must meet or exceed the state standards. This set of events suggested again that accountability policy making has to be iterative, professionally pragmatic, stakeholder sensitive, and use and propose educative criteria and processes.

For example, the processes used systematically involved stakeholders and educational experts, and gained feedback to draft policies from the public and elected representatives. There are plans to repeat the processes to establish secondary subject standards by 1999. It is also notable that the challenges from the Left and Right largely neutralized each other and helped generate support from the business sector. The accommodation of local control traditions, by providing multi-level structures and consultations, fostered positive perceptions about institutional change capacities. These perceptions, in turn, appeared to generate both 'the institutional and political support and societal attention necessary to foster future sustainability of the policies' (Cibulka and Derlin 1996b, pp. 28–29).

Again, the rules for successful policy making and implementation include participation by all stakeholders, the use of expertise, facilitating the competition between ideas, and building institutional and societal capacities. Cibulka and Derlin found common reasons for why accountability policy changes were sustained in these two cases, suggesting that specific criteria were appropriate at each of three stages leading towards policy adequacy (see Table 8.4).

It is interesting that this model requires political support at all stages. Cibulka and Derlin also considered the possibility of policy adaption as an alternative to policy adoption. In a situation where US school communities are increasingly being expected

Table 8.4: Cibulka and Derlin's stages and requirements of policy adequacy

Stage	Adoptability	Sustainability	Effectiveness
Relevant criteria for assessing adequacy at each stage	Credibility of design. Political support. Societal attention.	Feasibility of design. Implementation quality. Institutional capacity. Institutional support. Political support. Societal attention.	Credible results. Scientifically defensible results. Political support.

Table 8.5: Sashkin's six change agency models

Models	Consultant role	Trainer role	Researcher role
Adoptive: research, development and diffusion	Facilitating the adoption of a 'given' innovation.	No direct role although clients might learn how to obtain similar knowledge in future.	No direct role.
Adoptive: social interaction and diffusion	Facilitating the adoption of an innovation or new practice.	No direct role although clients may become aware of their role as training resources.	No direct role.
Adoptive: linkage	Creating linkages between knowledge source and user, including roles, structures and institutional resources.	Clients may acquire and pass on learning capacities such as information retrieval skills.	No direct role.
Adaptive: intervention theory and method	Advises not on content but on process e.g. data gathering, analysis and decision making.	Trains clients in new process skills and methods.	Trains clients in new research methods and skills during the change process.
Adaptive: planned change	Advises on content and process.	Trains clients in skills and understandings needed to create adaptive changes.	Evaluates the effects of changes and the training of clients in research skills.
Adaptive: action research	Advises on content, process and capacity building.	Develops clients' capacities for change design and change management.	Integrates evaluation, research, planning and professional development.

Table 8.6: An action research process for school community accountability policy research

Phase	Objective	Process	Key issues
1	Define accountability together.	Collective mapping of the situation.	Plural stakeholder views, interests and values.
2	Describe the policy challenge of accountability in context.	Joint specification of key aspects.	Antecedents, causes, effects, terms.
3	Review wider knowledge of policies and practices.	Research and search teams commissioned and report.	Alternative ideas, policies, trials and findings.
4	Gather and evaluate consequences of options.	Co-operative evaluation of the relative merits of options.	Competing explanations and solutions.
5	Deciding policies and planning practices.	Select strategies and joint planning.	Feasible? Beneficial? Plausible? Educational?
6	Action.	Teams are organized to accomplish specific tasks.	What? Why? Who? When? With? How? Evaluation?
7	Evaluate policy outcomes.	Collective review and report.	History of ideas, intended and unintended effects.
1a	Redefine accountability together ...	Joint formative evaluation and mapping of the situation ...	Newly revealed challenges, views, interests and values ...

to articulate policies within state and or national guidelines, and to become increasingly self-governing, in addition to being self-managing, opportunities for policy adaption must be increasing. This would apply particularly to charter schools. The practical implications with regard to change agency and policy research methodology are summarized in Table 8.5 (Sashkin 1974, pp. 212–213).

Table 8.6 suggests in greater detail how action research (Burns 1994, pp. 293–309) can be adapted to serve policy research purposes, particularly how it might be used to help produce school community accountability policies. The processes could also be modified to generate inter-level systemic, inter-institutional or inter-agency policies.

Summary

There was little substantive change to accountability policies in US public schooling in the 1990s. There were two surges of national political interest that were finessed by presidents and state governors using a rich rhetoric of symbolic reform. There was a marked intensification of scrutiny and reporting but little real change to purposes intended or served, to data and processes used to discharge obligations, or any real advance reported towards more moral, responsive and formative forms of accountability.

Comparative research (OECD 1995) has shown that US public education still relies heavily on standardized tests to monitor the progress of students and the quality of educational services provide by states and districts. The focus of accountability practices has shifted to the performance of schools. Federal programmes are using standards and curricular reforms to set clearer expectations. About 45 states appear to be dissatisfied with traditional testing and are trying to obtain greater coherence between programme standards and their curriculum, assessment and teacher development policies. Many have removed accumulated regulations in the hope that this will facilitate school-based reform.

The demands on educators have intensified. They are held accountable by their local school board or district for students' achievements, and made to feel accountable by political rhetoric to parents, the community and to the nation. There is no national system of school inspection. Most evaluation is focused on school districts, not on schools or the quality of teaching. It is argued that the sheer scale of US public education and the absence of reliable and acceptable ways of measuring schools' performance had made it difficult for policy actors to develop school-sensitive and performance-based accountability systems. When systems switched from a bureaucratic monitoring of inputs and processes to outcomes, the readily available test scores were adopted as convenient performance indicators for the evaluation of schools and districts, despite their low construct validity.

Test results continue to provide passports to higher education while criticism mounts. Tests are still often unrelated to curriculum, compromised by teachers and students and unable to measure high order skills and understandings. The Right attacks them for propagating low national standards of achievement. The Left attacks them for their 'high stakes' consequences and inequitable outcomes. In some states they are being phased out by more authentic forms of assessment that are based on the taught curriculum and emphasize skill mastery, understanding and application. Progress is slow.

The high mobility of students and teachers has made the calculation of the value added by each school virtually impossible. Equity-driven policy making has promoted OTL and other more educational policies. The traditionally more successful systems tend to highlight the professional competence of teachers. Systems which are dissatisfied with the performance of their schools are more likely to emphasize institutional reform. The status of teachers is therefore crucial to any reform, yet teachers have relatively poor pay and qualifications when compared to their colleagues in England and Wales or Australia. It is fair to conclude that while the profession is seen as part of the problem, it is yet to be widely regarded as part of the solution. The same can be said with regard to US school communities.

In the interim, US policy makers continue measuring outcomes for two major and two minor reasons:

- accountability; defined as a political and administrative impulse concerned with how well schools are performing, how their performances compare and to see if schools are meeting externally set standards,

- school improvement; since measurement may identify the areas best attended to,
- to monitor the implementation of reforms in terms of actual change, to judge the success of new policies, and
- to improve the general understanding of the factors that create successful schooling.

In general, it has proved difficult to reconcile the interests of experts, who seek fair and reliable assessment means of showing what worthwhile learning has occurred, and policy actors, who seek robust, inexpensive and easily quantifiable measures (McDonnell 1994). Evaluations of networking, reculturing and restructuring strategies have identified the need for greater levels of alignment between accountability purposes, policies, theories and mechanisms, to lessen objective contradictions, and to boost interactive links with other aspects of the policy cycle such as evaluation and planning. Learning needs to be monitored using data aggregated *from* the classroom, not *to* the classroom. The contribution that accountability networking, reculturing and restructuring is making to classroom, school and system improvement needs to be evaluated. Systems need to map teachers' understandings of expectations, their purposes, the level of objective coherence they experience, and their critical consumption of new ideas.

The conceptual impossibility of rigorously economic 'accountability for productivity' policies introduced the need for more educative forms of 'resource accountability'. Resources were related to the quality of the delivery of educational services and to the production of educational outcomes. The value of iterative systemic policy research that revisited quantitative, cross-sectional, longitudinal and site-based qualitative data in an action-research process was demonstrated. Action research was also recommended by equity-driven attempts to advance accountability policies, along with multiple indicators that set aside systems theory to attend to three crucial contextual issues; access to knowledge, press for achievement and professional conditions for teaching.

The evolution of accountability policies will need to be more sensitive to the political, economic and social development of district communities if they are to help them to develop and redevelop locally shared goals and understandings, and then translate those moving expectations into accountability policies, criteria and practical processes. Policy making was recommended as a therapeutic community project. More generally, the methodology of policy making was found to be a central part of the problem of accountability. The standard methodology of accountability policy making was found to exhibit high degrees of vicious circularity within the confines of logical empiricism.

Alternative policy-making methodologies were reviewed. One sought an appropriate mix of accountability types determined by level and by policy issue. Another suggested that by guaranteeing equitable access to resources, a state might encourage local initiatives, fairer distributions of teaching staff and more customized curriculum and support options. A third proposed student-centred imperatives to traverse levels in a coherent manner and for equal attention to be paid to the quality of learning, teaching, leadership and governance. It became evident that accountability criteria and processes are a major means by which the 'moral economy' of an organization is sustained or changed.

In the mid-1990s a second presidential politics of symbolic reform cycle began. It was shown that public perceptions of reform incoherence were due, in substantial part, to the absence of agreed criteria for judging the adequacy of such policies. The rules for successful policy making and implementation in such a setting appeared to include participation by all stakeholders, the use of expertise, facilitating competition between policy ideas, and building group, institutional, systemic and societal capacities.

It was concluded that US school communities are more and more being expected to govern local policies within state and national guidelines or charters, in addition to the expectation that they will be largely self-managing. Opportunities for policy adaption will probably expand considerably as opposed to degrees of policy adoption.

The practical implications with regard to change agency and policy research methodology were then examined. Action research was identified as a potentially appropriate means of creating accountability policies in community, agency and systemic settings. As the next chapter demonstrates, the emergent situation in the United States is not unlike the conditions encountered in 1992 in Tasmania's public education system.

9
Accountability Policy Research in Tasmania

Introduction
The context and methods of a policy research project conducted in Tasmania between 1992 and 1996 are reported in this chapter. The project aimed to produce educative accountability policy options, educative in the sense that they would identify processes and criteria that are generally believed to improve the quality of learning, teaching and leadership. The commissioning of the project did not include the issue of governance.

The research objectives were to identify accountability criteria and processes preferred by Tasmania's education stakeholders and locally-managed school communities, map the patterns of support, measure the intensity of support for each policy option and facilitate systemic policy making. The methods used attempted to take account of the advice provided in the chapters above concerning policy research. The deeper epistemological reasons for the methods selected and some of the insights gained by this approach to policy research are discussed in Chapter 10. The setting of the project is clarified first to contextualize the purposes and processes of the research.

The Policy Research Commission in Context
A memorandum was sent out in September 1992 by Graham Harrington (1992), Deputy Secretary (Education), to principals of all high, district high and primary schools, senior officers, chairpersons of schools councils and chairpersons of parents' associations. In it the Department of Education and the Arts (DEA) announced research intended to identify forms of accountability preferred by various interest groups. It was made clear that, 'in mapping forms of accountability, the project will address the criteria and the processes that are or should be used to identify, report on, and improve learning, teaching and leadership' (p. 1). The memorandum also specified that the findings would be 'reported in 1994 in a form which will assist school communities and the Department to review their accountability policies' (p.1). This concurrent approach to policy research and policy making has to be understood in a broader historical context of systemic policy review and development.

Attempts to improve accountability policies in the Tasmanian state education system have originated in local, state and national settings. They have been driven at various times in Tasmania's history by economic, political, administrative and educational concerns (Macpherson 1992c). For example, the general reforms to educational administration between July 1982 to May 1989 were driven by a political desire at state and national levels to introduce corporate management processes and by a state administrative agenda of generating efficiencies and greater effectiveness (Government of Tasmania 1981, 1982). Introduced or refined during that period were strategic planning, computerized accounting, information systems integrated with school-based programme budgeting, an integrated career structure and a senior management service. It was also a time of watershed curriculum development documents, such as *Secondary Education and the Future* (1984), which had sections written by the Minister of Education, the Hon Peter Rae. *Secondary Education and the Future* set the philosophical ground for *Our Children and the Future* (1991a), another still influential policy document. While there was a significant and progressive devolution of management responsibilities to schools and colleges and pioneer synthesis done on the concept of the 'self-managing school' (Caldwell and Spinks 1988), there was little movement of governance powers. As the end of the 1980s approached, and the resources available to education and the legitimacy of services began to fall again, pressures for a review of accountability policies rose. The Federal Government was one source of pressure.

Negotiations with the Commonwealth Government over resource agreements between 1986 and 1989 forced Tasmania to reconsider the introduction of performance indicators. There was substantial resistance encountered in Tasmania on educational grounds. Hocking and Langford (1990) reviewed the exploration of performance indicators in the DEA's Evaluation and Research Unit and reported that the concepts of 'formative evaluation' and the 'self-generated improvement of schools' were being deployed as a defence against the 'imposition' of an 'externally generated indicator-based accountability system' (p. 191). Performance indicators were reputedly seen as less threatening and more effective when developed in a context of teaching programme evaluation, planning and decision making in schools. The key problem they identified was that the 'parsimony of performance indicators fails to capture the full complexity of a given educational process' (p. 200). Performance indicators then played little role in accountability policy making until 1993 when they were re-introduced to help demonstrate student achievement.

Interest in the issue of accountability abated for a period when the Tasmanian Liberal Government fell at the May 1989 election. The need to economize on services became paramount. With the Labour-Green Accord in place, Labours' Michael Field formed a minority government to face an inherited financial crisis. The new Education Minister, the Hon Peter Patmore, commissioned Cresap Ltd to identify $18 million savings in the Education budget for the 1990–1991 year. Cresap's Final Report (1990, p. 26) attempted to justify the cuts by simply listing the 'inefficient areas' of system management as including planning, accountability, delegation, duplication, budgeting, management information systems, policy co-ordination, system performance audit, communications, human resource development, and procurement and supply arrangements. There was no research base cited with regard to these alleged inefficiencies. These claims were vigorously contested at the time to little effect. The net effects included schools having to 'down-scale' their staff by about 10%, schools devolving many administrative functions, all regional personnel being replaced by lean district offices, most systemic professional and curriculum development structures being cut, and the DEA having to reduce its capacities. It is fair to say that the meaning of 'self management' was elaborated during the process.

The politics of education in Tasmania were bitter in the early 1990s. Despite this bitterness, it was still found to be in everyone's interest to negotiate a system of staff appraisal. *Having* a staff appraisal system was a condition of receiving a Tasmanian Industrial Commission's salaries determination made in August 1990 that reflected the national context of award restructuring. The terms of the *Policy Statement: Staff Appraisal* (DEA 1991b) were negotiated by the DEA with the Tasmanian Teachers Federation and the Secondary Colleges Staff Association. The new practices were phased in from 1 January 1992. The dual purpose of the new appraisal system was to review and improve (a) teaching practices and (b) administrative and leadership services related to the quality of learning. Teachers and principals were expected to undergo appraisal once every four years, nominate an 'appraiser', 'critical friend' or 'mentor' (p. 3), negotiate processes and criteria, and document conclusions. The DEA was to provide training. Schools were to report progress by the end of 1996.

What happened fell well short of the agreed policy. The scheme was never formally mandated as DEA policy and implementation allowed to gradually run down. Data on criteria, processes and outcomes are yet to appear in the public domain. It was 'explained away' to me by senior DEA and Union officials in the mid-1990s as 'not fitting culturally' with other accountability policies. While some schools did develop staff appraisal schemes, and were still using them to educative effect in the mid-1990s, more and more came to realize that the staff appraisal policy had been assigned the status of a symbolic policy.

Other attempts to advance accountability policies in the Education portfolio were made in the early 1990s despite some deep political crises. Minister Patmore's rapid and tough decisions in the portfolio in search of economies eventually drew condemnation from the Greens. While he retained the confidence of his Premier, he resigned in 1991 to save the Labour-Green Accord and was transferred to another portfolio. Nevertheless, one of his initiatives, *School and College Councils: Interim Guidelines* (DEA 1991c), was published by the new Minister for Education and the Arts, the Hon Michael Aird. An unequivocal policy statement therein declared that the 'Government requires that every school and college establish a council' (p. 2).

The rationale for school councils included the need for broader participation in school decision making and policy making. The minimum role was to include the development of a school charter (a negotiated statement of values, beliefs, needs, goals and priorities), the approval of the school development plan, the approval of resource allocations that cohered with the charter and the development plan, and reviewing the school's charter and policies. The new policy (DEA 1991c) was also held to be a substantial devolution of governance powers. The proposed lines of accountability for the school council and the principal were, however, a recipe for systemic control and potential role confusion for school councils (p. 7):

The school council is accountable both to its school community and to the Minister of Education and the Arts through the Secretary of the Department and the District Superintendent.

As a Government employee, the principal is accountable to the Secretary for Education and the Arts through the District Superintendent.

In sum, the history of accountability policy making in the 1980s and early 1990s in Tasmania can be characterized as the moving outcome of the interplay between local, state and national agendas driven in different ways at various times by economic, political, administrative and educational factors. Despite cuts and symbolic policies, the decade culminated in a foreshadowed move towards localized accountability policies so that the devolution of governance might better match the pioneering development of 'self-managing schools'.

Tasmanian Accountability Policy Initiatives in the 1990s

In February 1992, Tasmania elected a Liberal (conservative) Government. It came into office with economically rational policies and immediately began to down-scale many ministries. Education was largely exempted from this process. Instead, the incoming Minister for Education and the Arts and Deputy Premier, the Hon John Beswick, initiated reviews of many policies in his portfolio. It became clear in the DEA that, while many schools had improved their planning capacities in recent years, there was also compelling evidence that evaluation processes were much less effective. The Minister then 'went soft' on the devolution of governance proposed by his predecessor, emphasizing instead a bureaucratic form of line management accountability.

In August 1993, the DEA published a new *Accountability Policy* for the Division of Education (DEA 1993a). It directed all sections and schools in the Division to establish processes of accountability that applied 'sound' planning, resource management, monitoring, reviewing and reporting practices, and used two basic criteria; 'the best possible education for all students with the most effective and efficient use of resources' (p. 1). Three operational principles were emphasized; reporting outcomes, a range of forms of evidence and the co-ordination of information from schools. While many of these terms were not defined, the new policy was fundamentally centralist; intended to ensure that 'coherent, consistent and realistic policies and practices are applied across the system' (p. 3).

Two implementation strategies were used. First, all sections and schools were directed to establish an 'accountability cycle' that had aims that were consistent with the DEA's strategic plan and policies concerning evaluation, reporting and school planning cycles. Second, each school's accountability responsibilities were spelt out in a series of policy documents; *Local School Leadership and Management* (DEA 1993b), *School Planning 1994* (DEA 1993c), *School Resource Package* (DEA 1993d) and the *Framework for Curriculum Provision K-12* (DEA 1993e). The role of school personnel was to implement these increasingly detailed statements and to consult with district superintendents whenever interpretation was required.

The accountability themes in these policies can be summarized. *Local School Leadership and Management* (DEA 1993b) made it clear that principals were expected to embed accountability processes and criteria in the annual planning cycle, provide accounts to the district office and the school community, as well as monitor, evaluate and report the outcomes of teaching, learning and resource allocation. In addition to these classic tools of bureaucracy, the presence of a systems model of organization became explicit (p. 8):

> if evaluation is for Central level, system-wide performance indicators will be used; each school should determine its own ... indicators for performance [which] need to be established for inputs, processes and outcomes ... [and] ... if a school fails consistently to produce outcomes which are commensurate with Central and local goals and priorities, the District will intervene in the management process.

Almost as an afterthought, *Local School Leadership and Management* added (p. 10) that 'the involvement of the School Council, Parents' and Friends' Association or any other appropriate parent body should be seen as integral to the local management and accountability of the school'. It is fair to record that most stakeholder groups remained marginal to all phases of the accountability policy making and implementation processes used.

A great deal of detailed advice arrived in schools to guide the administrator's hand. *The School Resource Package* (DEA 1993d) and the *Framework for Curriculum Provision K-12* (DEA 1993e) provided guidelines for the efficient and effective management of resources and indicators for demonstrating student achievement. *School Planning 1994* (DEA 1993c) explained how schools should satisfy accountability requirements within the planning cycle, meet system priorities, record particular information and provide evidence to a district superintendent to demonstrate that quality educational services were being provided to students. The main forms of evidence required were (a) a 'School Charter', a statement of vision and intent, (b) a 'long-term plan', a projection of priorities, activities and schedules, (c) a 'short-term plan' of programmes and objectives, and (d) an 'annual report' to the district office and the School Council, Parents and Friends or other representative body. There was a symbolic acknowledgement of other legitimate sources of policy ideas; all documents were meant to reflect the context of national, system, community and school needs and priorities. Other requirements for 1994 included the possible need to review of the state-wide sample testing of numeracy and reading, the need to provide information for system planning, to the Minister and for national reporting purposes, and the possible need to review reporting, pending the advice of the Tasmanian Education Council (TEC).

Minister Beswick had commissioned his independent policy advisory reference group, the TEC, to advise on parents' opinions concerning the nature and frequency of reports from schools and parents, the type of information parents preferred, and the extent to which reports on students should provide information about a student's performance compared with that of other students (TEC 1993, p.1). The TEC's survey gained responses from 2166 parents and another 21 extended responses from schools and school organizations. The diversity of views revealed how plural the Tasmanian education policy community was with regard to performance reporting and made it difficult for the TEC to generalize. Data were interpreted with care and subtlety.

In *Reporting to Parents*, the TEC confirmed that the most frequent requests from parents were for written reports once a term, formal parent/teacher interviews twice a year (essentially current general practice) and curriculum information sessions early in the year outlining the programme to be covered and identifying expected learning outcomes (comparatively rare in practice). General satisfaction was also recorded with recent initiatives, such as journals and folios, which had helped improve awareness of student learning. The perceived importance of early advice of educational or behavioural difficulties and collaboration between teachers and parents were also emphasized.

With regards to the data that parents actually wanted, the TEC reported that they valued accurate information on curriculum content, expected learning outcomes, their child's academic progress, their child's attitude, behaviour and social skills, their strengths and weaknesses and how they as parents could help their child learn. They also recorded (p. 8) that:

> there was overwhelming support for some type of comparative assessment and reporting. Parents, particularly in the primary sector, are keen to have some form of 'benchmark' by which to evaluate their child's educational development. They particularly stressed the need for having a statement of expected learning outcomes early in the school year, against which they could evaluate their child's progress during the year.

There was no evidence published about the demand for norm-referenced and standardized testing of numeracy and reading, then current practice when children were aged 10 and 14. The tests were known colloquially as 10N, 10R, 14N and 14R. On the other hand, the TEC went on to suggest that (a) parent education in the area of criterion-based assessment and (b) a central clarification of educational goals and operational guidelines, particularly for early childhood and primary education, would be seen as very helpful by parents.

To this latter end, and to the chagrin of some senior DEA officials, the TEC proposed a draft policy comprising 'educational objectives' and an 'educational framework' intended to ensure that the accountability procedures in Tasmanian schools were both flexible and effective. It is my personal view, confirmed by experience when I served on the Schools Board of Tasmania, that this chagrin showed that DEA officials preferred to be the sole conduit of policy advice to the Minister and used hierarchical practices to ensure that school personnel accounted solely to the DEA. The TEC took a very different view. *Reporting to Parents* (TEC 1993, p. 10) identified a major structural limitation to more effective accountability policies in schools:

If mechanisms such as this are going to work, they need to be under control of, and accountable to, a school-based authority. We believe that the best way for this to happen is through school councils. Therefore the Council is concerned that, in the proposed new *Education Act*, schools councils have not been made compulsory.

The Minister chose not to accept this advice or to act decisively on the other recommendations of the TEC Report. When *Interim Guidelines for School and College Councils* (DEA 1995) were derived from the recent *Education Act 1994*, school councils remained optional (p. 4) and were not given a serious role in accountability. The *Interim Guidelines* (p. 4) did oblige schools council to 'make school decision-making more accountable' but limited their role (p. 6, pp. 12–14) to 'developing the school charter, determining school policies and approving and monitoring school plans and school budgets' along with managing commercial operations such as hostels, farms, child care facilities and canteens. However, with regard to developing the school charter, plans and budgets, the principal and staff were awarded automatic leadership based on positional authority (p. 12). Each principal was given additional and over-riding obligations, responsibilities and powers to (a) implement DEA policies and instructions, (b) prepare plans, budgets and reports for the school, (c) manage the school including the welfare and professional development of teachers, and (d) implement any school policies developed by the school council. The *Interim Guidelines* also insisted (p. 14) that schools councils adhere to determinations made by the DEA Secretary concerning matters relating to discipline, curriculum, teaching practice, homework, assessment and reporting. Put simply, school communities were obliged to comply with ministerial reserve powers held and exercised personally by the most senior line manager in the bureaucracy or by his nominees; other line managers.

This also meant that Tasmanian schools councils did not acquire a democratic accountability role in the most recent legislation. The *Interim Guidelines* did not provide processes and criteria by which they might monitor, evaluate and improve 'school decision-making'. Instead they appeared to legitimate imbalances of power between immediate school community stakeholders and defined school council structure in little more than symbolic democratic rhetoric. While the possibility of some local exceptions that persisted, schools councils had little more that policy advisory powers, tended to work for and report to their principals, who in turn, reported to district superintendents. The continuing absence of mandatory requirements gave principals an informal veto on the establishment, the continued existence, the responsibilities and the operations of their school council. This raised the real possibility that the legislation under-cut the broader public legitimacy of schools councils and retarded the broader evolution of local governance capacities in Tasmanian public education.

How was this preference at state level for line management accountability rather than school community accountability justified? Minister Beswick and the DEA regularly expressed the view that local demand and professional support for school councils and available public expertise varied widely. In such circumstances, they argued, an evolutionary approach was more likely to be effective in the longer term, and less controversial in the short term, than the use of coercive strategies. A sole and relatively junior DEA official was appointed to support the formation and development of schools councils, and the Minister and his senior officials at central and district levels occasionally encouraged principals to take a pro-active role in the formation and development of schools councils. There was one much cited concession in the Act. The *Education Act* (1994) enabled any group to approach the Minister of Education with a proposal to set up a school council.

Tasmanian Accountability Structures and Processes

The recent advances in Tasmanian accountability policy making described above occurred in a context of steady state funding and relative political peace, although, as it is shown below, there was substantial national pressure building in the 1990s for the introduction of outcomes-based performance indicators and the testing of learning in order to better justify the investment of resources in state education. In some contrast, Tasmanian DEA policy reviews tended to focus more and more on educational issues, despite the presence of policy documents that detailed line management responsibilities. Another dilemma was that accountability processes and criteria concerning the quality of learning, teaching and leadership tended to be partitioned as policy issues and were clearly at very different stages of development. One implication here is that the research commissioned and reported here must be seen as part of an ongoing state effort to generate more comprehensive and coherent systemic accountability policies and to better co-ordinate the management of evaluation between schools, district offices and state levels. Another is that there was a broad awareness that accountability was shaping up as a litmus issue that would severely test the centralized nature of policy-making capacities of the DEA and the increasingly restive principals of 'self-managing' schools. In the time-honoured way of bureaucracy, a structural response was devised.

Minister Beswick and senior DEA officials consulted extensively with the teachers' unions, principals' associations and parents and friends' associations in 1992 and 1993 prior to announcing that an Educational Review Unit (ERU) would be established in Central Office. The Secretary of Education and Chief Executive Officer (CEO) (Davis 1993) explained its establishment in terms of the need for the central co-ordination of review processes, while noting that district offices would continue to plan and manage school and college reviews. An Assistant Secretary (Review), Dr Jan Baker, was appointed to lead the ERU and to report directly to the CEO in order to be structurally independent from service delivery being managed by the Deputy Secretary (Education). Since the Education Secretary was an architect by profession, this gave the Assistant Secretary (Review) near identical capacity to influence systemic evaluation and development policies as the Deputy Secretary (Education), hitherto the highest placed educator in the bureaucracy. Further, the key function added by the new ERU was reporting on the quality of the learning outcomes, supplementing the planning cycles at all levels. Performance indicators for the ERU were also provided (Davis 1993, p. 4):

An Educational Review Unit is only useful if it brings about an increase in the quality of the educational outcomes in schools. Increases in quality need to be demonstrated by processes of measurement which:
ensure that the curriculum ... is taught to standards whose criteria are defined ...
check to see if the children are receiving the core education required by the Government ...
investigate whether or not children are learning to agreed levels of achievement; and
direct teachers to strategies of intervention where assessment shows children's learning does not reach these levels.

What appears in such documents and what happens in practice can be two different things. In this case the ERU began work by monitoring and reporting on the implementation of policies; the integration of special education students, the outcomes of the educational outcomes of the K-12 health curriculum, schools' reports on the learning outcomes of K-8 mathematics, new assessment standards for K-3 literacy. Of immediate relevance to this research is that it nurtured a fresh policy discourse concerning accountability between senior officials, district superintendents and principals and between staff in schools.

This discourse increasingly focused on the co-ordination and development of the school review process. The first of six major functions of district offices was to 'review and monitor each school's progress in the implementation of an approved school plan and facilitate change consequential to the review' (Bowen District, DEA 1996, p. 1). A typical school review process began with an internal review considering indicators and preparing answers to 'focus questions' posed by the district superintendent. The questions were standardized across the state. This internal review of plans and achievements was often regarded as the most valuable component of the process, suggesting that it was tending to be seen less as part of an ongoing process of organizational action research and more as a means of satisfying external demands for accountability.

The school inspection panel, comprising the superintendent and a paired principal, typically interacted with the senior staff, programme managers, some school councillors, parents and students over a few days. Some review teams made classroom visits. Superintendents later provided reports in less than two pages setting out achievements in the three priority areas nominated and specifying agreements and recommendations for future directions.

The ERU also encouraged formative evaluation of the school review process itself. In the Bowen District (DEA 1996, p. 2), for example, consultations with principals and school communities found that the school review process stimulated valuable reflection, further planning and celebration. Four areas for improvement were also identified; (a) the questions, indicators and role of paired principals, (b) where appropriate, the need for follow up action and a second visit, (c) where appropriate, the need for staff release resources, and (d) the need for new priorities for review in 1996 (science, parental participation and school councils, plus a school-nominated area). These district evaluations of school review processes co-ordinated by the ERU culminated in revisions to the *School Plan Implementation Review 1996* (DEA 1996) so that it would better attend to the needs of 'under-performing schools' in 1997. I come back to this matter in Chapter 15.

On the other hand, the same formative evaluation processes had made it clear, by 1995, that some long-standing staff and school evaluation processes in use had serious technical limitations. The 'stand alone' staff appraisal scheme mentioned above, for example, was originally intended to review and improve teaching, administrative and leadership services related to the quality of learning. It failed to provide public criteria and data that could have served accountability policy review purposes.

Conversely, some accountability policies had high technical merit but remained distant from daily professional practices and client interests. The *Accountability Policy*, for example, located accountability processes within the planning cycle of each section and school. This presumed that these planning processes would translate (via formative evaluation) into improved professional practices, better learning and more effective forms of public accountability. Accruing evidence across districts suggested that such an assumption was widely doubted in schools. And while the *Local School Leadership* and *Management and School Planning 1994* policy documents and the new *Education Act* (1994) strongly emphasized the educational leadership role of principals, it was less clear how the forms of evaluation required to nurture and measure improvements in quality were to be designed and introduced. In sharp contrast, the new structural arrangements were blunt in their clarity. While the line accountability roles of principals, district superintendents, section managers and senior executives were sharpened and reinforced, roles that reflected classroom, school community, public and consumer interests and responsibilities were rendered more obscure and marginalized.

There was, for example, limited encouragement for school communities to build accountability capacities. Apart from the evolving school review activity, they were simply given goals, priorities, and accountability processes and criteria. *The School Resource Package* and the *Framework for Curriculum Provision K-12*, for example, provided the criteria and the processes to be used for evaluating the use of resources and students' learning achievements. There were no criteria or processes offered concerning the quality of teaching, leadership or governance services. They were, by formal omission, defined as inadmissible policy issues. District offices were unable to significantly broaden the agenda; they were obliged to provide line management and monitorial services in addition to advisory support. There also seemed little possibility that federal authorities would encourage accountability capacity building in school communities. Sadly, as the TEC pointed out, the legitimacy of state education was unlikely to advance in locally-managed schools while local governance structures were destined to play such a negligible role in accountability.

National Accountability Agendas in the 1990s

Many problems in Tasmanian public education have extra-state origins. As noted above, Tasmania has often been encouraged by Commonwealth agencies, using their resource power as leverage, to reconsider performance indicators and to provide even more detailed outcomes-related accounts of expenditure. Some brief examples illustrate the phenomenon. Meetings of State Premiers and the Commonwealth (federal) Prime Minister in 1995, at the Council of Australian Governments (COAG), agreed to publish performance indicators of effective and efficient schools. The parallels with the symbolic reforms of US presidential politics were striking.

COAG wanted data collected that would permit inter-state comparisons of school performance and to compare Education against performances in seven other portfolios. These demands were part of a wave. In the same year a federal Senate Select Committee collected evidence on how to improve the means by which states account for their use of Commonwealth funds. Further, a federal Joint House committee conducted a 'Public Accounts Inquiry' with a special focus on the administration of 'Special Purpose Funds', including those dispersed to support state schools. National testing of literacy were proposed in the Commonwealth Department of Prime Minister and Cabinet (1994) report, The Working Nation. The Commonwealth's Department of Employment, Education and Training (DEET) was asked to negotiate a national testing scheme with the states.

While most states accepted that federal authorities and COAG have bounded rights to devise accountability policies concerning the public education they fund, it was considered by many states, including Tasmania, to be quite another matter as to whether the purposes, criteria and processes they were proposing were as appropriate as they might be. The DEA took the view that this national policy context has consistently preferred summative, political, systemic and quantitative forms of public accountability, principally to help justify the use of Commonwealth resources, to create greater economies and to provide the federal government with greater steerage over state's education policies. The key problem to Tasmanian educators, it was argued in the DEA, is that this trend moves the focus of accountability criteria and processes away from formative, educational, local and qualitative forms of evaluation related to learning, teaching and leadership.

This criticism appears to be warranted. Key emergent criteria in the initiatives listed above include comparative indicators of efficiency and effectiveness of schools, tests scores in literacy, the comparative return on investment by state, the value adding capacity of schools and the economic returns to the individual (Yeatman 1990). Displaced by this national policy trend to commodification, vocationalism and quantification (Pusey 1991) were the criteria related to the enhancement of learning conditions, professional development and school improvement. It was therefore understandable that states have sought to resist some 'federal fishing expeditions', developed data and processes that might do 'double duty' at both school and national level while attempting to steer national accountability initiatives into serving more educational purposes.

Research Approach and Questions

In this context, and despite the presence of deeply embedded systems theory and line management legislation, policies and practices, the Tasmanian DEA provided public schools with two fundamental policy directions on accountability processes and criteria (Davis 1993, pp. 2–3). The first was that schools should account in the first instance to their communities as they contributed to systemic accountability. Second, such accountability should be couched primarily in terms of student learning outcomes. The practical research challenge identified by the DEA in 1992 was to identify accountability processes and criteria that might serve such educative ends.

Discussions with the senior officials in the DEA determined that the general two-part policy research task was to identify educative forms of accountability preferred by the immediately responsible groups in Tasmanian state education; teachers, principals, parents and educational administrators, and to further the development of state education policies. 'Educative forms of accountability' were defined as the processes and the criteria that should be used to collect data, report on and improve learning, teaching and leadership. Hence, two general research questions were used; (a) what processes (procedures, actions or methods) should be used to collect data, report on and improve students' learning, teachers' teaching and leaders' leadership, and (b) what criteria (standards, benchmarks or indicators) should be used to evaluate students' learning, teachers' teaching and leaders' leadership?

There were several often concurrent phases designed to add a systemic action research process to the DEA's internal policy review processes managed by the ERU; the qualitative description of the problem in school community and systemic settings, adding international resources to the Tasmanian policy process, the more systematic articulation of policy preferences by stakeholder groups, the inclusionary refinement of a survey instrument, quantitative data collection, the collective interpretation of qualitative and quantitative findings, and participating in ongoing school community and systemic accountability policy making and improvement processes. These processes are now briefly discussed.

The Initial Description of the Policy Problem

The first phase involved careful consultations with senior officials in the DEA and the Union, with leading office bearers in professional and parents' associations and with a widening circle of teachers and students. The first contact with Secretary Bruce Davis and Deputy Secretary Graham Harrington occurred in February 1992. They explained that the early devolution of management to Tasmanian public schools had been overtaken by the Cresap Report (1990), stripping out so much of the regional accountability infrastructure that it was proving much more difficult to report on the achievement of state and federal policies, in such areas as excellence and equity, to evaluate and plan the use of scarce resources in a sophisticated way, to boost the improvement of schools and to identify the schools deserving extra support in a reasonable and effective manner. They also noted that a small number of principals were confusing 'school self-management' with 'school autonomy', despite the change in official rhetoric to 'locally-managed schools'. This vocal minority was reputedly conceding neither the administrative authority of the DEA nor the need to prove publicly or improve the quality of schooling. Above all, the senior DEA officials were deeply concerned at the effects of the national accountability policy agenda noted above and wanted to find a more educative balance to Tasmania's accountability policies.

About eight months passed before I felt that it was appropriate to have an official memorandum issued by the DEA formally inviting stakeholders to participate. By then I had consulted with most of the senior DEA officials with functions related to accountability, most of the district superintendents appointed post-Cresap, many of the executive members of the Australian Education Union including their Tasmanian President, Ms Penny Cocker, the Presidents and many of the office bearers of the primary and secondary principals associations, and the office bearers of the Tasmanian Council of State School Parents' and Friends' Association, then led by Julie Roberts and Ivan Williams.

There were three key purposes to these focus group and interview-based consultations. First was to gain trust and support for the policy building project. Second was to negotiate an interactive process with each group for such participation. The third was to collect preliminary baseline data concerning concepts and views concerning educative accountability criteria and processes. All three purposes were achieved by August 1992. The weight of preliminary data gathered suggested that three dominant views prevailed, as summarized in Table 9.1.

Two problems immediately apparent were that the views appeared to be mutually exclusive, and, since they were buttressed by deep beliefs about other stakeholders' views, quite unlikely to change. Also evident were the similarities between these three perspectives and the models in competition identified by Elmore and Associates (1990) and summarized in Table 2.2. It will also be shown in later chapters that these three perspectives were the basis of myths held about the positions taken by other stakeholders on particular issues. For example, with regard to use of the standardized tests of reading and numeracy, with exceptions, teachers believed that they had been introduced by the DEA to evaluate the quality of teaching. DEA officials believed that parents strongly supported their continued use. Parents thought they were used by the DEA to monitor the performance of schools. To jump ahead momentarily, none of these myths were found to be sustained and the use of the tests was suspended. It was also shown that such powerful myths can suspend the growth of policy knowledge until research findings refute their validity.

Two tentative proposals follow that will be picked up at the end of this chapter. Policy knowledge appears to grow in people's heads; not through hypothesis testing and proof by data but by conjecture and refutation during interaction. Policy research might, therefore, usefully consider methods that identify the unrealized extent to which theories in use already overlap, despite the existence of what appear to be ironclad and mutually exclusive perspectives, enlarge current areas of overlap found, and create new areas of such touchstone. I take up this issue in Chapter 10.

To summarize to this point, by September 1992, the first phase of the research process had built support for the project in all stakeholder groups represented at systemic level, negotiated an interactive process of participation and mapped three theories of accountability that appeared to be mutually exclusive. The next phase was intended to provide a more detailed and qualitative description of accountability in

Table 9.1: Dominant perspectives on accountability policy in Tasmanian education 1992

Perspective	DEA officials	Teachers	Parents
Primary purposes	Prove and improve the quality and equity of schools and the system.	Demonstrate, reiterate and celebrate professionalism.	Publicly demonstrate the quality and fairness of services to clients.
Best target for reform	Curriculum, teaching, leadership, assessment	Professionalism of and evaluation in schools teachers and school leaders and resource policies.	The power relationships between providers and clients.
Conditions crucial for improvement	Teaching, learning and leadership to be based on DEA published policies.	Teachers and team leaders to develop their judgement and control their own work.	Principals and teachers to account directly to parents and their community.
Best source of criteria	Principles of line management.	Principles of professionalism.	Clients' experiences and consequences of outcomes.
Most appropriate accountability processes	System sets policies, priorities purposes and performance indicators. Use objective performance and outcomes data to use in the next planning round. Embed accountability in school planning and development strategies.	Professionals empowered to review school policies and priorities. Expert planning and co-operative teaching and learning. Collaborative staff action research reported to parents and system.	School councils to review outcomes and govern school policies, evaluation and development plans. Stakeholders participate in school planning and evaluation of learning, teaching, leadership and governance.

school community and systemic settings, articulating more systematically the policy preferences held by stakeholder groups in the hope that policy touchstone might be found, developed and created through dialogue.

The Detailed Description of The Problem

It was noted above that much of 1992 was given over to building trust, consultative processes and an understanding of basic perspectives. The balance of 1992, all of 1993 and a good part of 1994 was used refining the views of stakeholders, converting them into policy preferences and checking that this process both included all ideas and represented proposals accurately. Extensive use was made of the language of respondents. Despite the strongly supportive involvement of five research associates, namely Christine Barber (1992), Helen Davies (1993), John Ewington (1996), Margaret Taplin and Sue Wilson (1994), progress was affected by my having to unexpectedly provide departmental leadership services in the School of Education, University of Tasmania, from late 1992 until early 1994.

Focus-group workshops were used to gather and discuss qualitative data across the state 1992–1994. Whether the setting was a lunchtime workshop with the entire staff of a secondary school, an after-school meeting of senior primary school staff with their school council, an evening meeting with the parents and students of a primary school, part of the meeting of the state Executive of the Australian Education Union, small group discussions with students and teachers in a secondary school, an afternoon and evening meeting of all interested stakeholders in a district high school community, an evening or weekend gathering of parents' representatives in Hobart or Launceston, district gatherings of all primary and secondary school principals, or policy research seminars for senior DEA district or central staff, the data-gathering processes were very similar.

The purposes of the research were explained as an attempt to develop educative accountability policies, that is, accountability processes and criteria intended to evaluate, report on and improve the quality of learning, teaching and leadership. Processes were defined as procedures, actions or methods used to collect and report data. Criteria were defined as standards, benchmarks or indicators used to evaluate data. Leadership was defined as the sum of leadership services given to a school community, including the leadership provided by senior staff, teachers, students and community members.

A routine was developed and used in all settings to trigger the release of qualitative data. Participants were asked to suggest *processes* that should be used to find out about, report and improve students' learning in schools. A preliminary list of about three methods was recorded on a large blackboard, a white board or butchers paper. The second question asked participants to identify *processes* that should be used to find out, report and improve teachers' teaching in schools. A second brief list of about three procedures was recorded below the first. The third question asked participants to suggest *processes* that should be used to find out about, report and improve leaders' leadership, with the third list of actions being added. It was quickly found that any provision of examples biased subsequent responses and the practice was discontinued. At no point were suggestions refused or modified. All data were accepted.

Participants were then asked to suggest *criteria* that should be used to evaluate students' learning. A few suggestions were listed to the right of the short list of learning processes proposed earlier. Similarly, brief lists of desirable *criteria* concerning teachers' teaching and leaders' leadership were recorded. At this point the framework was laid over the brief lists and clarified. Free-response schedules (see Table 9.2) were then distributed to all participants with a request that they copy the items already on display that they agreed with, form social groups, and, through discussion, develop and record other processes and criteria that they would personally recommend.

Table 9.2: Open-response qualitative data schedule 1992–1994

I Recommend	**Processes** Procedures, actions or methods	**Criteria** Standards, benchmarks or indicators
To report, improve and evaluate **Students' learning**		
To report, improve and evaluate **Teachers' teaching**		
To report, improve and evaluate **Leaders' leadership**		

Groups typically worked well over the hour at this task and then always stayed on for a plenary process that summarized the data in each of the six cells in Table 9.2. These data gathering processes also provided many opportunities for clarification, verification, comparisons and validation. Through feedback they were found to be reliable over time and between comparable cohorts. The interactive processes were designed to satisfy the criteria for the conduct of 'naturalistic' research (Crowther and Gibson 1990, p. 46) summarized in Table 9.3.

Once the views of stakeholder groups represented at state level had been gathered, a one eighth stratified sample of all Tasmanian school communities were selected randomly for the collection of qualitative data from professionals. This excluded schools that served regions, such as special schools and secondary colleges (Years 11–12 only). There were 209 primary, district high and high schools in Tasmania serving local communities in 1992–1994. The sample of schools was structured to be representative of size, region, socio-economic status (as measured by the DEA's Educational Needs Index) and relative isolation from urban settings. Twenty-eight schools were selected; five high schools, four district high schools and 19 primary (elementary) schools. In each school the principal was asked to respond to the open-response schedule and to invite a sample of five other teachers to take part, the sample to traverse subject area and experience in secondary schools or grade level, and experience in primary schools. All responses were returned in confidence and rendered untraceable.

Forty-two percent of the open-response schedules distributed to principals and teachers were returned and deemed usable. Telephone follow-up could not find any untoward explanations for non-response. Content analysis was then used to identify categories that were mutually exclusive so that the coding of responses in each of the

Table 9.3: Crowther and Gibson's criteria for the conduct of naturalistic research

Scientific criteria	Naturalistic criteria
Internal validity	*Credibility.* Do research subjects find the researcher's views of his or her subjects' responses clarified by prolonged probing during interviews? Have the researchers returned to the data source for confirmation of categories produced during data analysis?
External validity	*Transferability.* Does the researcher, in interpreting and reporting data, make significant quantities of 'thick description' public?
Reliability	*Dependability.* Are the procedures like 'investigator triangulation' employed to ensure that the researcher's values, beliefs and actions can be subjected to close scrutiny and questioning?
Objectivity	*Confirmability.* As conclusions are derived, are original data re-examined to ascertain consistency or to stimulate further analysis? Is there a clear 'audit' trail between raw data and conclusions?

six cells became a clerical task. The rationale and process used to identify each of 73 different proposals across all six cells is described in detail by Wilson (1994, pp. 60–70).

These proposals were then converted into Likert items and randomly distributed in a pilot questionnaire trialed with 12 teachers. Many minor improvements were made and the 73-item instrument was then administered to a second stratified sample of professionals in 28 schools using the same procedures. The response rate jumped to 68%. Very few improvements could be made to the 73-item instrument on the basis of their responses.

At this point another 61 items were added from the parallel process of gathering and synthesizing proposals (and testing them as items) from all other stakeholders. All proposals from all stakeholders were included and validated by iterative and consultative processes. Since some editing was required to avoid excessive demands on respondents, some slight loss of reliability might have occurred.

The final Accountability Policy Questionnaire (APQ) had 134 items. It was shown by Wilson (1994, pp. 74–78) that there were five possible sources of bias; the semantics of the open-ended format, the structure of some questions, the face validity of some items, the sampling procedures, and the non-face-to-face nature of surveying. Conversely, the iterative and qualitative methods of category building, the constant comparison of concepts and proposals, the step-wise trialing of items and instrumentation, and the two-year long interactive process of consultation generated policy proposals that appeared to represent the totality of recommendations from all stakeholders.

Measurement of Support for Policy Preferences

It was noted above that policy preferences were gathered and processed as qualitative data by interviews, focus-group workshops and feedback sessions. Stratified and opportunistic samples of parents, teachers, principals and system administrators were used. The qualitative data were used to develop and refine an instrument which was then used to measure the intensity of support in each group. In this section I explain how quantitative data were collected and instances of non-response analysed.

The APQ was administered by mail in August 1994. Again a one-eighth sample of 28 school communities was selected randomly, excluding those school communities that had provided qualitative data or assisted with questionnaire trials. The sample was once again structured to be proportionately representative of type, size, rurality, educational needs of students and isolation. In each selected school, the principal was asked to invite the ten parents and ten teachers most interested in educational policy making to participate. Given their actual and potential responsibilities concerning school and systemic accountabilities, all primary, district high and secondary principals, all district DEA personnel, and all central DEA personnel with schools-related functions were surveyed.

The samples, responses and response rates by group and subgroup are shown in Table 9.4. The sampling regime ensured that all types of schools were well represented and that district high school parents, teachers and principals were slightly over represented. This was seen as warranted since district high school populations are small yet relatively diverse. District high schools also tend to be in relatively conservative and rural locations, more often facing questions of viability, and more closely connected with their communities than urban schools. Teaching, learning and leadership are, therefore, more transparent in district high schools, the schools of Tasmania that tend to have the least experienced leaders and staff.

An analysis of non-response to the APQ was conducted by telephone interviews of four primary, one district high and two secondary principals, after they had made

Table 9.4: Responses to the Accountability Policy Questionnaire, August 1994

Group/subgroup	Sample size	Usable responses	Response (%)
Primary school teachers	190	79	41.6
District High School Teachers	40	28	70.0
Secondary School Teachers	50	25	50.0
All Teachers	280	132	47.0
Primary School Principals	150	89	59.3
District High School Principals	25	15	60.0
Secondary School Principals	34	21	61.7
All Principals	209	125	59.8
Primary School Parents	190	96	50.5
District High School Parents	40	28	70.0
Secondary School Parents	50	24	48.0
All Parents	280	148	52.8
District DEA Personnel	14	10	71.4
Central DEA Personnel	31	16	51.6
All DEA Personnel	45	26	57.8
Totals	814	431	52.9

discreet inquiries among parents and teachers. There were many, reasonable and unique explanations for individuals not responding to the APQ reported. One shared reason given was fatigue and low morale among teachers induced by 'innovation overload'. Another shared reason was that some teachers found the concepts of professionalism and accountability ideologically incompatible, particularly in primary schools and when related to the quality of teaching. A third shared reason given for not responding to the questionnaire was the length of the instrument. An early decision to use an inclusionary approach meant that support for 134 policy proposals had to be measured. And while non-responses to items in usable questionnaires were typically about or less that four (1%), one item was not responded to by 15 respondents. It was judged to be ambiguous and all responses to it were set aside. These problems aside, it was the considered view of the DEA's project reference group and other stakeholders' leaders, that the systematic articulation of policy preferences had been achieved.

The Ongoing Review of Wider Knowledge

A concurrent literature search was used to identify the nature of accountability policies in English and Welsh, North American and Australian systems of 'self-managing' government schools, placed in the public domain and presented to the public at a Honours research seminar (Barber 1992). This study found that policy makers all three settings were concerned with the increasing costs of public education, doubts about standards and professionalism, the need for more decentralized and participative forms of accountability, the falling credibility of tests and the strengthening demand for multiple indicators. It concluded that Tasmanian system of public schools needed a new suite of coherent policies that encouraged educators to demonstrate and improve the quality of schools, to show and develop their professionalism, and to allow a broader range of stakeholders real roles in order to boost the legitimacy of public policies in education. There was a good deal of interest but little immediate response to the public reporting of these findings and recommendations.

Between 1992 and 1996, drafts of a number of papers were distributed for comment prior to publication. The first was a justification for the methodology of the policy research project (Macpherson 1996b). The second contested the view that accountability is a reasonably 'politically incorrect' concern among educators (Macpherson 1996a). The third summarized some of the persistent dilemmas in accountability policy making (Macpherson 1994a). The fourth provided interim policy findings (Macpherson 1996c). The fifth brought particular competencies, performance indicators and the need for professional development to the attention of principals (Macpherson and Taplin 1995). The sixth asked Tasmanian educational administrators to take stock of the extent to which they were providing educative accountability (Macpherson 1996d). These papers were intended to alert stakeholder leaders to the broader literature and ideas relating to accountability policies and policy making. Simultaneously doctoral research was investigating and providing occasional progress reports concerning parents' perceptions of the effectiveness of Tasmanian public schools (Ewington 1996).

As the project developed, participating groups and schools regularly gained access to new information concerning accountability policy preferences. The preliminary patterns of responses from each stakeholder group was distributed with permission to all other stakeholder groups. Each school community and stakeholder group received feedback within the month on the categories derived from the data they had contributed. All interim analyses of trial data went back to the contributors. In each case the data was delivered personally, to assist with interpretation, or was accompanied by an offer for such an interpretation, if so requested. It might be noted that Tasmania is reputed to have the highest proportion of practising teachers in Australia with masters level qualifications. Many professional and stakeholder groups made it clear that they were well able to manage their own interpretation of data, and moreover, occasionally suggested improvements to research methods. On the other hand, the weight of casual and regular feedback suggests that most school communities did not use the research findings on accountability in a systematic manner.

With regard to the analysis, feedback and interpretation of the responses to the APQ, measures and classifications of support for policy options were negotiated. The percentages of respondents in each group and subgroup expressing strong agreement (SA), agreement (A), not sure (NS), disagreement (D) and strong disagreement (SD) were calculated. The responses SA, A, NS, D and D were assigned the values 1–5 to permit means, modes and standard deviations to be calculated. 'Support' in a group or subgroup was defined by stakeholder leaders as more that 70% indicating SA or A. This rule of interpretation was agreed to indicate 'political significance'. It had far greater fecundity in this policy community than 'statistical significance'. Similarly, when the total percentage strongly agreeing and agreeing with a policy option was 30 to 70, support was deemed to be 'ambivalent'. Where less that 30% of a group agreed or strongly agreed to a policy proposal, the item was held to be 'unsupported'. Where a policy option was simultaneously supported, regarded with ambivalence and unsupported by different groups, the different profiles of support were ranked by means and displayed. Despite the prevailing consensus valuing 'political significance', the differences between each stakeholders' means and all respondents' means on all items were tested for statistical significance using the t test. For example, the concordance or differences between principals and all other groups were easily explained using this method along with implications; educative performance indicators, key competencies and issues requiring professional development (Macpherson and Taplin 1995).

Collective Interpretation of Findings and Ongoing Application

Interpretation occurred simultaneously at three levels from late 1994; central (DEA and stakeholders), in districts, and in school communities. The preliminary research findings were presented in a report to the Secretary of the DEA on 21 October 1994 at a seminar for senior executives (Macpherson 1994a). The report reviewed the project's purposes and processes, explained the instrumentation used, provided detailed analysis of the extent to which each stakeholder group and sub-group supported each policy option, and summarized the patterns of support, ambivalence and non-support. The meeting agreed unanimously that:

- similar presentations should be made as soon as possible to the Union's and Parent's executives by arrangement as well as to district superintendents, principals' association meetings and state conferences and to school communities on request,
- the ERU and the School of Education of the University of Tasmania would co-host a day-long policy seminar on campus for stakeholder executives, all DEA district and central staff with functions related to accountability, and representatives from principals' association, in April 1995,
- summary and display materials were to be prepared by the ERU to convey the essence of the findings to school communities in a manner that would help them review their accountability policies,
- further participation by the research team in accountability policy review at central, district and school level may occur thereafter by request, and
- the findings and processes of the project may be reported to professional and academic audiences and that they were to be related to other state, national and international policy research initiatives.

The decisions were conveyed to all respondents and schools with an interim summary of the main findings that was later published (Macpherson 1996c). While the ERU did not distribute display materials for schools, it moved straight into the organization of the seminar for state educational stakeholder representatives. Held on the 30 April 1995, it was organized to explore the implications for system accountability policies and policy making. The state executives of the major interest groups attended. The public domain components of this seminar were recorded and transcribed for text analysis. The way that participants handled the refinement of policy ideas is discussed in coming chapters.

The implications for the pre-service education of teachers and in-service professional development were considered by the University of Tasmania and the DEA. The competencies of principals proposed by the national principals' associations were reviewed in the light of the findings at an inter-state workshop hosted by the DEA, 29–30 March 1995. More detailed analyses were similarly presented to all other stakeholder groups in 1995 and 1996. By 1996, seminars given to the district superintendents and Treasury officials indicated that they were applying the findings to the formative evaluation of schools and systemic service delivery. It was then considered time to gather the project's materials, findings and implications into one text and make them available to interested wider audiences.

Before leaving the research process to focus on epistemological issues, it is important to note that the three-year co-operative policy research process was commissioned in January 1992 by those who manage the delivery of educational services in the DEA (Harrington 1992), prior to the establishment of the ERU. While strong interest continues from this sector of the DEA, the third and ongoing phase is being managed by the ERU. Three questions will be of continuing interest. To what extent is the ERU able to sustain policy review and development concerning accountability in school communities with limited local government structures in schools? What are the benefits and challenges that attend the continuing co-operation between delivery and audit functions in the DEA over the design and review of accountability policies? What impact will the policy process managed by the ERU for the DEA have in state, interstate and national settings?

Summary

In this chapter I have described in some detail the practical methods used to explore the key research issue; the accountability processes and criteria that should be used to collect data, report, evaluate and improve the quality of students' learning, teachers' teaching and leaders' leadership. The selection of methods was informed by the problems encountered in the practical policy context of Tasmania and policy research elsewhere. The methods were added to systemic policy review processes facilitated by the ERU and were, in substance, systemic action research processes.

It was demonstrated that accountability policy research in Tasmania had been moderated but not determined by historical, economic, political, administrative, ideological or educational concerns.

Accountability policy content at the end of the 1980s foreshadowed more localized accountability policies in order to obtain greater coherence between the advanced devolution of administrative powers and responsibilities and the reluctant devolution of governance. The incoming Liberal Minister preferred a bureaucratic form of line management accountability and subsequent policy documents in the 1990s embedded the processes of accountability in planning, resource management, monitoring, reviewing and reporting functions. Operational principles stressed proving rather than improving education; reporting outcomes, a range of forms of evidence and the co-ordination of information from schools. The new policies were fundamentally centralist and tried to standardize policies and practices across the system. Schools were directed to establish an 'accountability cycle' that cohered with the DEA's strategic plans and evaluation activities. School personnel had to implement these detailed statements in consultation with district superintendents. These classic tools of bureaucracy were underpinned by a systems model of organization. School communities were obliged to comply with the powers exercised personally by the most senior line manager in the bureaucracy or his nominees.

Tasmanian schools councils did not acquire a serious accountability role in the most recent legislation. Their structures were defined in symbolic democratic rhetoric. The legislation added little to the broader public legitimacy of schools councils and retarded the evolution of local governance capacities. Despite this, there was a growing awareness that accountability was a litmus process of democracy and a systemic strategy needed to both prove and improve the quality of schools. A structural response was devised. The ERU was established to co-ordinate the school and college review processes to be planned and managed by district offices. On the other, the legitimacy of state education and capacity building continued to be restricted by the negligible role given to accountability in limited local governance.

National policy communities have shown even less interest in school community capacity building, calling instead for hard-edged performance indicators and detailed outcomes-related accounts of expenditure. The national trends to commodification, vocationalism and quantification have contrasted with a sustained state interest in the enhancement of learning conditions, professional development and school improvement. In this context the DEA decided that research should identify educative forms of accountability preferred by the immediately responsible groups and further the development of state education policies.

Two research questions were used; (a) what processes (procedures, actions or methods) should be used to collect data, report on and improve students' learning, teachers' teaching and leaders' leadership, and (b) what criteria (standards, benchmarks or indicators) should be used to evaluate students' learning, teachers' teaching and leaders' leadership?

There were several policy research phases designed that added an action research process to the DEA's internal policy review processes managed by the ERU. The 'problem' of accountability was described in qualitative terms by stakeholders and school communities. International resources were added to the process. All policy preferences articulated by school community and stakeholder groups were refined across groups and converted into Likert items. The APQ then measured the extent to which each stakeholder group, including administrators at all levels, supported each policy proposal. A series of policy seminars facilitated the collective interpretation and application of findings in central, district and school settings. Meetings with district superintendents and Treasury officials in 1996 related the findings to the formative evaluation of schools and the quality of systemic service delivery systems.

Finally, it was noted in passing that the selection of methods tried to take account of the problems encountered in policy research elsewhere as well as the practical contingencies in the unique setting. The methods were selected to supplement existing systemic policy processes, especially the work of the ERU, and exhibited most of the features of action research. This combination of strategies reflected two tentative proposals made in passing about the growth of policy knowledge. First, it was noted that policy knowledge seems to grow not so much by hypothesis testing and through the provision of empirical proof, but by an interactive process of conjecture and refutation. Second, this implied that policy research might usefully proceed by (a) identifying the unrealized extent to which theories in use already overlap, despite the existence of apparently mutually exclusive perspectives, (b) through negotiation, developing the areas of overlap found, and (c) through conjecture and refutation, creating new areas of touchstone. These are very important matters and deserve further discussion. The epistemological aspects of the research methods used are discussed in the next chapter.

10
Epistemological Reflections on Policy Research

Introduction

This chapter comprises epistemological reflections on the selection of research methods described in Chapter 9. A central concern is coherence between the policy research methods used and the developing understanding of how knowledge actually grows, as understood in the field of educational administration. Chapters 3–8 above illustrated the considerable extent to which accountability policies are both the medium and outcome of policy processes; processes through which views of valuable knowledge (and historical commitments to how it might be demonstrated and improved) have been continually reconciled with changing priorities, local and national events and emerging perspectives. What has not been well understood in the field of educational administration is how particular theories of knowledge have impacted on the evolution of such policies and embedded serious limitations in practices. This chapter reflects on the part that epistemology has played in the development of policy-research methodology. The discussion begins with the persistent existence of, and the policy impasse apparently sustained by, competing theories of accountability.

Policy Gridlock and Epistemological Advice

Elmore's three competing knowledge systems introduced in Chapter 1 were noted again in Chapter 5, when both Halstead and Radnor *et al.* used mutually exclusive categories in England and Wales. Again, in Chapter 9, they were found to be a characteristic of the preliminary data collected in Tasmania. This suggests that accountability policies are a by-product of ongoing competition between three fundamentally different theories of accountability; theories driven by technical, client or professional assumptions. These theories carry with them very different assumptions about the nature of knowledge.

The technical perspective emphasized applying scientifically validated knowledge to the 'core technologies' of schooling, effective implementation, mandated systemic priorities, and the reliable measurement of actual achievements. Given this logical and empirical view of knowledge growth, and therefore how the *acquisition* of knowledge was to be demonstrated, accountability policy and practices required clarity of purposes, precise performance indicators and objective performance data. The research reviewed in Chapters 3-8 demonstrated that this perspective has dominated accountability policy making in England and Wales and in the United States for many decades.

It was also shown that the second consumerist perspective promoted contractual accountability between providers (administrators and teachers) and clients (parents, students and the community). This perspective has valued political, market and managerial mechanisms to prove and improve quality and choice. It has been strongly promoted in recent reforms on both sides of the Atlantic. While it accepts the empirical realities of (say) resource levels, it regards policy knowledge as a subjective and intersubjective construction, and thus capable of recurrent mediation and responsiveness to economic conditions and political action.

The third professional perspective held that accountability for school improvement is most effective when a by-product of professional development, autonomy, respect, resources and expertise. This knowledge system promoted collegiality, collaborative planning, and co-operative teaching and learning. The purpose and nature of policy knowledge is, therefore, critical to the liberation of people from determinist patterns of relationships and structures; hierarchy, class, gender, age and race. Knowledge is used for, and therefore conceived in terms of, socially-critical purposes.

The first and intensely practical problem here is policy gridlock. Gridlock can be given a genealogy in each setting by using a sociology of policy knowledge but without necessarily unlocking and moving on from revealing or the contested theories of knowledge involved. The epistemological challenge is to explain how gridlock works and to recommend policy research methods that will (a) unlock the frozen policy production processes, (b) do it in a way that yields supported policy options, and (c) indicate the relative trustworthiness of polity options. Nevertheless, the most evident feature of the three accounts, their representation of incompatible interests, is likely to attract proposals concerned with the reconciliation of interests in a context of differentiated power. The crucial point here is that there are epistemic limits to how far rational, economic and political analysis of interests can reach.

Assuming that we set aside trivial procedures and tyrannical decision methods that eliminate vetoes on implementation, Pareto's Immeasurability theorem (see Chapter 8) indicates that a new policy cannot be determined rationally in terms of utility. Similarly, Arrow's Impossibility theorem (see Chapter 8) advises that analytic methods cannot determine what is to be preferred. Taken together, these conditions imply that unless overlap between theories is developed by their guardians, and are expressed as shared practical policy preferences, no policy change will actually be achievable in policy actors' heads and hearts. The point I come back to is that a rational, economic and political analysis of policy gridlock is technically unable to unravel the role played by the theories of knowledge in use. Philosophical tools are needed to unpack the mutually exclusive nature of the knowledge systems involved.

Current practices reproduce the problem. When logical empiricism is used to arbitrate the three sets of claims, only one theory can be right. This theory insists that there can only be one set of claims that will best match the facts of the policy matter. However, it is also a fact that, in a plural society, vitally important non-empirical aspects of policies get to be defined in subjective and normative ways. And while the plurality of definitions of the 'real world' can be further increased by the interplay of interests, values and issues, they can also be reduced when epistemically critical processes help policy actors understand better and attend to the relativity of their views as knowledge claims (Walker and Associates 1987).

This is to suggest that a less ideologically committed, a more pragmatic and an epistemically critical approach to the articulation and testing of policy proposals might be more appropriate in public education than continuing to use logical positivism. What appears to be overdue in the settings reviewed above is an accountability policy process that accepts plural interests and multiple definitions of the situation, finds workable blends of values and theories, and which iteratively develops what are seen as the most appropriate ways of proving and improving the quality of schooling. An example was Kogan's call for multiple zones of accountability policy and legitimacy (in Chapter 4).

To answer such a call requires the clarification of rules for a pragmatic and epistemically critical approach to the production of policy. The first step would involve accepting many types of data, each on its own terms. This, in turn, would imply the need for a rigorous process that could identify common ground between the claims. This, as Evers and Lakomski (1991) argued, requires a coherence test of empirical, subjective and normative evidence, a process which has the added advantage of 'directing the process of conjecture and refutation towards the *growth* of knowledge' (p. 37) [original emphasis]. Lakatos (1970) argued that the knowledge compiled through such a process should be regarded as 'touchstone' and be used to provide additional criteria for the admission of further claims. This approach also suggests that policy research should be designed to enable guardians to (a) state their own initial position, (b) together, deconstruct all stakeholder positions as theories of the situation, (c) examine the extent to which they share values and interests, (d) consider the best of relevant knowledge from external sources, and then, (e) together, reconstruct theory and policy on the 'touchstone', so that practical trials may begin.

From an early stage in the project discussed in Chapter 9, it became clear that such a process of developing policy knowledge also had the advantage of being derived from, and thus cohering with, people's idealized view (and sometimes practical experience) of policy making. Stakeholder-sensitive action research was also regarded as attractive by policy actors for its capacity to mediate contested knowledge claims in three crucial ways. It allowed stakeholders to recurrently reconcile interests and values in the policy communities of Tasmania. It helped them both solve problems and discover new challenges to do with the quality of learning, teaching and leadership. It promoted and broadened the learning capacities of local and central policy communities.

On the other hand, it has to be acknowledged that the best way of arbitrating underlying theories in competition is a hotly contested matter in the field of educational administration, especially in the practices of educational administration conceived holistically as policy making, policy implementation and a moral art (Hodgkinson 1978, 1981, 1983, 1991, Duignan and Macpherson 1992, pp. 171–185). The advice given to and developed in the field of educational administration about the quality of knowledge may be broadly classified as logical positivism or logical empiricism, humanist subjectivism, critical subjectivism and non-foundational naturalistic coherentism. While it is also recognized that there are other options available in the philosophy of science, the relative influence of the main imports and the absence of alternatives is only now being systematically evaluated in the field (Park 1995a, b, 1997).

In the next section it will be shown that logical positivism stands apart from the others as being distinctively positivist; it assumes that data are objective observations about the way the 'real world' of administration really is. The other three are post-positivist in assumption; they take the markedly different view that observations are inevitably theory laden, and since the same data can support any number of theories, data always underdetermine theories. One implication for policy researchers is that data are of limited use in deciding which theory is best when the options have been clarified. Although there is the remote possibility that an exclusive epistemology could generate the most trustworthy policies over time, the far greater probability is that the policy research and accountability practices traditionally based solely on logical empiricism are flawed.

Positivist Policy Research

Logical positivism or logical empiricism has a long and honoured tradition in the field of educational administration and still deeply influences policy research. The recurrent patterns to the evolution of policies reviewed in Chapter 3–8 above reflect the deep belief in policy makers that better accountability policies will be discovered by a dispassionate search through the measurable facts of inputs, processes and outputs. US presidents and state governors and British governments have bet their political futures on the belief that hypothesized policy effects predicted by theories of accountability should be verified by collecting objective evidence. Factual student test data have been the preferred indicator in the United States, and when combined with school inspection reports, similarly favoured in England and Wales. Accountability policy making in these settings has long been assumed to be a positivist decision science: a science that can rely on facts for the verification of theories already expressed as policies, in law and as hierarchical structures. If this approach to accountability policy making can be shown to have used an inadequate epistemology, its outcomes (policy content, legal conditions and governance and administrative structures) will be significantly less worthy of public trust. A review of theory development in the field of educational administration suggests that this is very likely to be the case and that it could be a major reason why reform policies have enjoyed such little success and come to be associated with recurrent legitimacy crises.

After a period in the early decades of this century, when the search for better administrative technique was the all-consuming interest, a number of scholars initiated the 'New Movement' or the 'Theory Movement' (Halpin 1958, Culbertson 1981, Griffiths 1983, 1988). The theories considered most trustworthy at that time were those refined using hypothetical-deductive research methods. Since education was regarded primarily as a social system, the main theories of educational administration were derived logically or by experiment from the behavioural sciences. The leading theorists of the day tended to use the version of 'good science' developed in the 1920 and 1930s by the Vienna Circle of logical positivists. This version of science relied heavily on a verification theory of meaning. A knowledge claim stated as a sentence was considered to be trustworthy if and only if it could be proven analytically (using mathematics or logic), or if it was verifiable and falsifiable by observation or experience (Stroud 1992, p. 264). Unfortunately, this also meant that positivists could not deal with the immeasurable aspects of administration, such as ethics, politics, culture or emotion, or accommodate plural views of reality. Claims that backed up into non-empirical realms could not be verified or falsified using empirical evidence. Logical empiricism deals only with the 'facts of the matter' in the one 'real world'. How was this watershed accomplished?

Nobel Peace Prize winner Herbert Simon (1976) provided a systematic application of logical empiricism in his seminal text, *Administrative Behavior*, first published in 1945. It was used to sweep aside anecdotal theory in the field of educational administration. It was enormously influential. Faced with the limited rationality of human beings who were making decisions about how best to be organized, Simon decided to set aside ethical issues and built a positivist theory of administration using verifiable facts. Similarly, Dan Griffiths (1957), who was trained in mathematics and science, used logical positivism to develop his 'administrative science' that cohered with Parsons' (1951, 1966) systems theory, although he later conceded that the approach was mistaken and reflected 'the scientific ideology of the times' (1983, p. 203). Hence, although logical empiricism eventually collapsed under attack from philosophers of science (Popper 1972, Quine 1951, 1960, 1961, Kuhn 1957, 1970, Feyerabend 1975), it bequeathed a rich if flawed epistemological legacy to educational administration. The legacy was evident in the chapters above; especially where policy actors exhibited an ideological commitment to logical empiricism in the face of recurrent and mounting evidence that their theories and practices did not work. Post-positivist views of policy research are now examined to suggest that empirical, subjective *and* normative types of data are required to theorize and develop accountability policy.

Post-positivist Policy Research

If Simon and Griffiths provided the first epistemological watershed in educational administration, by introducing logical positivism, Thom Greenfield provided the second in 1974 when he introduced a humanist and constructivist variant of subjectivism. He attacked the pervasive effects of logical positivism, proposing instead that people and their values be placed central in organizational theory, and that their multiple definitions of the 'real world' in education should be accepted and regarded as social constructions (Greenfield 1975, 1980, 1988). By the lights of his radical constructivism, accountability criteria can be regarded as the standards, benchmarks or indicators invented by well intentioned people engaged in the process of making judgements and constructing a preferred future. They are an expression of how people see their world, how they value education and believe that it can be demonstrated and improved. Theories are not built by compiling verified facts, he argued, but by wilful people who invent and develop a seamless web of facts, values and norms to justify their actions. He set out new moral, metaphysical and epistemological conditions for policy research (1978, p. 12):

> These subjectivist assumptions mean that organisation theory must not only describe the process people use in constructing reality; it must somehow *be* that reality with all the possibilities that the human mind reads into experience [original emphasis].

There were special features to Greenfield's epistemology. His 'humanist science' embodied a deep respect for people, and unlike most administrative theory at the time, cherished the sanctity of others' minds (Greenfield and Ribbins 1993, p. 250). For example, his theory defined the building of an accountability policy as a social, moral and metaphysical construction. People would assemble ideas on this aspect of being organized for educational purposes in a way that was valuable and meaningful to them. Greenfield's humanist science also assumed that people are reflective; policy actors will inevitably evaluate and change the processes and criteria they employ. It also presumed that people habitually use objective, subjective and normative data, and a changing mix of qualitative and quantitative processes of analysis. Clearly, humanist subjectivism was incompatible with the idea that the growth of knowledge should be mediated solely by empirical verification, a central idea of logical positivism.

How did humanist subjectivism propose to arbitrate knowledge claims in policy making? Greenfield argued (1979a, p. 171, 1979b) that the 'theory ladenness of observation' meant that theory entails evidence, rather than empirical evidence entailing theory as believed by logical positivists. This meant turning policy research processes back to front. Theories in use, including purposes and values in practice, had to be evaluated rigorously in order to determine fresh data collection and analysis methods and appropriate applications. Rather than collecting data in order to develop and test given policies favoured by policy elites, the focus shifted on to the quality and the improvement of the theory used to drive the collection and use of data. An immediate implication for accountability policy research was that the scope had to be far broader than student achievement and needed to include stakeholders' theories of teaching, leadership and governance, and be mindful of the specific historical context of complex social, political, ideological and economic dimensions (Rizvi 1985).

Similarly, since data underdetermines theory, Greenfield concluded (1975, p. 84) that theory choice entails moral decisions. This implied that accountability policies and practices are to be regarded primarily as moral actions and that they should be subjected to formative evaluation, not restricted to a decontextualized process of objectively measuring and reporting the achievement of standards prior to (say) allocating blame or praise.

Greenfield was less clear why moral choice entailed adoption of Hodgkinson's (1991) theory of values. He never developed his own theory of value. What was clear, nevertheless, was that the fact-value and theory-observation distinctions, so essential to logical positivism, began to dissolve. And with them went the positivist foundations of the behavioural science and systems theories that had long dominated the field of educational administration (Hoy and Miskel 1978). This also meant that the products of positivist policy research were no longer as worthy of trust as once supposed and suggests that policy actors' assertive and heroic attempts at reform detailed in earlier chapters were probably misconceived.

Ironically, given the degree to which trust was invested in subjective data, Greenfield's humanist subjectivism provided little guidance on how to evaluate intersubjective knowledge (Lakomski 1987). This problem was taken up by Willower (1979) who set aside the positivism-subjectivism dualism on the grounds that it failed to represent the array of epistemological options available. This was a reasonable, informed and far-reaching argument. Post-positivism had been described elsewhere (Boyd 1991) as comprising subjectivism or constructivism, constructive empiricism and scientific realism. Indeed, Griffiths (1986, p. 139) adopted constructive empiricism en route to multi perspectivism, although later generously conceding that Greenfield had 'jolted me into examining the assumptions underlying my theoretical position and caused me to move to theoretical pluralism' (1995, p. 152). Willower's more systematic attention to the problems of positivism produced a blend of naturalism, empiricism, instrumentalism and pragmatism (1981, 1986, 1988, 1993, 1994). His research programme developed four general theses of immediate relevance to the policy research reported here:

- a broad range of methods should be used and cohere with research purposes,
- while observation is theory laden, it did not follow that knowledge is the product of subjectivity alone,
- the warranted assertability of claims should be raised through open, self- correcting and intersubjective verification, and
- a Deweyan (1916, 1938, 1963) blend of humanist values and pragmatic science should inform theory choice.

At this point it is important to consider another important variant of subjectivism that emerged in the 1980s, critical subjectivism, and relate it to the challenges of policy research.

Socially Critical Subjectivism

Greenfield's work was built on by Richard Bates (1980, 1982, 1983, 1988, 1990) and other socially critical subjectivists (Watkins 1985, 1986, Foster 1986a, b, Rizvi 1986). They showed, for example, that positivism had helped portray educational administration as a masculine enterprise (Kenway 1990, Yates 1990, Blackmore, 1992, pp. 25–46) conducted in a mono-cultural manner (Rizvi 1990, Rizvi *et al.* 1990). Grace (1995, p. 2) summarized one of the current agendas, namely 'socially critical leadership':

> For those who wish to resist the assertion that strong and effective school leadership is inevitably the property of one person (or of a small, elite group) and therefore a continuing manifestation of necessary social hierarchy, the critical study of educational leadership becomes essential ... [and if] ... educational leadership can be shared, transforming, empowering and democratic enterprise, how is this to be achieved?

This way of understanding defined policy research and leadership as cultural and political methods of defining social reality and organizational purposes, and then achieving those ends. Such reculturing and political processes could either force people to accept the values embedded in the methods used, or help them to review and develop the values in use. Given these two possibilities, policy research and leadership should be evaluated, it was argued, for their capacity to create 'psychic prisons' or 'emancipatory conditions'. This form of analysis, for example, revealed and successfully contested a form of institutional racism faced by the indigenous Maori of New Zealand (Walker 1985, p. 79).

The strongly sociological interest in knowledge growth soon identified the way that administrative processes act in both surgical and genetic ways. They can change both the structures of daily relationships and the deep assumptions that govern the longer term evolution of structures (Giddens 1979, 1984, pp. 16–28). Socially critical action research (Kemmis and McTaggert 1988) revealed how the 'moral economy' of an organization is reproduced. While these agendas are often very attractive to socially aware and politically active educators, there are limits to socially critical subjectivism that are less well appreciated.

The first problem was that such research adopted Habermas' (1971) early proposal that knowledge could be partitioned according to basic interests served. The first partition included empirical-analytic knowledge. Such knowledge was associated by early socially critical subjectivists with a concern for control and manipulation. The second partition encompassed historical or hermeneutic knowledge, which was linked with a liberal, progressive and practical interest in generating understandings and more open communication. The third form of knowledge was deemed 'socially critical' since it emphasized dialectic and collaborative processes needed to serve an emancipatory interest in social justice.

A number of questionable foundational assumptions were used to justify these partitions. It was assumed that this partitioning of knowledge would emancipate people and resolve comparative interests. It was also assumed that these partitions cohered with a trustworthy theory of knowledge. It was believed that the proposed re-partitioning of knowledge according to lights of a particular ideology was more valuable than de-partitioning knowledge in search of coherence and solutions to problems. None of these assumptions seem to have endured.

The socially-critical perspective did resonate with a concern for the disadvantaged in society and the suspicion that administration plays a part in sustaining the broader inequities in society. The ready acceptance was, however, not as epistemically critical as it might have been. For example, the socially critical perspective developed out of Hegel's exploration of contradictions, not practice, and his idealistic body of doctrines or social theories. Education policies and theories, according to Marxist analyses (e.g. Bowles and Gintis 1976), were to be advanced by identifying contradictions, reconciling contradictions, and then uncovering fresh contradictions. The New Sociology of Education (Young 1971, Young and Whitty 1977) introduced into educational administration by Richard Bates, for example, suggested a subjectivist epistemology to interpret educational processes, a neo-Marxist structural analysis of class power and control, and evaluated policy and theory claims in terms of emancipation and human betterment. Similarly, Bates' synthesis of the ideas from the Frankfurt School of Social Theory assumed that policy knowledge in education reflected three crises faced by modern states; rationality, legitimacy and motivation (e.g. Habermas 1970, 1978, 1979). There was a tendency, therefore, to assume, sometimes with justification, that policies and practice should cease exploiting an underclass, often indigenous people or teachers, and promote instead a normative order of struggle, emancipation and liberation (e.g. Apple 1982).

This position appeared to soften as the normative order of educational administration changed. It is often assumed today that unfair social structures should be remodelled and that inappropriately skewed distributions of power should be questioned as part of the educational process. It is often assumed that administrators should use collaborative administrative methods that better share cultural capital and give a higher priority to equity. It is often assumed that greater equity will be achieved by conducting a shared analysis of interests and helping the disadvantaged acquire greater communicative competence. It is often assumed that leaders should develop participative and collaborative structures such as co-governance and co-management. These assumptions, however, are far from the reality of practice, and from possibly becoming common practice in many settings.

Critical research typically requires subjective and normative data concerning the historical context in order to evaluate how economic, political, social, cultural and ideological conditions help sustain inappropriate theories in use in policies and practices. An epistemically critical review would also identify the part played by the normative theory behind such data collection. It is also vital to note that the more compelling contributions of Habermas (1970, 1971, 1978, 1979) do not require a Marxist 'working class' to act as the agent of revolutionary change (Bender 1970) but caring individuals and groups who critically analyse policy discourses that originate in the superstructure of society and then generate imaginative interventions (Rockmore *et al.* 1981, Berrell and Macpherson 1995).

This is not a revolutionary but a pragmatic and evolutionary approach. It affirms John Dewey's (1963) view that what constitutes an educated person is an issue that must be answered by communities, if they wish to remain communities. Epistemically critical reflection is required to connect deliberation about ethics, compassion, social justice and democracy to strategic evaluation and planning in educational systems and institutions so that learners and educators can refresh their understanding of civilized living.

Such imaginative problem solving works within current governance policies to untangle the distorted communications between institutions, structures and groups, and if need be, to reform government. It invents interventions in order to eliminate unnecessary modes of authority, exploitation, alienation and repression using both empirical and non-empirical data (Ritzer 1975, 1981, 1991). It reconstructs modes of governance from within. Indeed, Vivian Robinson (1993, 1994, p. 73) has suggested moving the focus from socially critical values to problem solving as a valuable end in its own right, and accommodating different powers and values in a politics of collaboration and critical dialogue.

To summarize this section, socially critical subjectivism shares with humanist subjectivism the view that knowledge is a social construction. It alone promotes the view that educational administration should aim to achieve greater equity and social justice through the agency of public education. A serious limit to its epistemology is its willingness to repartition knowledge to serve pre-set commitments, although this is often the source of its inspiration.

Another limit has been its use of a radical language that tends to polarize people instead of gaining their understanding and attracting support. On the other hand, we have not been educated to hear alternative voices. It was only in recent times that our field realized that it had to reconstruct the post-graduate curriculum of educational administration (Griffiths 1993, pp. 52–53), essentially by adding cultural, educational, organizational, methodological, political and ethical dimensions to the passing on of technique; leadership processes and management functions.

This implies that we should regard the virus-like effects of positivism as a learned incapacity. Similarly, we must accept that the proposed antidotes of humanist and socially-critical subjectivism turned out to be carrying another form of virus that gave privileges to foundational items of belief. Humanist subjectivism often valued the sanctity of people's minds to the extent that it could not discriminate between theories. Socially critical subjectivism often used pre-set agendas to automatically discriminate between theories. On the other hand, a growing epistemic awareness has led to blends of concerns and approaches (Bates 1994).

Despite these advances, the continuing presence of epistemic incapacities explains why the field of educational administration stumbled so easily into the 'paradigm trap', although it might have learned enough from the experience to avoid the 'trap' of postmodernism.

The Paradigm and Post-modern/Post-structuralist Traps

The 'paradigm trap' was set inadvertently in organizational research when a typology of research paradigms was clarified by Morgan (1980) and Burrell and Morgan (1981). They identified two profoundly different philosophical orientations in organizational science to social reality and values. They argued persuasively that these perspectives had long silently mobilized bias in problem solving, research methodologies and in the growth of knowledge about organization.

Their first question echoed Greenfield; Is the social reality of organization objective or subjective in nature? Their second question echoed Bates; Should research and other practices in organizational problem solving aim to be reformist or regulative? When the answers to these questions were arranged on two axes, the four quadrants created were deemed to reflect four basic research perspectives, investigative paradigms or puzzle-solving mindsets.

Each quadrant was shown to have its own metaphors in administration and research; metaphors that were used to define the relevance of prior knowledge, the appropriateness of research questions and most useful data. The four quadrants were labelled 'structural functionalism', 'interpretivism', 'radical humanism', and 'radical structuralism'. They were soon regarded as discrete research paradigms (McCarthey 1988, Guba 1990, Ribbins 1985).

Even though Kuhn had explained that paradigms provided the relationship between evidence and theory, our field's relatively weak auto-immune system failed to trigger a critical examination of our theorists' theories of knowledge (Campbell *et al.* 1987, pp. 208–210). One of the confusing factors was that the term 'paradigm' had at least 21 different meanings (Masterman 1970, p. 61-65). Hence, Kuhn's suggestion that 'justification was relative to paradigm' was too quickly interpreted as an affirmation of positivism, and somewhat patronizingly, as a good reason for being a little more tolerant of non-positivist approaches as they 'struggled' to establish their credentials (Lincoln 1984).

The epistemic trap was sprung when we were found that we could not arbitrate between paradigm-specific knowledge claims (Willower 1988, p. 741). It was clearly preposterous that all technically competent research processes and outcomes had to be accepted as being of equal value. Some so-called 'research-based theories' of educational administration in the 1980s had no boundary between belief and knowledge (Culbertson 1988, pp. 18–24).

The hook in selecting a paradigm that cohered with a research purpose was that it was tantamount to arguing that the justification of the knowledge about to be created was automatically provided by the paradigm and by the arbitrary selection and definition of 'the problem'. Although the logic involved was almost ridiculously circular, a new way of thinking then appeared that promised to dissolve the problem; postmodernism or post-structuralism. The precise meanings of these terms are elusive. As Lather (1991, p. 34) explained:

> postmodernism/poststructuralism is the code name for the crisis of confidence in Western conceptual systems. It is a produce [*sic*] of the uprising of the marginalised, the revolution in communication technology, the fissures of a global multinational hyper-capitalism, and our sense of the limits of Enlightenment rationality, all creating a conjunction that shifts our sense of who we are and what is possible.

Following Rorty (1979, 1989), post-modernism came to the view that 'good science' comprised no more that a number of alternative stories about knowledge production. Like most socially-critical research, post-modernism challenged the legitimacy of the dominant social discourses in organizations, and how they relate to 'master discourses' about society (Lyotard 1986). Unlike most socially-critical research, however, postmodernism has no detectable epistemology, and indeed, dismissed the possibility of choosing between theories. As Tarnas (1991, p. 398) observed; 'The postmodern human exists in a universe whose significance is at once utterly open and without warrantable foundation'.

The post-modern approach to theory choice encouraged researchers to use their favourite 'paradigm' as a philosopher's touchstone to justify the relationship between their evidence and their theory. Two results were that post-modern analysis tended to stay at the surface structures of society and organization and allowed extreme versions of subjectivism and relativism. It made no provision for the growth of knowledge outside of a paradigm. Apart from these fatal practical flaws it had no answer to the self-referential test; why adopt post-modernism as an epistemology when it denies the possibility of an epistemology?

Two solutions were, therefore, required; one that reconstituted 'paradigm' and another that promised alternatives to the circularity of paradigmatic thinking and the chaos of post-modernism. By comparing how educational sociology and educational administration have responded to the 'paradigm trap', paradigms were rehabilitated (Berrell and Macpherson 1995) as representing broad, organizing and competing theories of knowledge. While this tidied up the concept of paradigms to a degree, and proposed a reformulation of research methodologies in educational sociology, the second step required an epistemological justification for post-paradigmatic research that celebrates the growth of knowledge. It was provided by Colin Evers and Gabriele Lakomski when they drew a line between foundational and non-foundational theories of knowledge. It was the third watershed in the production of knowledge in our field.

Foundationalism and Non-Foundationalism

Evers and Lakomski (1991) argued that positivism and subjectivism share two epistemic flaws. They partition the domain of knowledge. Their attempts to validate the partitions regress to foundational claims, claims that turn out to be arbitrary and unjustified. Earlier, when foundational claims about the divisions to knowledge had been traced to their origins (Evers 1979, 1984, 1987a, 1988, Walker and Evers 1982, 1984, Lakomski 1987), they had been found to be the preloved articles of faith of a 'discipline' or to be an expression of ideological commitment, rather than intrinsically secure and universally reliable truths or axioms. The implications for policy researchers and leaders were substantial and are still being worked out.

Escaping from the problem of vicious regress means that policy researchers should neither partition knowledge nor appeal to foundational premises when identifying options. They should also set aside the positivist logic; linking premises, hypotheses, data and findings as a way of 'proving' new ideas. Instead, they are advised to start with how participants and stakeholders see the 'facts of the situation', and to use all of these perceptions and interpretations to map the current theories being used to define 'the problem'.

This approach should also accommodate personal and group theories, sensed data, beliefs, known solutions, international research experiences, and hypotheses, as well as ideas from less traditional sources, such as paradox, dialectic, intuitions and speculations (Quine and Ullian 1978, Hesse 1974, 1980). The intention would be to build a multi-dimensional and multi-perspectival network of ideas about the problem-context-solution in an holistic and iterative manner (Walker and Associates 1987, p. 15). Leaders who have used the early stages of action research to build a collective appreciation of policy options will recognize the nature of the process recommended.

The second implication relates to theory building. From the outset, the network of ideas used as a conceptual frame of reference has to be regarded as a provisional assembly. It has to be evaluated regularly as components are added or adjusted in a holistic way. There also has to be a constant search for coherence between theoretical, empirical and value ideas, without giving special status to a particular type of claim or data. Each idea needs to be examined in terms of the theory it represents, to raise the level of internal coherence. It has to be checked against other data, practical experience and other theories of reality, values and knowledge to raise the level of external coherence. This also means that policy proposals have to be checked out against empirical and subjective evidence, the broader base of societal and international knowledge, and the assumptions embedded in the most effective known practical solutions. The point here is that each policy option is justified by the extent that it logically integrates with and supports others.

The third implication concerns theory choice. This pragmatic approach generates the standards for judging between theories as it proceeds. As the problem-context-solution set becomes more comprehensive, more logical and more practical, the set reflects an increasing degree of overlap between the theories held by all participants. This theory overlap or touchstone also contains the criteria that have been developed during the coherence testing. These criteria are then used to make theory choices. As Walker and Associates (1987, p. 16, original emphasis) explained:

> We extract *common standards* from the overlapping accounts of shared problems, or we adopt them from other shared areas of the theoretical frameworks of participants. By examining the actual content of touchstone, we discover what values and procedures each of the competing theories is committed to in common with the others, and ask which of the theories comes out best in view of these shared values and procedures.

The fourth implication relates to the respect to be given to different types of information. Non-foundational epistemology gives no privileges to foundational items of knowledge outside the realm of theory. Items of knowledge such as objectified facts, subjective experience, ideological commitments, as well as perceptions of the context such as historical inequities and economic conditions, are all treated respectfully as theory; as provisional and theoretical constructions. This accepts that evidence is riddled with the theoretical assumptions used during its creation, and that beliefs have degrees of empirical validity and subjective potency. There are no neat distinctions between facts, perceptions and values made. It refuses to believe that 'data can speak for itself'. As Willower (1988, p. 742) put it, it is 'an epistemology that recognises the fallibility of science and seeks warranted assertability ... , not certainty'.

A fifth implication of non-foundational coherentism is that it insists on a broad and contextualized view of a problem. It gradually selects the most coherent explanation of the situation, its context and effective solutions. When the approach is given an educative purpose, such as taking reflective action in natural settings, it does use the empirical world 'out there' to 'naturalize' policy claims prior to application. However, while non-foundational pragmatism subjects policies to 'the empirical test' (Robinson 1994, p. 74), it controls the virus of positivism by using a holistic test of justification and a coherence test of theory to select a problem-context-solution set (Quine and Ullian 1978, Ch. 6).

Evaluations of Non-foundational Naturalistic Coherentism

An immediate limit to Evers and Lakomski's proposals is that they constitute a normative theory about methodology in policy research. Their research programme is still unfolding. The methodological implications of their position are yet to be fully developed systematically and tested in practice. A few of the still rare applications available in educational administration are now reviewed to indicate potential value and possible limitations.

Aspin *et al.* (1994), for example, aimed to identify better theories of quality schooling to underpin educational and administrative policies. They collected qualitative data from a broad range of stakeholders and tested their theories against other bodies of knowledge and theory. Then, by putting the ideas through a process of constant comparison and critical reference to different theoretical perspectives, to force theory competition and correction, they searched for touchstone. The touchstone of 'quality schooling' was found to have two core values; the development of autonomous and self-motivated individuals capable of independent judgement, and, developing such agency in networks of mutual relationships that comprise the economic, social, cultural and political aspects of community.

A key epistemic finding was that the research was unconstrained by the ideological boundaries of positivism and foundationalist subjectivism. It had been able to redefine 'effective schooling' by collecting plural values in an inclusionary manner. After systematic evaluation on a number of such post-positivist research projects, Judith Chapman (1995, p. 6) came to the view that the potency of post-positivist methodology lay in its recurrent evaluation of theories-in-use, the theories in people's heads that were determining and predicting improvements. When researchers worked with practitioners to critically evaluate and constantly compare the theories-in-use, they were able to indicate which theories would be better for specific purposes in context.

There is good reason to believe that epistemic understanding will be further advanced by a number of doctoral research programmes. Anna Ciccarelli (1996), for example, has just finished relating Habermasian critical theory and Deweyan pragmatism to Bob

Young's (1995, 1996) and Stephen Crump's (1992, 1995) research programmes and has proposed a critical-pragmatist methodology for organizational analysis. Her innovative approach includes problem solving and critical dialogue between researchers and practitioners at all levels to address the 'interactive dimensions of structure/individual/interaction and the concomitant problems of power, agency and communication' (p. 304). Park Sun Hyung's (1995a–c, 1997a, b) epistemic evaluation of the theory in our field has reviewed coherentist theories of knowledge, traced the genealogy of Australian naturalism, evaluated Greenfield's subjectivism and discussed potential alternatives to naturalistic coherentism. Alan Pritchard (1996) has decided to set aside the attractions of a socially critical epistemology in favour of naturalistic coherentism to develop a new QA policy for Western Australia. He took this approach to avoid either pre-setting a normative ideal, justifying the ideal by dialectical analysis and appeal to authoritative persons, or collecting and using of data biased in favour of achieving the pre-set ideals. He wanted to retain a more open research agenda than that permitted by radical humanism, employ language and strategies more likely to attract the support of systemic and institutional policy makers, and keep open the possibility of systemic position holders serving educative ends.

My own path to naturalistic coherentism dates from the early 1980s at Monash University when Judith Chapman, Peter Gronn and Colin Evers helped me identify some of the limitations of the humanist subjectivist methodology I had derived from Thom Greenfield's theory (Macpherson 1984, 1988). Later, at the University of New England, Patrick Duignan and I became very concerned about the new technicist orthodoxy of 'self-management' sweeping Australia, and devised an inter-state research methodology intended to generate a 'practical theory' of 'educative leadership'. It involved Colin Evers, Jim Walker, Jeff Northfield, David Pettitt and Ian Hind, and Fazal Rizvi as lead theorists, teams of exemplary practitioners nominated by three states, and used what are now regarded as features of a non-foundational, naturalistic and coherentist methodology.

The project was driven by two questions. How should leaders in education decide what is important? How will they know that they are morally right, when they act? It demonstrated the importance of coherence between three conditions; (a) leaders at all levels in systems providing processes that create, promote and apply knowledge about how best to be organized, (b) feedback being accepted as crucial to the growth of knowledge, and (c) educational organizations being regarded as moral cultures and receiving proactive moral leadership (Duignan and Macpherson 1993, Macpherson 1992a, 1993a). There was strong support found for Argyris and Schön's (1978) concept of a 'learning organization'.

Since 1992 Pat Duignan and Narottam Bhindi (1995) have been building a theory of authentic leadership, 'authentic' defined as having high congruence between values and actions. They have used many aspects of naturalistic coherentism to identify twin imperatives of servitude and stewardship (p. 4):

> As a result of the analysis of the responses from the interviews, open-ended surveys, research workshops, and the literature review, a number of themes related

to the research questions have emerged, illuminating the concept of authentic leadership. There are: (1) cynicism about leaders and leadership, (2) need for an authentic appreciation of self; (3) need for authentic relationships; (4) need for authentic learning; (5) need for authentic governance and organisation; and (6) the special nature of authentic leadership.

Since 1992, as explained in Chapter 9, I have been helping to review accountability policies for the public school system in Tasmania. Two years were spent collecting ideas and policy proposals; accountability processes and criteria that would collect data, report on and improve the quality of learning, teaching and leadership. The third year was spent refining theories and proposals, and measuring support for each policy proposal. Two years were spent interpreting and applying the findings at system and site level, despite the context of budget cuts, restructures, industrial disputation and political turbulence.

Throughout the policy-research process I have been able to watch how participating stakeholders created and applied policy options. This enabled me to conduct a preliminary evaluation of Evers and Lakomski's naturalistic coherentism. Apart from periods when ego, altruism and creativity ruled momentarily, there was consistent evidence that stakeholders' knowledge of accountability did comprise temporary subjective data, reified objective data and normative types of data, and that attempts were made to honour each on their own epistemic terms. Most people did not partition knowledge or appeal to foundational premises when using the measures of support and ideas from elsewhere to revise personal theories of educative accountability.

Stakeholder groups did construct a multi-dimensional, provisional and inter-perspectival network of ideas that was evaluated regularly for its coherence and problem solving capacity as components were added or adjusted in a holistic way. Supported proposals were compared with wider and external theories of 'a good education'. The emergence of touchstone policy put pressure on participants to move from away using just 'the facts of the matter' towards using internal and external coherence tests.

While the empirical components of touchstone were used to 'naturalize' claims, a substantial number of other criteria were used, much as suggested by Churchland (1985) and Evers and Lakomski. Prudence was evident in incremental rather than radical elaboration's of past beliefs. Modesty was seen in limited risk taking. Simplicity was obvious when refining processes and criteria. The possibility of generalization was valued when considering wider applications. Refutation ended debates over disputed proposals, not proof by data. Consistency checks were used to monitor the relationships between proposals. Potency underpinned a concern to make a difference to the quality of learning, teaching and leadership. Explanatory unity was seen in efforts to relate all proposals to the touchstone set of supported options.

There were limits found to Evers and Lakomski's ideas. The first is that behaviours seen were sometimes different than the epistemic 'best practices' they suggested. There were moments in policy forums when ego, altruism and creativity played a decisive role as criteria for theory or policy choice.

They could also have underplayed the role of touchstone. Touchstone is a metaphor drawn from mining where ore is rubbed on a piece of schist to test for the presence of gold or silver. The metaphor was quickly adopted but soon had four distinct meanings; (a) finding the common ground, (b) the common ground between theories, (c) the epistemic standards intrinsic to the common ground, and (d) the process used to test other claims for the presence of the common, significant and valued standards. It also had deeper effects not inconsistent with Cicarrelli's proposals.

The teasing out of touchstone modified how people in the setting communicated, used their powers and thought about their service. Most interesting, when accountability policies were 'implemented' not by a separate follow-up strategy but through participation in construction, 'touchstone' served as structuration (Giddens 1979). It provided the conditions and processes that governed the use and development of structure.

Touchstone had two other interesting properties. Apart from specifying technically effective accountability processes and evaluation criteria, the development of touchstone simultaneously generated moral conditions for leadership practice. Moral knowledge was integrated with other forms of knowledge during its production. For example, when the stakeholders at a state policy seminar decided that the 'common ground' should become system accountability policy, and be used to refine practices, they stressed the importance of co-operative policy review processes at all levels. This 'importance' was loaded not with rational or empirical meanings but with moral imperatives concerning relationships and structures. The dynamics of building touchstone had created a moral code that applied equally to leadership and followership.

The other special property of touchstone was that, as it developed, participants grew increasingly reluctant to permit epistemic privileges. This was most evident when the principals, parents and union participants present at the state policy seminar used moral consequentialism to argue that the system's and school's policy gate-keepers should be held accountable for the provision of not consultative but collaborative and participative policy making and implementation practices. In sum, they had become epistemically aware that policy and moral knowledge grow together.

Summary

In this chapter I have suggested that the growth of policy knowledge will require epistemically critical research methods that challenge policy gridlock. Accountability policy gridlock could be a consequence of the positivism used by policy élites being resisted by other stakeholders who use subjectivist and other post-positivist epistemologies.

The development of epistemically critical research methodology has been retarded in the field of educational administration. The field has experienced three epistemological watersheds; positivism, subjectivism (humanist and socially critical), and nonfoundationalism. Having learned from falling into the 'paradigm trap', and thus remaining sceptical of post-modernism/post-structuralism, the field now seems to be the cautiously evaluating naturalistic coherentism.

The policy research methods reported in Chapter 9 provided a preliminary evaluation of naturalistic coherentism. Consistently strong support was found for the normative theory proposed by Evers and Lakomski. Mindful of the methodological limits of the research, three caveats and an extension to their proposals are tentatively suggested. Ego, altruism and creativity were given epistemic privileges by participants. The function of touchstone in theory building could have been understated. It appears to be central to structuration. While they could have been artefacts of the derivative methodology used, these are two matters of considerable practical and theoretical significance and further more direct research is warranted.

In the earlier chapters it is clear that accountability policy research has tended to use systems theory, an objectivist view of social reality and logical empiricism in order to identify performance indicators of learning. It is also clear that policy knowledge production concerning accountability needs to take a more holistic and causally interdependent view of teaching, learning and leadership services and to begin with the theories in use.

There is also a significantly underdeveloped moral aspect to accountability policy research. The accepting of responsibility in education obliges leaders to provide reasonable and coherent forms of accountability in a broad context of formative evaluation, educative reporting relationships and politically sensitive planning. Accountability policies need to be developed in an demonstrably educative manner. An appropriate methodology might consider using a consequentialist moral theory, try to reconcile stakeholder perspectives through their active engagement in action research, and employ a non-foundationalist account of knowledge production.

Unlike the logical empiricism that has long dominated policy research, the methods described in Chapter 9 did not seek a rational and functional theory of appropriate behaviours based on objective facts but a practical set of policy options that best accommodate empirical, subjective and normative data available, that is, until an even more comprehensive and more coherent account of educative accountability develops. While it lacked the positivist precision of past policies and many of the neo-centralist aspects of policy making in the past, the approach used exhibited the features of naturalistic coherentism. It:

- used falsification rather than proof to advance policy options,
- incorporated ethics, values, politics and economic dimensions in a holistic and inclusionary manner, and
- used an evolutionary process where touchstone and refutation gradually developed a new web of belief in a state policy community.

It is tentatively concluded that epistemically critical research is needed to create greater legitimacy in policies and practices in a complex democracy. They give practical expression to the metavalues of organization. They formalize collective and personal obligations to stakeholders. They are vital in *educational* institutions and systems as they have the unique feature of being expected to both reproduce and transform societies' values. It follows that clarifying and improving policies in such settings requires forms of policy research that iteratively produce educative policy proposals. The extent to which this approach produced a practical theory or policy of educative accountability for Tasmania is an issue addressed in the chapters ahead.

11
Accountability Policy Preferences in Tasmania

Introduction

In this chapter I present the first analysis of the data collected by the APQ. It was part of the preliminary research findings presented to senior DEA officials 21 October 1994 (Macpherson 1994a). The report was organized as requested to display policy-options in each of the six areas researched; processes and criteria to do with students' learning, teachers' teaching and leaders' leadership. In each table, the mean was calculated after the values 1–5 had been assigned to the responses SA–SD. The percentage that strongly agreed or agreed to each proposal is shown by %SA+A.

Five points must be stressed. The findings were initially reported as distributional data only, using benchmarks of 'political significance' to define proposals as 'supported', 'ambivalent support' and 'unsupported', all as negotiated. The preliminary interpretations provided in this chapter were those agreed at this first meeting with senior DEA officials. It was considered important for analysis in this form of systemic action research to be paced by their needs and their understandings.

The immediate concerns at that meeting were to 'take stock of the project', implement any 'obviously useful findings and improvements', and to design appropriate 'follow-up' analysis and policy processes. Subsequent chapters incorporate interpretations provided by the senior executives of all stakeholder groups and those negotiated at state and local policy forums, and those generated later when the findings were related to history and context, as well as to theories, research findings, policies and practices noted elsewhere.

Accountability Processes and Students' Learning

Table 11.1 presents the 22 policy options provided by all stakeholder groups ranked by %SA+A within three categories; the items found to be 'supported' by all stakeholders (i.e. > 70% SA+A), the items found to 'unsupported' (i.e. < 39% SA+A), and the items with ambivalence expressed by at least one stakeholder group (<70% and >30% SA+A). The term 'P and Fs' refers to schools' Parents and Friends Associations,

Table 11.1: Support for processes to collect data, report on and improve students' learning

Option	Supported, i.e. in all groups >70% SA+A	Mean	SD	% SA+A
1.1	Parent/teacher interviews.	1.46	0.61	97.2
1.2	Teachers evaluating and planning their lessons thoroughly.	1.58	0.59	96.4
1.3	Teachers' observations.	1.71	0.52	97.9
1.4	Conferencing between teacher and student.	1.74	0.63	93.2
1.5	The sampling of student work (e.g. folios).	1.77	0.69	91.1
1.6	Teachers evaluating and planning their programs systematically.	1.77	0.69	91.8
1.7	Parent/ teacher/ student discussions.	1.80	0.71	91.9
1.8	Teachers clearly identifying intended outcomes for each student.	1.80	0.74	89.3
1.9	Parents given many different opportunities to provide their views and to have teaching and curricular policies explained.	1.82	0.68	90.0
1.10	Reports of student's learning through newsletters, school magazines, displays and positive public relations.	1.84	0.87	87.3
1.11	Reports with clear and accurate descriptions of learning.	1.89	0.70	86.8
1.12	Student's own self-assessment.	1.92	0.89	82.5
1.13	Teacher-designed tests (to measure mastery, to diagnose individual understanding).	1.93	0.67	87.7
1.14	Formative evaluation related to teaching objectives.	2.17	0.72	73.6
	Ambivalence in at least one group, i.e. SA+A <70% and >30%			
1.15	Teachers using written checklists and running records to monitor student progress.	1.93	0.85	82.7
1.16	Reports from other support staff (such as guidance, welfare, speech and health specialists).	1.99	0.79	83.4
1.17	Parents being given the goals of the curriculum, intended outcomes and individualized expectations at the beginning of each year.	2.12	0.97	73.5
1.18	Reports with marks or grades.	2.89	1.37	47.6
1.19	State-wide, norm-referenced, standardized tests of literacy and numeracy.	2.70	1.18	54.2
1.20	Peer appraisal.	3.05	1.05	35.1
1.21	P and Fs/School Councils review, discuss and report learning.	3.09	1.09	33.4
	Proposals unsupported, i.e. in all groups SA+A < 30%			
1.22	Reports that allow parents to compare their child with others.	3.89	1.17	17.0

which tend to have fund raising and canteen management roles, rather than school policy advice; the main role of Tasmanian schools councils.

During the workshops and meetings used to collect qualitative data, a number of groups expressed their beliefs about the preferences of other groups. Teachers' and parents' leaders often asserted that 'parents want reports with grades and marks'. Parents representatives regularly claimed that 'teachers are against providing them'. Many DEA personnel tended to claim that 'parents want grades and marks' and that 'most teachers are reluctant but willing to oblige'. Proposal 1.18, that 'one of the processes used to report on students' learning should be reports with marks or grades', drew a complex range of responses by group and subgroup. They are summarized in Table 11.2 and ranked by means.

This proposal was supported by district high parents (89.2% SA+A), secondary parents (87.5%), secondary principals (76.2%) and secondary teachers (72.0%). It was unsupported by primary principals (13.5% SA+A) and primary teachers (24.1%). These findings were interpreted by DEA officials to mean that support was polarized on primary–secondary lines, essentially because primary school professionals sustained more interactive relationships with parents where secondary professionals felt obliged to provide more quantitative and comparative indicators of success in the coming end-of-school examinations that determined life-chances. Other groups were, to various degrees, ambivalent. Above all, it was regarded as particularly notable that none of the assumptions noted above that group leaders held about other's views were sustained by the distribution of responses. They were redefined as myths.

Table 11.2: Distribution of responses to the proposal that one of the processes used to report students' learning should be reports with marks or grades.

Respondents	SA(%)	A(%)	NS(%)	D(%)	SD(%)	Mean	Mode	SD
Sec. parents	14(58.3)	7(29.2)	2(8.3)	0(0)	1(4.2)	1.63	1	0.97
Dist. parents	16(57.1)	9(32.1)	0(0)	2(7.1)	1(3.6)	1.68	1	1.06
Sec. principals	7(33.3)	9(42.9)	1(4.8)	3(14.3)	1(4.8)	2.14	2	1.20
All parents	55(37.2)	51(34.5)	11(7.4)	21(14.2)	10(6.8)	2.19	1	1.26
Sec. teachers	5(20)	13(52)	1(4)	5(20)	1(4)	2.36	2	1.15
Central DEA	1(6.7)	9(60)	3(20)	1(6.7)	1(6.7)	2.47	2	0.99
Prim. parents	25(26)	35(36.5)	9(9.4)	19(19.8)	8(8.3)	2.48	2	1.30
All DEA pers	1(4)	13(52)	6(24)	3(12)	2(8)	2.68	2	1.03
Totals	81(18.8)	124(28.8)	53(12.3)	107(24.9)	65(15.1)	2.89	2	1.37
Dist. teachers	3(10.7)	11(39.3)	3(10.7)	7(25.0)	4(14.3)	2.93	2	1.30
District DEA	0(0)	4(40)	3(30)	2(20)	1(10)	3.00	2	1.05
Dist. principals	0(0)	6(40)	2(13.3)	7(46.7)	0(0)	3.07	4	0.96
All teachers	14(10.6)	37(28)	21(15.9)	39(29.5)	21(15.9)	3.12	4	1.28
Prim. teachers	6(7.6)	13(16.5)	17(21.5)	27(34.2)	16(20.3)	3.43	4	1.21
All principals	11(8.8)	23(18.4)	15(12)	44(35.2)	32(25.6)	3.50	4	1.29
Prim. principals	4(4.5)	8(9)	12(13.5)	34(38.2)	31(34.8)	3.90	4	1.12

Similarly, proposal 1.19, specifically 'that one of the processes that should be used to report on students learning should be the use of state-wide, norm-referenced, standardized tests of literacy and numeracy', had a rich mythology masking a complex set of positions. Table 11.3 summarizes the complexities involved.

Proposal 1.19 was supported by all parents (73.1% SA+A), particularly by district high parents (84.6%), and by district DEA personnel (80%). It was unsupported by district high teachers (28.6%) with district high principals (33.4%) and secondary principals (33.4%) tending to be unsupportive. Primary teachers (40.5%), secondary teachers (48.0%), primary principals (50.0%), central DEA (56.3) were ambivalent. These data were taken by senior DEA staff to mean that the proposal was more likely to be controversial in settings where district high teachers, principals and parents meet, and when teachers' and parents' representatives debate this issue.

Proposal 1.20 (Table 11.1), that 'one of the processes that should be used to report on students learning should be peer appraisal', was regarded with ambivalence by all groups with primary parents (12.6% SA+A) and district high parents (22.2%) unsupportive. Similarly, proposal 1.21, that 'P and Fs/School Councils should review, discuss and report on student learning' was regarded with ambivalence by all groups, with district DEA personnel (20.0% SA+A), all teachers (22.7%) and primary principals (23.5%) unsupportive.

Proposal 1.22 was heard regularly from small and vehement groups of parents and some central DEA personnel during the gathering of qualitative data. It was contended that 'one of the processes that should be used to report on students learning should

Table 11.3: Distribution of responses to the proposal that one of the processes that should be used to report on students' learning should be the use of state-wide, norm-referenced, standardized tests of literacy and numeracy.

Respondents	SA(%)	A(%)	NS(%)	D(%)	SD(%)	Mean	Mode	SD
Sec. parents	3(12.5)	15(62.5)	4(16.7)	2(8.3)	0(0)	2.21	2	0.78
All parents	34(22.1)	74(51)	16(11)	14(9.7)	9(6.2)	2.27	2	1.10
District DEA	0(0)	8(80)	1(10)	1(10)	0(0)	2.30	2	0.68
Prim. parents	22(21.1)	46(48.4)	12(12.6)	9(9.5)	8(8.4)	2.36	2	1.17
All DEA	4(15.4)	13(50)	5(19.2)	3(11.5)	1(3.8)	2.39	2	1.02
Central DEA	4(25)	5(31.3)	4(25)	2(12.5)	1(6.3)	2.44	2	1.21
Totals	56(13.1)	175(41.1)	73(17.1)	86(20.2)	36(8.5)	2.70	2	1.18
Prim. principals	10(11.4)	34(38.6)	15(17)	20(22.7)	9(10.2)	2.82	2	1.21
Sec. teachers	2(8)	10(40)	5(20)	6(24)	2(8)	2.84	2	1.14
All principals	12(9.7)	44(33.5)	23(18.5)	34(27.4)	11(8.2)	2.90	2	1.17
Prim. teachers	5(6.3)	27(34.2)	17(21.5)	20(25.3)	9(11.4)	3.01	2	1.16
All teachers	8(6.1)	44(33.6)	29(22)	85(26.5)	15(11.4)	3.04	2	1.14
Sec. principals	1(4.8)	6(28.6)	5(23.8)	8(28.1)	1(4.8)	3.10	4	1.04
Dist. principals	1(6.7)	4(26.7)	3(20)	6(40)	1(6.7)	3.13	4	1.13
Dist. teachers	1(3.6)	7(25)	7(25)	9(32.1)	4(14.3)	3.29	4	1.12

Table 11.4: Distribution of responses to the proposal that one of the processes that should be used to report on students' learning should be reports that allow parents to compare their child with others

Respondents	SA(%)	A(%)	NS(%)	D(%)	SD(%)	Mean	Mode	SD
Central DEA	2(12.5)	6(37.5)	2(12.5)	5(31.3)	1(6.3)	2.81	2	1.22
All DEA	2(7.7)	7(26.9)	4(15.4)	10(38.5)	3(11.5)	3.19	4	1.20
Dist. parents	3(10.7)	7(25)	2(7.1)	10(35.7)	6(21.4)	3.32	4	1.36
Sec. parents	2(8.3)	4(16.7)	3(12.5)	10(41.7)	5(20.8)	3.50	4	1.25
Sec. principals	2(9.5)	3(14.3)	0(0)	12(57.1)	4(19)	3.62	4	1.24
All parents	10(6.8)	24(16.3)	13(8.8)	59(40.1)	41(27.9)	3.66	4	1.24
Prim. parents	5(5.3)	13(13.7)	8(8.4)	39(41.1)	30(31.6)	3.80	4	1.18
District DEA	0(0)	1(10)	2(20)	5(50)	2(20)	3.80	4	0.92
Totals	19(4.4)	54(12.6)	39(9.1)	157(36.7)	159(37.1)	3.89	5	1.17
Dist. principals	0(0)	3(20)	1(6.7)	5(33.3)	6(40)	3.93	5	1.16
All principals	6(4.8)	14(11.2)	7(5.6)	45(36)	53(42.4)	4.00	5	1.17
Prim. principals	4(4.5)	7(9.0)	6(6.7)	20(31.8)	43(48.3)	4.10	5	1.15
Prim. teachers	0(0)	7(8.9)	9(11.4)	26(32.9)	35(44.3)	4.16	5	0.96
All teachers	1(0.8)	9(6.8)	15(11.4)	43(32.6)	62(47.0)	4.20	5	0.95
Sec. teachers	1(4)	0(0)	4(16)	7(28)	13(52)	4.24	5	1.01
Dist. teachers	0(0)	2(7.1)	2(7.1)	10(35.7)	14(50)	4.29	5	0.90

be reports that allow parents to compare their child with others'. Table 11.4 summarizes the patterns of measured support.

The most evident feature of Table 11.4 is that while district high parents (35.7% SA+A) were ambivalent, and there was a bloc of particularly strong support among central DEA personnel (50% SA+A), all other groups were unsupportive (i.e. <30% SA+A). This was interpreted as revealing the presence of a myth concerning client demand. The generally high standard deviations were also taken to indicate that the proposal was likely to be controversial within many groups. Given the overall patterns, specifically that 73.8% of all surveyed disagreed or strongly disagreed with this proposal, it was quickly decided that it would continue to be difficult to sustain support for this policy in locally managed Tasmanian state schools.

As indicated in Chapter 9, the use of 10N, 10R, 14N and 14R standardized tests was quietly suspended within a month of these findings being given to senior DEA officials. A state project team was then assembled and asked to devise a method of monitoring learning in ways that would provide both qualitative and quantitative data, and, give teachers a central role in the process. This policy process is referred to again in Chapter 15.

Accountability Criteria for Evaluating Students' Learning

Table 11.5 provides the 13 policy options suggested by all groups and ranked by means. Options 2.1–2.5 were supported by teachers, principals, parents and system

Table 11.5: Support for criteria for evaluating students' learning

Option	Supported, i.e. in all groups SA+A >70%	Mean	SD	% SA+A
2.1	Measures of individual progress.	1.77	0.63	93.2
2.2	Results of objective assessment should be used to plan improvements to students' learning.	1.85	0.57	92.7
2.3	The attitude of the student to school, teachers, peers, learning and homework.	1.90	0.70	89.4
2.4	Measures of students' self-esteem and life skills.	1.94	0.72	84.4
2.5	Performance indicators developed within schools by teachers.	1.95	0.71	85.4
	Ambivalence in at least one group, SA+A <70% and >30%			
2.6	Judgements by teachers.	1.99	0.77	84.2
2.7	The performance indicators outlined in state and national policy documents (e.g. Frameworks, Profiles).	2.13	0.74	77.8
2.8	Indicators developed jointly by parent/teacher/student.	2.17	0.96	72.7
2.9	Indicators from research literature should be included in the criteria used to plan improvements to students' learning.	2.23	0.62	72.0
2.10	Performance indicators developed by teachers at subject moderation meetings.	2.24	0.75	71.4
2.11	Criteria developed by research in the classroom.	2.24	0.74	66.4
2.12	Rates of participation by students (attendance and retention).	2.54	0.94	57.5
2.13	Parental expectations.	2.96	1.03	37.6

administrators. The level of support for options 2.6–2.13 was classified as ambivalent using the criteria described above.

Again, special attention was paid to items that nearly gained across-the-board support. Proposal 2.6, that 'judgements by teachers should be used to evaluate students' learning' was supported by all groups except secondary parents (65.2% SA+A) who exhibited a small degree of ambivalence. Similarly, all groups were supportive of proposal 2.7, that is, 'the evaluation of students' learning should use the performance indicators outlined in state and national policy documents (e.g. Frameworks, Profiles)', except secondary parents (56.5% SA+A) and primary parents (62.8%) who were ambivalent. Given the relative recency of national curriculum profiles, this finding was taken to mean that the newly designed package of materials and a public information campaign then under way should also target school communities. This interpretation was acted upon.

The items that attracted some ambivalence were examined closely. The use of 'indicators developed jointly by parent/teacher/student', i.e. proposal 2.8, was supported by all DEA personnel (80.7% SA+A), primary parents (84.1%), secondary parents (83.3%), district high principals (73.3%) and primary teachers (74.7%). Different degrees of ambivalence were evident in the responses by primary principals

(67.8%), district high parents (67.3%), secondary principals (65%), district high teachers (64.3%) and secondary teachers (44%). The outlier response by secondary teachers is discussed in a later chapter.

Proposal 2.9, that 'indicators from research literature should be included in the criteria used to plan improvements to students' learning', was supported by all groups except all teachers (62.2% SA+A) who indicated some ambivalence. Likewise, all groups supported proposal 2.10, the use of 'performance indicators developed by teachers at subject moderation meetings', except primary principals (58%), primary teachers (60.8%) and primary parents (65.2%) who were ambivalent. This was regarded as an unsurprising finding since primary personnel had little experience of moderation. On the other hand, the unexpected level of support from secondary and district high teachers suggests was interpreted to mean that an equivalent to moderation in primary schools could be a very effective professional development strategy with regard to teaching practices.

'Criteria developed by research in the classroom', proposal 2.11, was supported by all DEA personnel (80.8%), all parents (75.2%), district high principals (78.5%) and secondary teachers (74%), with the other groups ambivalent. This was interpreted as being a policy option that would probably attract strong support when the benefits were more widely appreciated. All parents (71%) supported proposal 2.12, that is, 'rates of participation by students (attendance and retention) should be used to indicate students' learning', with all other groups ambivalent. The view was taken that while it was a 'too crude' proxy of learning, however convenient, and that it should really be disregarded for such a purpose, it would be retained for another reason; it would help keep parents accountable for enforcing attendance.

Proposal 2.13, that 'parental expectations should be used as criteria to evaluate students' learning', was unsupported by secondary teachers (8% SA+A) with all other groups expressing ambivalence. I come back to the nature of parents' preferences in a later chapter where they are interpreted in wider contexts.

Accountability Processes and Teachers' Teaching

Table 11.6 provides the 31 policy options suggested by all groups. They are ranked by their means. Options 3.1–3.8 were supported by all groups. The level of support for options 3.9–3.27 was classified as ambivalent.

Once again, the attention of DEA officials focused on items that nearly drew general support. There was a reflex action to search for touchstone. Proposal 3.9, that 'one of the accountability processes that should be used to collect and report data and to improve teachers' teaching should include negotiating new goals for professional development', was supported by all groups except by secondary parents (65.2% SA+A) who were ambivalent. All groups supported 'the appraisal of student outcomes' except district high teachers (67.9%) and secondary teachers (68%) who indicated minor degrees of ambivalence. 'The documentation of best practices', proposal 3.11, was supported by all groups except by primary parents (57.4%), primary teachers (63.3%) and secondary parents (65.2%) who were ambivalent. It was assumed that discussion would lead to support for these items.

Table 11.6: Support for processes to collect data, and report on and improve teachers' teaching

Option	Supported, i.e. in all groups SA+A >70%	Mean	SD	% SA+A
3.1	Discussion between colleagues.	1.38	0.56	98.4
3.2	Planned development of teachers.	1.49	0.66	95.8
3.3	Training and support to identify and cope with students 'at risk'.	1.54	0.58	97.7
3.4	Teacher appraisals should be reported to the individual teacher.	1.64	0.63	92.7
3.5	Co-operative learning between colleagues (e.g. mentoring).	1.80	0.64	92.0
3.6	Self-evaluation.	1.82	0.79	89.7
3.7	Encourage teachers to read and do research.	1.86	0.76	88.0
3.8	Newly appointed teachers should be given a transition program.	1.86	0.84	80.4
	Ambivalence in at least one group, SA+A <70% and >30%			
3.9	Negotiating new goals for professional development.	1.92	0.73	88.3
3.10	Feedback and appraisal by peers.	2.00	0.82	81.5
3.11	Documentation of best practices.	2.06	0.75	78.1
3.12	Appraisal of student outcomes.	2.09	0.75	80.3
3.13	An appraisal of planning.	2.16	0.72	81.6
3.14	The school review process.	2.22	0.87	66.4
3.15	Appraisal discussed between the individual and senior staff.	2.28	0.98	70.4
3.16	The planned development of classrooms.	2.28	0.78	65.3
3.17	Feedback and appraisal by more senior school colleagues.	2.44	0.86	65.7
3.18	Summative report on standardized test results should be reported to the individual teacher.	2.47	0.96	63.4
3.19	Quality of teaching should be reported to district or system level for promotion and school development purposes.	2.55	0.98	55.4
3.20	Parents to be given opportunities to develop as co-teachers.	2.58	0.96	55.2
3.21	Feedback and appraisal by students.	2.59	0.98	65.9
3.22	Feedback and appraisal by parents.	2.60	1.05	52.8
3.23	Assessment of the teacher's contribution to school planning.	2.72	0.98	52.3
3.24	Parents should be given opportunities to be consulted and/or co-plan teaching programs.	2.88	1.09	42.3
3.25	Feedback and appraisal by an independent expert.	2.89	1.11	36.0
3.26	Feedback and appraisal by the P and F/School Council.	3.11	1.06	32.0
3.27	General outcomes of teacher appraisals should be reported to school colleagues, as part of the professional development program.	3.20	1.03	30.6
	Unsupported, i.e. SA+A in all groups < 30%			
3.28	Schools to network more effectively with the teachers' union.	2.96	0.93	28.8
3.29	P and F/School Council discuss teacher and classroom development strategies	3.23	1.11	29.0
3.30	The selection of teachers should be more localized.	3.35	1.06	23.9
3.31	General outcomes of teacher appraisals to be reported to school parents as part of the school planning and development program.	3.69	1.11	19.8

Proposal 3.13 suggested that 'one of the accountability processes that should be used to collect and report data and to improve teachers' teaching should be an appraisal of planning'. It was supported by all groups except primary teachers (64.6% SA+A) who were slightly ambivalent. Putting the school review process to similar purpose, proposal 3.14, was greeted with ambivalence by all parents (61.9%) and all teachers (59.8%) but supported by all DEA personnel (88.5%) and all principals (73.2%). Proposal 3.23, that 'an assessment of the teacher's contribution to school planning should be used to help improve teachers' teaching' was regarded with ambivalence by all groups. The idea that 'parents should be given opportunities to be consulted and/or co-plan teaching programs', proposal 3.24, was regarded with ambivalence by all groups except for primary teachers (26.6%) and secondary teachers (28.0%) who were unsupportive. An emerging theme was noted by DEA officials at this point; exclusivist professionalism. It was also noted by teachers and is addressed in Chapter 13.

The 'planned development of classrooms' was proposed as means of improving the quality of teachers teaching. This proposal, 3.16, drew support from secondary parents (83.4% SA+A), district high principals (73.4%) and secondary teachers (72%). All other groups were ambivalent. As indicated in Chapters 9 and 10, this distribution of responses was quickly interpreted as an example of an unfamiliar yet obviously creative idea that seemed to improve its level of support with discussion.

Proposal 3.17, that 'feedback and appraisal by more senior school colleagues should be used', was supported by secondary principals (85.7), district high school parents (85.2%), central DEA personnel (81.3%), secondary parents (79.2%), district high teachers (71.4%) and marginally by primary principals (69.7%), with all other groups ambivalent. The general interpretation was that leadership was an expectation of but not exclusively the duty of those who hold positions of responsibility.

Proposals 3.18–3.20 were greeted with general ambivalence with some groups supportive. Most, for example, were ambivalent about the idea that 'a summative report on standardized test results should be reported to the individual teacher', with district high parents (81.5% SA+A) and secondary parents (83.3%) supportive. The suggestion that 'the quality of teaching should be reported to district or system level for promotion and school development purposes' also drew mixed responses, with secondary parents (87.5%) only supportive. All groups were similarly ambivalent about the suggestion that 'parents should to be given opportunities to develop as co-teachers' except district high parents (70.4%) and DEA personnel (70%) who were supportive. These findings were thought to be too controversial to be useful by DEA officials.

Five proposals tested which feedback and appraisal processes the respondents believed should be used to collect data, report on and improve teachers' teaching. 'Feedback and appraisal by peers', proposal 3.10, was supported by all groups except by primary parents (66.3% SA+A) who were ambivalent. 'Feedback and appraisal by students' was greeted with ambivalence by all groups less all DEA personnel (84.6%) and district high school principals (85.7%) who were supportive. 'Feedback and appraisal by parents' was also regarded with ambivalence except for primary parents (82.7%) who supported this proposal. 'Feedback and appraisal by an independent expert', proposal 3.25, was greeted with ambivalence by all groups except for primary teachers (22.8%), secondary principals (23.8%), primary principals (25.8%) and district high teachers (28.6%) who were all unsupportive. 'Feedback and appraisal by the P and F/School Council', proposal 3.26, was regarded with ambivalence by all parents (51.4%) and all DEA personnel (34.6%) or as insupportable by all teachers (18.9%) and all principals (22.4%). This array of responses was interpreted as meaning that feedback policies and practices were worthy of immediate attention and sustained development at site and district level, particularly as part of the school review process.

Proposal 3.27, that 'the general outcomes of teacher appraisals should be reported to school colleagues as part of the professional development program', was regarded with ambivalence by all parents (38.9% SA+A), all DEA personnel (36.0%), district high principals (35.7%) and secondary principals (33.3%) with all other groups unsupportive. The suggestion that 'an appraisal should be discussed between the individual and senior staff' was supported by all parents (76.0%), primary principals (72.8%) and district high teachers (82.2%). All other groups were ambivalent. The view taken by senior DEA staff was that teacher appraisal policy was a worthy idea, but given the history of the mechanism as largely symbolic policy and practice in Tasmania, unlikely to regain more than a small foothold.

Accountability Criteria for Evaluating Teachers' Teaching

Eighteen policy options were suggested by all groups, ranked by the means of subgroup and group responses and are provided in Table 11.7. Options 4.1–4.11 were supported. The level of support for options 4.12–4.18 was classified as ambivalent.

Again, items that nearly attracted general support were closely examined. Proposal 4.12, that 'student progress should be used as a criterion to evaluate teachers' teaching', was supported by all groups except district high school teachers (57.3% SA+A) who were ambivalent. The special challenges of district high schools were addressed briefly. It was agreed that student progress was, by definition, the most accurate indicator of learning, it did not follow that the quality of teachers' teaching was an accurate predictor of learning or that accountability for learning should be solely attributed to the quality of teaching. Given the dangers of over attribution, the view was taken that this criterion should only be used in conjunction with and moderated by other touchstone items.

Using 'communication skills with stakeholders' as a criterion was supported by all principals (87.2%) and all DEA personnel (80.7%) with all teachers (56.8%) and all parents (54.2%) ambivalent. Putting 'the competencies outlined by departmental/system job descriptions' to the same purpose was supported by secondary teachers (80%), secondary principals (80%), district high parents (75.1%) and all DEA personnel (88%) with all other groups were ambivalent.

'Using the effectiveness of teachers' written records and plans' as a criterion to evaluate teachers' teaching, proposal 4.15, was supported by primary principals (77.5% SA+A), secondary principals (72.4%), all parents (74.1%) and district DEA personnel (90%) with all others groups ambivalent. All groups were ambivalent about using two other criteria when evaluating the quality of teachers' teaching; 'the extent of leadership services performed by teachers within the school', and, 'teachers' participa-

ACCOUNTABILITY POLICY PERFERENCES IN TASMANIA 203

Table 11.7: Support for criteria for evaluating teachers' teaching

Option	Supported, i.e. in all groups SA+A >70%	Mean	SD	% SA+A
4.1	Classroom environment.	1.60	0.66	95.3
4.2	Interpersonal communications within the classroom.	1.70	0.61	94.7
4.3	Behaviour management skills.	1.79	0.67	91.5
4.4	Organizational skills.	1.79	0.71	92.0
4.5	Teachers' knowledge (of subject matter and child/adolescent development).	1.86	0.68	90.1
4.6	Teachers' attitude towards students, parents and colleagues.	1.97	0.75	85.1
4.7	How well work is set, monitored and marked.	2.00	0.84	83.5
4.8	Effectiveness in implementing school and curriculum policies.	2.02	0.63	85.7
4.9	Instructional expertise.	2.03	0.66	82.3
4.10	Willingness to engage in continuing professional development.	2.08	1.03	77.9
4.11	The attitude of children (e.g. enthusiasm).	2.10	0.88	76.5
	Ambivalence in at least one group, SA+A <70% and >30%			
4.12	Student progress.	1.99	0.93	81.0
4.13	Communication skills with stakeholders.	2.23	0.76	67.0
4.14	Competencies outlined by departmental/system job descriptions.	2.40	0.87	63.6
4.15	The effectiveness of teachers' written records and plans.	2.46	0.97	65.8
4.16	The extent of leadership services performed by teachers within the school.	2.66	0.95	52.0
4.17	Teachers' participation in school and community activities.	2.66	0.95	52.0
4.18	Students' achievement levels as listed in the K-12 Framework.	2.83	1.06	45.6

tion in school and community activities'. Proposal 4.18, that 'students' achievement levels as listed in the K-12 Framework' be used as a criterion for evaluating teaching was supported by secondary parents (79.2%) with all other groups ambivalent. These findings were regarded as potentially useful by DEA officials.

Accountability Processes relating to Leaders' Leadership

Twenty-six accountability processes were suggested by all groups to collect and report data and to improve leaders' leadership. They were ranked by means and are presented in Table 11.8. Options 5.1–5.5 attracted support from teachers, principals, parents and system administrators. The level of support for options 5.6–5.20 was considered to be ambivalent. Policy options 5.21–5.26 were found to be unsupported.

Table 11.8: Support for processes to collect data, report on and improve leaders' leadership

Option	Supported, i.e. in all groups SA+A >70%	Mean	SD	% SA+A
5.1	An appraisal of the support and feedback given to staff.	1.78	0.62	91.8
5.2	Evaluate the coherence between the school vision, the development plan and outcomes.	1.84	0.65	90.3
5.3	An appraisal of the extent to which they both provide and generate a school vision.	1.89	0.67	87.7
5.4	Improved by skill development programs, especially in the areas of governance and management.	1.90	0.70	87.1
5.5	Improved by using feedback from staff.	2.06	0.63	84.3
	Ambivalence, in at least one group SA+A <70% and >30%			
5.6	Self-appraisal.	1.96	0.87	83.3
5.7	Improved by peer networks that share reflections on the challenges of practice.	1.97	0.72	79.7
5.8	An appraisal of how well they report to the parents and the community.	2.11	0.62	82.1
5.9	Survey of the school climate.	2.17	0.74	74.4
5.10	An appraisal of policy-making strategies used.	2.18	0.67	74.7
5.11	An appraisal by school colleagues.	2.24	0.82	74.0
5.12	Peer appraisal.	2.25	0.83	70.7
5.13	Improved by using feedback from parents and students.	2.26	0.80	71.7
5.14	Improved by using a mentoring process.	2.28	0.82	60.5
5.15	An appraisal of the quality of external liaison.	2.40	0.79	59.8
5.16	Parental, professional and DEA involvement in the selection of principals.	2.42	1.05	60.3
5.17	An appraisal by the P and F/School Council.	2.76	1.13	47.2
5.18	Appraisals should be reported to the DEA.	2.76	0.97	43.3
5.19	An appraisal by the DEA.	2.78	1.02	45.5
5.20	Appraisals should be reported to the individual and to school colleagues as part of the professional development program.	2.99	0.97	34.2
	Unsupported, i.e. in all groups SA+A < 30%			
5.21	Improved by overseas exchanges.	3.02	0.93	28.8
5.22	An appraisal by the community.	3.13	0.92	25.7
5.23	Improved by having fixed term and negotiated performance contracts.	3.33	1.23	27.4
5.24	The selection of leaders should be more localized.	3.35	0.97	18.1
5.25	The P and F/School Council should set the policies about the preferred nature of leadership services.	3.35	1.02	21.9
5.26	Appraisals should be reported to parents as part of the school development program.	3.37	1.03	21.3

There were a number of appraisal processes suggested for collecting data, reporting on and improving leaders' leadership that drew ambivalent degrees of support. Again this was interpreted as meaning that current appraisal policies lacked substance and legitimacy. Uncertainty related first to who should conduct appraisals. Leaders' 'self-appraisal' was supported by all groups except district high parents (60.7% SA+A) and secondary parents (50%) who were ambivalent. All groups supported 'an appraisal by school colleagues' except all parents (64.1%) who were marginally ambivalent. 'Peer appraisal' was supported by all groups except primary parents (51%) and secondary parents (60.6%) who were ambivalent. All groups were ambivalent, none supportive or unsupportive, of 'an appraisal by the P and F/School Council'. 'An appraisal by the DEA' was regarded by all groups with ambivalence except secondary teachers (24%) who were unsupportive and district high principals (73.4%) who were, conversely, supportive. All groups were unsupportive of 'an appraisal by the community' except primary parents (32.3%) and all DEA personnel 34.6%) who were ambivalent.

Uncertainty was also considered to be related to the substance of appraisals. An appraisal of 'how well leaders report to the parents and the community', for example, was supported as a process by all groups except primary teachers (65% SA+A) who exhibited some ambivalence. An appraisal of the policy-making strategies used by leaders was supported by all groups except primary parents (67.4%) and secondary parents (65.2%) who were marginally ambivalent. The proposal to appraise the quality of external liaison drew ambivalent support from all groups except district high parents (75%) and district DEA personnel (90%) who were supportive.

Some uncertainty was interpreted as being related to the reporting of appraisals. Proposal 5.18, that 'appraisals of leadership services should be reported to the DEA', was supported by secondary parents (70.9% SA+A) while district high teachers (28.6%), district high principals (26.7%) and district DEA personnel (10%) were unsupportive. All other groups were ambivalent. All groups were ambivalent about the suggestion that 'appraisals should be reported to the individual and to school colleagues as part of the professional development programme' except primary teachers (27.9%), primary principals (27.6%) and central DEA personnel (20%) who were unsupportive.

There were also ambivalent views acknowledged concerning improvement strategies. 'Peer networks to share reflections on the challenges of practice' was supported by all groups except some primary teachers (67.1% SA+A). 'Surveying the school climate' was supported by all groups with the exception of district high teachers (53.5%), secondary teachers (64%), district high principals (60%) and primary parents 63.3%) who were, to various degrees, ambivalent. 'Using feedback from parents and students' was supported by all groups except all teachers (55.3%) and district high school principals (66.7%) who exhibited degrees of ambivalence. 'Improving leadership by using a mentoring process' was a proposal all groups supported, except primary teachers (49.3%), district high teachers (60.8%) and all parents (45.9%) who were ambivalent. 'Overseas exchanges' to improve leadership were unsupported by all groups, except all principals (52.8%) and district DEA personnel (40%), who were ambivalent.

Proposed appointment processes and service conditions evinced mixed views. Responses to proposal 5.16, which called for 'parental, professional and departmental involvement in the selection of principals' were complex but generally supportive or ambivalent. Teachers views were bimodal and ambivalent; while 47.7% agreed or strongly agreed, 22.7% were unsure with 28.8% indicating that they disagreed or strongly disagreed with the proposal. Strongest support came from secondary principals (79% SA+A), district principals (86.7%), district DEA personnel (80%) and district parents (77.7%). The greatest ambivalence was expressed by secondary teachers (32% SA+A), primary teachers (48.1%) and district teachers (60.7%). The detail is evident in Table 11.9.

Responses to the proposal that 'the selection of leaders should be more localized' showed that all parents (28.7% SA+A) were ambivalent with all other groups unsupportive. All groups were unsupportive of the suggestion that 'the P and F/School Council should set the policies about the preferred nature of leadership services in schools'. All teachers (24.3%) and all principals (14.6%) were unsupportive while all parents (38.7%) and all DEA personnel (38.4%) were ambivalent about the suggestion that 'leadership would be improved by having fixed term and negotiated performance contracts'.

Accountability Criteria for Evaluating Leaders' Leadership

A total of 22 accountability criteria were suggested to evaluate leaders' leadership services. They are provided in Table 11.10 and are ranked by the means of responses.

Table 11.9: Distribution of responses to the proposal that, to improve the quality of leadership, there should be parental, professional and departmental involvement in the selection of principals.

Respondents	SA(%)	A(%)	NS(%)	D(%)	SD(%)	Mean	Mode	SD
Sec. principals	6(28.6)	11(52.4)	4(19.0)	0(0)	0(0)	1.91	2	0.70
Dist. principals	4(26.7)	9(60)	1(6.7)	1(6.7)	0(0)	1.93	2	0.80
Dist. DEA	3(30)	5(50)	1(10)	1(10)	0(0)	2.00	2	0.94
Dist. parents	9(33.3)	12(44.4)	3(11.1)	3(11.1)	0(0)	2.00	2	0.96
Sec. parents	7(29.2)	9(37.5)	7(29.2)	1(4.2)	0(0)	2.08	2	0.88
All parents	34(23.6)	63(43.8)	27(18.8)	18(12.5)	2(1.4)	2.24	2	1.00
All DEA	5(19.2)	10(38.5)	9(34.6)	2(7.7)	0(0)	2.31	2	0.88
All principals	23(18.7)	58(47.2)	23(18.7)	15(12.2)	4(3.3)	2.34	2	1.02
Prim. parents	18(19.4)	42(45.2)	17(18.3)	14(15.1)	2(2.2)	2.36	2	1.03
Totals	79(18.6)	177(41.7)	89(21)	68(16)	11(2.6)	2.42	2	1.05
Central DEA	2(12.5)	5(31.3)	8(50)	1(6.3)	0(0)	2.50	3	0.82
Prim. principals	13(14.9)	38(43.7)	18(20.7)	14(16.1)	4(4.6)	2.52	2	1.08
Dist. teachers	6(21.4)	11(39.3)	4(14.3)	4(14.3)	3(10.7)	2.54	2	1.29
Prim. teachers	9(11.4)	29(36.7)	17(21.5)	21(26.6)	2(2.5)	2.72	2	1.07
All teachers	17(12.9)	46(34.8)	30(22.7)	33(25)	5(3.8)	2.72	2	1.10
Sec. teachers	2(8)	6(24)	9(36)	8(32)	0(0)	2.92	3	0.95

Table 11.10: Support for criteria to evaluate leaders' leadership

Option	Supported, i.e. in all groups SA+A >70%	Mean	SD	% SA+A
6.1	Capacity to hear and care for others.	1.56	0.57	96.8
6.2	The openness and climate/tone of the school.	1.76	0.73	88.8
6.3	Evidence of student and teacher morale and motivation.	1.77	0.72	88.8
6.4	Ability to plan outcomes and achieve priorities.	1.78	0.57	94.2
6.5	Capacity to make and implement policy.	1.86	0.62	90.9
6.6	Management and organizational skills, especially in evaluation, budgeting and governance.	1.89	0.68	88.3
6.7	The extent to which staff support their leaders.	1.93	0.75	84.5
6.8	The extent to which collaborative decision making is used.	1.93	0.75	83.4
6.9	The extent to which creativity and productivity are valued in the school.	2.04	0.70	79.5
6.10	The quality of internal and external communications.	2.10	0.74	79.9
	Ambivalence, in at least one group SA+A <70% and >30			
6.11	Evidence of learning by staff and students.	2.16	0.84	73.6
6.12	The extent of professional development within the school.	2.16	0.89	74.9
6.13	Evidence of the quality of teaching by the staff.	2.21	0.86	74.6
6.14	Capacities as learners and researchers.	2.29	0.81	68.6
6.15	Indicators from research literature used to plan improvements to leadership services.	2.32	0.71	61.1
6.16	Performance indicators in the guidelines provided by the DEA.	2.36	0.74	62.9
6.17	Recommendations from school reviews.	2.46	0.76	55.9
6.18	The expectations of the community.	2.46	0.85	61.0
6.19	The extent to which parents support their school leaders.	2.54	0.91	54.4
6.20	The quality of the physical environment.	2.53	0.92	58.8
6.21	Indicators adapted from national competency standards for managers.	2.69	0.77	40.6
6.22	Leaders' relevant qualifications.	2.75	1.06	51.7

All groups of teachers, principals, parents and system administrators supported Options 6.1–6.10. The level of support for options 6.11–6.22 was considered to be ambivalent.

As above, initial interpretation by DEA officials focused on items that nearly attracted full support. The patterns of support in some groups was taken to mean that concerns existed about the types of evidence proposed. For example, using 'evidence of learning by staff and students' was supported by all groups except all teachers (63.6% SA+A) who exhibited some minor ambivalence. All groups supported 'the extent of professional development within the school' being used as an indicator of leadership with the exceptions of all teachers (66.6%), primary parents (67.4%) and district high parents (66.7%) who shared minor ambivalence. Similarly, 'evidence of the quality of teaching by the staff' was supported as a criterion by all groups although district high teachers (53.6%), secondary teachers (56%) and district high principals (60%) were ambivalent. Leaders' 'capacities as learners and researchers' was also considered an appropriate criterion by all groups with all teachers (55.3%), secondary principals (61.9%) and primary parents (69.9%) exhibiting some ambivalence. Proposal 6.17, 'to develop criteria from the recommendations of school reviews' was greeted with ambivalence by all groups except district DEA personnel (80%), who supported the option. Managing school reviews was and remains one of the major roles of district superintendents.

Criteria from external-to-school sources attracted less support across the board. This was held to indicate another theme; the closedness of some schools to ideas and innovation. Proposal 6.15, that 'indicators from research literature should be used to plan improvements to leadership services', was supported by secondary parents (75% SA+A) and district DEA personnel (70%), but with all other groups ambivalent. The proposal to 'develop performance indicators from the guidelines provided by the DEA' was supported by all DEA personnel (80%), district high school parents (81.5%) and district high school principals (73.3%) but with all other groups ambivalent. Adapting 'indicators from national competency standards for managers' was not supported by secondary teachers (24% SA+A), primary principals (28%) and secondary principals (23.8%). district DEA personnel (70%), however, were supportive with all other groups ambivalent. These ideas were not seen as viable proposals by senior DEA staff.

Using 'the expectations of the community' as evaluative criteria was supported by district high school parents (74.1% SA+A) with all other groups ambivalent. All groups were ambivalent about evaluating leadership services in terms of 'the extent to which parents support their school leaders'. All groups were also ambivalent about 'using the quality of the physical environment to evaluate leadership services' except district DEA personnel (80%) who expressed support. The proposal to employ leaders' 'relevant qualifications' as a criterion drew ambivalent responses from all groups except secondary parents (75%) who supported this option. These suggestions were also regarded as being in the 'too hard basket'.

Two General Issues

Two general issues were raised in a number of forums. The first was the claim that, in order to improve the quality of teaching, learning and leadership, the number of acting and temporary appointments should be reduced. Second was the view that to improve the quality of teaching, learning and leadership, a formal complaints procedure should be negotiated by school communities within state guidelines that protect all legitimate interests with due process and prevent anyone from being victimized. Responses to the first of these two proposals are summarized in Table 11.11.

ACCOUNTABILITY POLICY PERFERENCES IN TASMANIA

Table 11.11: Distribution of responses to the proposal that, in order to improve the quality of teaching, learning and leadership, the number of acting and temporary appointments should be reduced

Respondents	SA(%)	A(%)	NS(%)	D(%)	SD(%)	Mean	Mode	SD
Dist. parents	13(48.1)	8(29.6)	4(14.8)	2(7.4)	0(0)	1.82	1	0.96
Prim. parents	49(52.7)	20(21.5)	15(16.1)	7(7.5)	2(2.2)	1.85	1	1.08
Dist. teachers	13(46.4)	10(35.7)	1(3.6)	2(7.1)	2(7.1)	1.93	1	1.22
All parents	68(47.2)	37(25.7)	23(16.0)	10(6.9)	6(4.2)	1.95	1	1.14
All teachers	54(40.9)	32(24.6)	21(16.2)	13(10.0)	10(7.7)	2.18	1	1.29
Prim. teacher	1(39.2)	17(21.5)	16(20.3)	6(7.6)	7(8.9)	2.23	1	1.31
Sec. teachers	10(40)	5(20)	4(16)	5(20)	1(4)	2.28	1	1.31
Totals	160(37.9)	99(23.5)	74(17.5)	57(13.5)	32(7.6)	2.29	1	1.30
Prim. principals	25(28.1)	26(29.2)	17(19.1)	12(13.5)	8(9.0)	2.46	2	1.29
Sec. parents	6(25.0)	9(37.5)	4(16.7)	1(4.2)	4(16.7)	2.50	2	1.38
Sec. principals	8(38.1)	3(14.3)	4(19.0)	3(14.3)	3(14.3)	2.52	1	1.50
All principals	35(28.5)	29(23.6)	23(18.7)	24(19.5)	12(9.8)	2.59	1	1.34
Central DEA	3(20)	0(0)	5(35.3)	6(40)	1(6.7)	3.13	4	1.25
All DEA	3(12)	1(4)	7(28)	10(40)	4(16)	3.44	4	1.19
Dist. principals	2(14.3)	0(0)	2(14.3)	9(64.3)	1(7.1)	3.50	4	1.16
District DEA	0(0)	1(10)	2(20)	4(40)	3(30)	3.90	4	0.99

Views on this proposal were quickly interpreted by DEA senior staff as being polarized. All DEA Personnel (16% SA+A) were not supportive. District high teachers (80.1%) were supportive with primary teachers (60.7%) and secondary teachers (60%) exhibiting some ambivalence. While all parents (72.9%) were supportive, especially district high parents (97.7%) and primary parents (74.2%), there was some ambivalence amongst secondary parents (62.5%). primary principals (57.3%) and secondary school principals (52.4%) were ambivalent with district high school principals (14.3%) not supportive of this proposal. The issue was regarded as also being in the 'too hard basket' until such time as political will changed. And the matter was partially resolved by political interventions in late 1995 not unrelated to the state elections held in early 1996. The Minister announced a number of new permanent positions and a new transfer policy. It was then assumed that the findings above were obsolete.

There was less nervousness with regard to the suggestion that a 'formal complaints procedure should be negotiated by school communities within state guidelines that protect all legitimate interests with due process and prevent anyone from being victimized'. It was soon regarded by DEA officials as 'a good idea whose time had nearly come'. All groups were either ambivalent or supportive. All teachers (59.8%) exhibited degrees of ambivalence, that is, secondary teachers (64%), primary teachers (60.8%) and district high teachers (53.6%). All principals (68%) exhibited a wider range of ambivalence; secondary principals (66.7%), primary principals (60.8%) and district high school principals (43.3%). All parents (80.6%) supported the proposal; primary parents (79.6%), district high parents (82.4%) and secondary parents (82.3%). District

Table 11.12: Distribution of responses to the proposal that, to improve the quality of teaching, learning and leadership, a formal complaints procedure should be negotiated by school communities within state guidelines that protect all legitimate interests with due process and to prevent anyone from being victimized.

Respondents	SA(%)	A(%)	NS(%)	D(%)	SD(%)	Mean	Mode	SD
Dist. parents	9(33.3)	13(48.1)	5(18.5)	0(0)	0(0)	1.85	2	0.72
All parents	41(28.5)	75(52.1)	22(15.3)	5(3.5)	1(0.7)	1.96	2	0.80
Prim. parents	26(28.0)	48(51.6)	15(16.1)	3(3.2)	1(1.1)	1.98	2	0.82
District DEA	3(30)	4(40)	3(30)	0(0)	0(0)	2.00	2	0.82
Sec. parents	6(25)	14(58.3)	2(8.3)	2(8.3)	0(0)	2.00	2	0.83
Totals	98(23.1)	196(46.2)	83(19.6)	38(9)	9(2.1)	2.21	2	0.97
Prim. principals	22(24.7)	41(46.1)	13(14.6)	9(10.1)	4(4.5)	2.24	2	1.08
All principals	29(23.2)	56(44.8)	21(16.8)	14(11.2)	5(4.0)	2.28	2	1.07
Sec. teachers	5(20)	11(44)	6(24)	2(8)	1(4)	2.32	2	1.03
All DEA	5(20.8)	9(37.5)	7(29.2)	3(12.5)	0(0)	2.33	2	0.96
Sec. principals	5(23.8)	9(42.9)	3(14.3)	3(14.3)	1(4.8)	2.33	2	1.15
Prim. teachers	13(16.5)	35(44.3)	21(26.6)	7(8.9)	2(2.5)	2.36	2	0.95
All teachers	23(17.4)	56(42.4)	33(25)	16(12.1)	3(2.3)	2.39	2	0.99
Dist. principals	2(13.3)	6(40.0)	5(33.3)	2(13.3)	0(0)	2.47	2	0.92
Dist. teachers	5(17.9)	10(35.7)	6(21.4)	7(25)	0(0)	2.54	2	1.07
Central DEA	2(14.3)	5(35.7)	4(28.6)	3(21.4)	0(0)	2.57	2	1.02

DEA personnel (70%) supported the proposal while DEA central personnel (50%) were ambivalent. The details are evident in Table 11.12.

The DEA subsequently encouraged the primary principals' initiative to draft new policy, and to draw their colleagues into the process, especially staff in district high schools. As noted above, the district high schools face special challenges, such as sometimes critical and conservative communities, high levels of professional transparency and isolation, and relatively inexperienced and defensive staff and leadership.

Interim Conclusions

The findings presented above were regarded by senior DEA staff as an immediately useful map of the patterns of support, ambivalence or non-support by the respondents at a particular point in time. They were not seen as explanations or interpretations of the situation. Interpretation and application were deemed by the officials to be matters for the state education policy-community of Tasmania and for school policy-communities. They also took the view that students' and second order interest groups in state education were to be consulted as part of school community policy-making. Decisions were immediately taken to share the findings with school policy-communities and other states, to invite all stakeholder executives to a policy-forum, and to advance the analysis of the data by comparing each stakeholder group responses to all views and by comparing findings with alternative views. These analyses are reported in the coming chapters.

What was also agreed at the first meeting was that the broad patterns of support and some implications were clear in ways not apparent before. Instances were noted where myths had been shattered. There was also a felt desire to assemble the policy-options that attracted more that 70% strong agreement or agreement from all groups and subgroups. These findings were regarded as a map of the policy-touchstone available concerning accountability processes and criteria, processes and criteria that are believed by respondents to be positively and causally related to the quality of learning, teaching and leadership. The discussion of the broad patterns were, in essence, a search for general theories that explained, as one senior official put it, how 'educative accountability actually works'. An equally interesting point agreed was that this 'common ground' was regarded as substantial and that it should be used to help develop a more integrated and comprehensive system accountability policy-and to review practices and policies at each school community site.

It was noted above that the proposals that came very close to attracting universal support were given a great deal of attention, attention that tried to find grounds that would permit them being added to the touchstone group. The senior officials found it helpful to have ambivalent groups and the degree of their ambivalence identified. They soon assumed that discussions with identified groups exhibiting relatively minor degrees of ambivalence could lead to specific policy-options being added to the touchstone options. Table 11.13, for example, is a summary of the touchstone and near-touchstone policy-proposals concerned with students' learning.

Table 11.14 summarizes the touchstone and near-touchstone policy-proposals concerned with teachers' teaching.

Table 11.15 summarizes the touchstone and near-touchstone policy-proposals concerned with leaders' leadership.

It was noted above that the project's reference group of senior officials in the DEA reviewed these initial findings in October 1994. These findings and additional commissioned analyses were presented to a seminar of state educational stakeholder representatives in early 1995 to elicit responses and to deepen the exploration of the implications for system accountability policies. These materials (Macpherson 1995) were also related to other state, national and international initiatives and compared to ideas from other sources. The analyses and interpretations generated through this interactive process are reported in the coming chapters.

Table 11.13: Touchstone and near-touchstone proposals concerned with students' learning

Touchstone processes to collect data, report on, evaluate and improve students' learning	Touchstone criteria for evaluating students' learning
Parent/teacher interviews.	Measures of individual progress.
Teachers evaluating and planning their lessons thoroughly.	Results of objective assessment used to plan improvements to students' learning.
Teachers' observations.	Attitudes of students to school, teachers, peers,
Conferencing between teacher and student.	Measures of students' self-esteem and life skills
Sampling student work (e.g. folios). learning and homework.	Performance indicators developed within schools by teachers.
Teachers evaluating and planning their programs systematically.	
Parent/teacher/student discussions.	
Teachers clearly identifying intended outcomes for each student.	
Parents given many different opportunities to provide views and teaching and curriculum policies explained.	
Reports of student's learning through Newsletters, school magazines, displays and positive PR.	
Reports with clear and accurate descriptions of learning.	
Student's own self-assessment.	
Teacher-designed tests (to measure mastery, to diagnose individual understanding).	
Formative evaluation related to teaching objectives.	

Near-touchstone processes, with ambivalent groups identified	Near-touchstone criteria, with ambivalent groups identified
Teachers use written checklists and running records to monitor student progress, district principals (53.4).	Judgements by teachers, sec. parents (65.2).
Use reports from other support staff (e.g. guidance, welfare, speech and health specialists), central DEA (56.3), sec. principals (65.1).	Performance indicators in state and national policies (e.g. Frameworks, Profiles), sec. parents (56.5), prim. parents (62.8).
Parents given curriculum goals, outcomes and individual expectations at year start, primary teachers (57) primary principals (65.1).	Indicators jointly developed by parents/teachers/students, primary principals (67.8), district parents (67.3), secondary principals (65), district teachers (64.3), secondary teachers (44).
	Indicators from research included as criteria used to plan improvements to students' learning, all teachers (62.2).
	Performance indicators developed by teachers at subject moderation meetings, primary principals (58), primary teachers (60.8), primary parents (65.2).

Table 11.14: Touchstone and near-touchstone proposals concerned with teachers' teaching

Touchstone processes to collect data, report on, evaluate and improve teachers' learning	Touchstone criteria for evaluating teachers' learning
Discussion between colleagues. Planned development of teachers. Training and support to identify and cope with students 'at risk'. Teacher appraisals reported to the individual teacher. Co-operative learning between colleagues (e.g. mentoring). Self-evaluation. Encourage teachers to read and do research. Give newly appointed teachers a transition program.	Classroom environment. Interpersonal communications within the classroom. Behaviour management skills. Organizational skills. Teachers' knowledge (of subject matter and child/adolescent development). Teachers' attitude towards students, parents and colleagues. How well work is set, monitored and marked. Effectiveness in implementing school and curriculum policies. Instructional expertise. Willingness to engage in continuing professional development. Attitude of children (e.g. enthusiasm).

Touchstone processes, with ambivalent groups identified	Near-touchstone criteria, with ambivalent groups identified
Negotiating new goals for professional development, sec. parents (65.2). Feedback and appraisal by peers, prim. parents (66.3). Documentation of best practices, prim. parents (57.4), prim. teachers (63.3), sec. parents (65.2). Appraisal of student outcomes, district teachers (67.9), sec. teachers (68). An appraisal of planning, prim. teachers (64.6).	Student progress, district, teachers (57.3).

Table 11.15: Touchstone and near-touchstone proposals concerned with leaders' leadership

Touchstone processes to collect data, report on, evaluate and improve leaders' leadership	Touchstone criteria for evaluating leaders' leadership
Appraise the support and feedback given to staff.	Capacity to hear and care for others.
Evaluate the coherence between the school vision, the development plan and outcomes.	Openness and climate/tone of the school.
Appraise the extent to which they provide and generate a school	Evidence of student and teacher vision, morale and motivation.
Improve by skill development programs, especially in governance and management.	Ability to plan outcomes and achieve priorities.
Improve by using feedback from staff.	Capacity to make and implement policy.
	Management and organizational skills, especially in evaluation, budgeting and governance.
	Extent to which staff support their leaders.
	Extent to which collaborative decision making is used.
	Extent to which creativity and productivity are valued in the school.
	Quality of internal and external communications.

Touchstone processes, with ambivalent groups identified	Near-touchstone criteria, with ambivalent groups identified
Self-appraisal, district parents (60.7), sec. parents (50).	Evidence of learning by staff and students, all teachers (63.6).
Improved by peer networks that share reflections on the challenges of practice, prim. teachers (67.1).	Extent of professional development within the school, all teachers (66.6), prim. parents (67.4), district parents (66.7).
An appraisal of how well they report to the parents and the community, prim. teachers (65).	Evidence of the quality of teaching by the staff, district teachers (53.6), sec. teachers (56), district principals (60).
Survey of the school climate, district teachers (53.5), sec. teachers (64), district principals 60), prim. parents (63.3).	
An appraisal of policy-making strategies used, prim. parents (67.4), sec. parents (65.2).	
An appraisal by school colleagues, all parents (64.1).	
Peer appraisal, prim. parents (51), sec. parents (60.6).	
Improved by using feedback from parents and students, all teachers (55.3), district principals (66.7).	

12

Tasmanian Parents' Policy Preferences and Implications

Introduction

When the Tasmanian state government devolved many administrative responsibilities to schools in the 1980s and 1990s it was assumed that they would develop fresh management and development capacities and deliver significant economies. This decentralization was part of an attempt to replace bureaucracy with corporate management, replace regional support structures with district personnel who focused on school development, and to displace hierarchy with more supportive collegial networks. In Chapter 9 I showed how the principle of public accountability was redefined in Tasmanian public education as a local issue to be resolved largely through more explicit and consultative planning.

In this chapter I use the interpretations of Tasmanian parents to the findings to suggest that they actually prefer a far more educative and communitarian approach to accountability, and that this view is broadly shared with other key stakeholders; teachers, principals and state government officials. And slightly more so that other stakeholders, parents prefer greater subsidiarity, pluriformity and complimentarity in their schools and education system, rather than neo-centralist and 'self-managing' corporate managerialism, uniformity and comparability. The discussion begins with the orthodoxy of 'self-management'.

The 'Self-managing' School

Most public schools in Tasmania exhibit the characteristics of a 'self-managing school,' not unlike SBM in the United States and LMS in England and Wales discussed in earlier chapters. The introduction of this approach in Australia has been boosted by the confluence of economic, political and ideological forces in the late 1980s. The original formulation of the 'self-managing school' was shown in Chapter 2 to be derived from early US school effectiveness literature and then developed with international consultancies and scholarship. It was also shown that Australian research into school 'self-management' has drawn attention to the dangers of an uncritical faith in corporate managerialism, such as the displacement of educational metavalues like quality pedagogy, democracy and social equity.

The evidence reviewed in Chapters 6–8 above supports the same worrying conclusions. For example, it will be recalled that Wohlstetter and Odden (1992) showed that SBM has become pervasive in the United States without systematic accountability structures. They also found little real delegation of authority, a primary concern for teacher morale and satisfaction, and that the links between SBM and student learning have remained obscure. More recent research (Wohlstetter *et al.* 1995) has shown that none of the 'school charter' laws passed in eleven states by the end of 1994 to tighten accountabilities have linked district support, school improvement and classroom development. The point here is that SBM could be a policy myth that defines educators as solely accountable for student learning while ingratiating neo-centralism in policy making and educational administration, particularly in the management of financial contraction, while the state withdraws from public education.

Similarly, it was demonstrated in Chapters 3–5 above that LMS has transformed the way that schools are managed in England and Wales. It has given expression to a New Right myth of greater educational choice, essentially by imposing new political, managerial and market mechanisms. An array of technical and philosophical problems have been encountered. Yet, despite considerable pressure from the centre, schools are using their self-governing powers to vote repeatedly against 'opting out' of Local Education Authority (LEA) control. Many school communities are refusing to set aside co-operative networks in LEAs in favour of competition. Governors are supporting educators more and more, even to the extent of defying the national government with deficit budgets.

Many English and Welsh schools are developing more educative evaluation and development strategies to supplement the formal and blunt accountability mechanisms of OFSTED's school inspections and standard student assessment tasks. Accumulating evidence also indicates that school governing bodies are providing supportive and advisory services while moving steadily towards the adoption of their legislated role of local public accountability.

It can, therefore, be speculated that the New Right's neo-centralist attempt in England and Wales to create a politics of choice and local contractual accountability is being challenged by a politics of subsidiarity that values responsive accountability. This speculation is examined below with regard to Tasmania when terms have been defined.

Politics of Choice and Subsidiarity

It will be recalled from Chapter 5 that 'contractual accountability' refers to the answerability of educators while 'responsive accountability' is about taking into account the requirements of all interested parties when making educational policy and operational decisions (Halstead 1994, p. 149). 'Subsidiarity' is an organizational principle that holds that 'decisions should be made at the lowest possible level' (Casey 1993, p. 173). The principle of subsidiarity (McBrien 1980, p. 1044):

was first formally articulated by Pope Pius XI in his encyclical *Quadrogessimo Anno* (1931): 'It is a fundamental principle of social philosophy, fixed and unchangeable, that one should not withdraw from individuals and commit to the community what they can accomplish by their own enterprise and industry. So, too, it is an injustice and at the same time a grave evil and a disturbance of right order, to transfer to a larger and higher collectivity functions which can be performed and provided by lesser and subordinate bodies' (cited by John XXII's *Mater et Magistera* 1961, para 53).

Today the subsidiarity principle is taken to imply that 'Any collectivity, before it usurps the power vested in the local body, must show cause why it can discharge that function better, more efficiently, more humanely, more skilfully' (Beare 1995, p. 147). The principle of subsidiarity is antithetical to the neo-centralism driving standardized forms of school 'self-management', SBM and LMS. It has two corollary principles; pluriformity and complimentarity. Pluriformity is the encouragement, development and celebration of diverse problem-solving structures. Complimentarity values collegiality and co-operative action between diverse member units for the greater common good.

To illustrate, the South Australian Commission for Catholic Schools (1987) recognized that, in its role of overall policy formulation, it had to be:

sensitive to the special character of those schools foundered and directed by a Religious Institute ... Conscious of the special charisma of each Religious Institute as a gift to the Church, the Commission shall endeavour to foster that special expression of the elements of Catholic Education which flows from that charisma. In doing this it will be preserving that special pluriformity that has been characteristic of, and very special to, the history of the Catholic school.

Of immediate interest here is the extent to which parents and other stakeholders in the 'home' of 'self-managing' schools, Tasmania, actually prefer contractual or responsive forms of accountability, and the principles of subsidiarity, pluriformity and complimentarity instead of 'self-management,' uniform structures and comparing the performance of learners, teachers and leaders in a context of neo-centralism.

This proposal is not irrelevant in wider settings. The two peak national parent's bodies of Australia, the Australian Council of State School Organisations (ACCSO) and the Australian Parents Council (APC), have recently collaborated to articulate their joint perspective on assessment and reporting (ACCSO/APC 1996). Extensive consultations in affiliated state organizations had identified six key parental needs (p. 5):

to feel welcome and comfortable in their children's school, and confident in offering suggestions and comments; opportunities and encouragement to share knowledge of their children and their children's experience of school with their children's teachers; to realise a partnership with teachers for the children's learning at school; to ensure and be assured that their children achieve optimum levels of literacy and numeracy; written reports covering all facets of their children's progress at school and which describe a relationship to the progress of children their age, and; exit reports encompassing the range of their children's academic and co-curricular achievements and participation at school.

ACCSO/APC (1996, p. 6) concluded that 12 principles should underpin effective, just, equitable and ethically defensible assessment and reporting procedures intended to provide balanced, comprehensive and valid information:

1. Parents are entitled to continuing, quality information regarding their children's education through a variety of reporting mechanisms.
2. Any form of assessment should be integral to the curriculum and designed to inform, support and improve learning outcomes.
3. Assessment and reporting processes should make provision for parent and student input about teaching and learning.
4. Parents and their organizations must have an active role in developing and implementing assessment and reporting policies and processes at the school, the system, the state and the nation.
5. Schools, systems and governments, state and federal, must make explicit and public the purposes for which they wish to collect assessment data.
6. Assessment data must not be used for the purpose of establishing and publishing competitive judgements about schools/systems/states or territories.
7. Parents must be informed by all those who seek such data about student performance, of the uses to which such information will be put.
8. Data collected from students in school should be used in accordance with its stated purposes. Any subsequent uses should be specifically negotiated.
9. Individual student assessments are confidential to the student, his/her parents and appropriate school staff.
10. Parents have the right to withdraw their children from specific system, state-wide and national testing.
11. Assessment data for state-wide or national purposes should be collected by statistically valid, light sampling procedures only.
12. Appropriate appeal mechanisms should be established and made public to protect the rights of students and parents in matters of student assessment and reporting at the school, state and national level.

The case made by ACCSO/APC for improving the assessment and reporting of student learning defined accountability as part of (and subsequent to) formative evaluation, promoted a mutually respectful partnership between stakeholders including the joint interpretation of data, argued for the development of trustworthy databases and benchmarks of achievement, and gave primacy to the interests of learners, parents and responsive professionals.

The sophistication of the ACCSO/APC case was impressive. It was situated in a context of social and cultural change, labour market and technological change, rising demand for participatory policy making and decision making, saliency of educational 'outcomes' and the potential 'narrowing' of curriculum and testing. The case urged caution, ongoing professional development and equal attention to inputs, process and outcomes. It is perhaps not an accident that there are five features common to the TEC survey findings (examined in Chapter 9) and the ACCSO/APC policy recommendations;

(a) the need for coherent systemic policies supported by stakeholders that are to be applied sensitively at classroom, school and system levels,
(b) the need for transparent, educative, fair, sensitive and rigorous processes,
(c) the need for appropriate, explicit and comprehensive criteria,
(d) system accountability obligations remain subordinate or additional to those to be discharged in classrooms and schools, and, therefore, that
(e) systemic processes and criteria are to be derived from or to cohere with rather than determine classroom and school accountability practices.

This set of demands supports the notion that parents of public school children might prefer organizational subsidiarity, pluriformity and complimentarity rather than neo-centralist accountability structures characterized by corporate managerialism, uniformity and comparability. The idea is now examined as parents' accountability policy preferences are compared to all stakeholders' preferences.

Parents and Stakeholders' Policy Preferences Concerning Students' Learning

Table 12.1 compares support for methods proposed for collecting data, reporting on and improving students' learning. The statistical significance of differences between means was established using t tests and when $p < 0.05$.

The four most evident features of Table 12.1 are:

(a) the extent of available touchstone concerning processes,
(b) the identification of 'best practices',
(c) that accounting for student learning is to be conducted and contextualized not at school or system level but in the classroom, and
(d) that the preferred accountability processes related to learning should focus largely on improving relationships and be part of collaborative action research in classrooms.

The statistically significant differences between parents' and other's mean responses to supported items was set aside given the high levels of political support. The unexpected variance in support of state-wide, norm-referenced, standardized tests of literacy and numeracy was related to other technical concerns and, as indicated in Chapter 9, soon led to the suspension of such testing. The expected variance of support for peer appraisal, involving Parents and Friends (P and F) or Schools Councils and the use of marks and grades in reports confirmed that these proposals remain controversial. The finding that less than one in four parents wanted to compare their

Table 12.1: Support for proposed processes to collect data, report on and improve students' learning

Processes proposed by all stakeholders	Parents' views Mean	Parents' views %	All groups' views Mean	All groups' views %	Differences of means t	Differences of means p <
Supported						
Parent/teacher interviews	1.42	97.2	1.46	97.2	0.42	ns
Teachers evaluate and plan lessons thoroughly	1.57	96.5	1.58	96.4	2.48	0.02
Teachers written checklists and running records	1.67	89.1	1.93	82.7	3.61	0.001
Conferencing between teacher and student	1.81	93.1	1.74	93.2	1.21	ns
Parent input and policy explanations	1.72	92.8	1.82	90.0	1.58	ns
The sampling of student work (e.g. folios)	1.73	92.5	1.77	91.1	0.64	ns
Reports—clear and accurate descriptions of learning	1.73	92.2	1.89	86.8	2.69	0.01
Parent/teacher/student discussions	1.76	91.6	1.80	91.9	0.64	ns
Teachers identify outcomes for each student	1.72	91.7	1.80	89.3	1.18	ns
Teachers' observations	1.78	98.0	1.71	97.9	1.48	ns
Reporting through publications and public relations	1.85	87.5	1.84	87.3	0.12	ns
Parents given goals, expected outcomes, and individual expectations at the beginning of each year	1.90	84.0	2.12	73.5	2.44	0.02
Support-staff reports; guidance, welfare, speech, health	1.93	84.2	1.99	83.4	0.77	ns
Teachers evaluate and plan programs systematically	1.94	86.2	1.77	91.8	0.17	ns
Teacher-designed mastery and diagnostic tests	2.10	82.5	1.93	87.7	1.14	ns
Ambivalent support						
Student's own self-assessment	2.16	63.5	1.92	82.5	2.63	0.01
Reports with marks or grades	2.19	71.7	2.89	47.6	5.70	0.001
Formative evaluation related to teaching	2.27	63.7	2.17	73.6	1.54	ns
State-wide, norm-referenced, standardized tests of literacy and numeracy	2.27	72.1	2.70	54.2	4.03	0.001
P and Fs/Schools Councils review, discuss and report learning	2.80	46.5	3.09	33.4	9.28	0.001
Peer appraisal	3.48	18.1	3.05	35.1	4.49	0.001
Unsupported						
Reports allow parents to compare child with others	3.66	23.1	3.89	17.0	1.97	0.05

Table 12.2: Support for proposed criteria for evaluating students' learning

Processes proposed by all stakeholders	Parents' views Mean	Parents' views %	All groups' views Mean	All groups' views %	Differences of means t	Differences of means $p <$
Supported						
Measures of individual progress	1.72	97.2	1.77	93.2	0.86	ns
Student attitudes to school, teachers, peers, learning and homework	1.79	93.8	1.90	89.4	1.72	ns
Measures of students' self-esteem and life skills	1.80	88.2	1.94	84.4	2.04	0.05
Results of objective assessment used in plans	1.86	94.5	1.85	92.7	0.19	ns
Indicators developed jointly by parent, teacher, student	1.97	70.1	2.13	77.8	2.29	0.05
Performance indicators developed within schools by teachers	2.06	81.3	1.95	85.4	1.63	ns
Indicators from research literature used in planning	2.15	78.7	2.17	72.7	1.51	ns
Criteria developed by research in the classroom	2.15	75.2	2.24	71.4	1.40	ns
Performance indicators developed by teachers through subject moderation	2.16	76.2	2.23	72.0	1.18	ns
Judgements by teachers	2.25	75.8	1.99	84.2	3.44	0.001
Ambivalent support						
Performance indicators in state and national policy documents	2.29	65.3	1.99	84.2	2.20	0.05
Student participation rates (attendance, retention)	2.30	71.0	2.24	66.4	2.79	0.01
Parental expectations	2.84	45.6	2.96	37.6	1.16	ns

child's learning to others destroyed a myth to the contrary long held by some other stakeholders.

Table 12.2 compares levels of support for all criteria suggested by stakeholders for evaluating students' learning.

Table 12.2 exhibits strong agreement between stakeholders over which learning evaluation criteria should be used. It is equally evident that all stakeholders believed that measurement should occur in the classroom and that a broad range of indicators of action research and improving classroom relationships should be used. Again, little was drawn from statistically significant differences except to indicate some reluctance by parents to rely solely on teachers' judgements. All stakeholders, moreover, acknowledged the legitimacy of external criteria such as national profiles of learning, state performance indicators, research findings and moderation standards. There was some ambivalence over using participation rates as a proxy for learning and considerable doubt in all stakeholder groups over using parental expectations, a point revisited below.

Table 12.3: Support for Proposed Processes for Collecting Data, Reporting on and Improving Teachers' Teaching

Processes proposed by all stakeholders	Parents' views Mean	Parents' views %	All groups' views Mean	All groups' views %	Differences of means t	Differences of means $p <$
Supported						
Discussion between colleagues	1.57	95.9	1.38	98.4	3.10	0.001
Training and support to identify and cope with 'at-risk' students	1.56	96.5	1.54	97.7	0.52	ns
Planned development of teachers	1.72	92.4	1.49	95.8	3.66	0.001
Report teacher appraisals to the individual teacher	1.73	89.5	1.64	92.7	1.48	ns
Encourage teachers to read and do research	1.76	94.2	1.86	88.0	1.58	ns
Transition program for newly appointed teachers	1.89	79.6	1.86	80.4	0.37	ns
Co-operative learning between colleagues (e.g. mentoring)	1.99	87.6	1.80	92.0	3.35	0.001
Appraisal of student outcomes	2.04	84.2	2.09	80.3	0.78	ns
Self-evaluation	2.10	81.4	1.82	89.7	3.49	0.001
Individual and senior staff discuss appraisals	2.14	76.0	2.28	70.4	1.69	ns
An appraisal of planning	2.17	81.2	2.16	81.6	0.16	ns
Documentation of best practices	2.18	73.1	2.06	78.1	1.75	ns
Feedback and appraisal by peers	2.20	72.8	2.00	81.5	2.43	0.02
Negotiating new goals for professional development	2.23	75.8	1.92	88.3	4.32	0.001
Ambivalent support						
The planned development of classrooms	2.16	72.5	2.28	65.3	1.63	ns
Feedback and appraisal by parents	2.21	70.3	2.60	52.8	3.87	0.001
Report teaching quality to the DEA for promotion and school development purposes	2.24	69.8	2.55	55.4	4.03	0.001
Standardized test results go back to the individual teacher	2.27	73.4	2.47	63.4	2.37	0.02
The school review process	2.29	61.9	2.22	66.4	0.86	ns
Feedback and appraisal by more senior school colleagues	2.34	71.3	2.44	65.7	1.37	ns

Processes proposed by all stakeholders	Parents' views Mean	Parents' views %	All groups' views Mean	All groups' views %	Differences of means t	Differences of means $p <$
Supported						
Opportunity for parents to develop as co-teachers	2.42	62.9	2.58	55.2	1.82	ns
Opportunities for parents to consult and co-plan teaching programs	2.58	54.9	2.88	42.3	2.85	0.01
Feedback and appraisal by students	2.63	54.8	2.59	65.9	0.42	ns
Assess teacher's contribution to school planning	2.72	54.2	2.72	52.3	0.00	ns
Feedback and appraisal by an independent expert	2.52	51.0	2.89	36.0	3.43	0.001
Feedback and appraisal by the P and F/School Council	2.69	51.4	3.11	32.0	3.94	.001
P and F/School Council discuss teacher and classroom development	2.87	43.0	3.23	29.0	3.31	ns
Reporting of teacher appraisals to colleagues as part of professional development	2.93	38.9	3.20	30.6	2.94	0.01
Network more effectively with the teachers' union	2.93	30.1	2.96	28.8	0.33	ns
General reporting of teacher appraisals to parents as part of school planning and development	3.12	36.8	3.69	19.8	5.11	0.001
Unsupported						
The selection of teachers should be more localized	3.15	27.4	3.35	23.9	2.04	0.05

Parents' and Stakeholders' Policy Preferences concerning Teachers' Teaching

It will recalled that the APQ also measured support for accountability processes and criteria concerned with the quality of teaching and leadership. Table 12.3 summarizes the levels of support for all processes suggested by stakeholders to collect data, report on and improve teachers' teaching.

When compared to the levels of agreement exhibited in Tables 12.1 and 12.2, Table 12.3 suggests that there was comparatively less touchstone available to stakeholders concerning teaching accountability processes. This suggested that 'the quality of teaching' was a less salient policy issue than 'the quality of teaching when defined in terms of accountability'. On the other hand, the policy proposals supported by all stakeholders suggested that accounting for and improving the quality of teaching was not seen as a classroom or system issue but as a school responsibility, with school defined as 'a community of professionals'. The focus was on the quality of professional appraisal and feedback, professional relationships and co-development, and collegial action research and planning.

On the other hand, some parents in policy forums were doubtful about the efficacy of methods that rely so heavily on local intra-professionalism, and all stakeholders doubted the capacity of the Tasmanian school review process to improve teaching. Since the school review process was one of the last surviving methods of providing feedback on the quality of pedagogy in Tasmania, this was regarded by most stakeholder executives as a very serious situation and referred to the ERU and district superintendents for attention. I return to this matter in Chapter 15.

The considerable number of proposals supported by parents yet doubted by other stakeholders were then examined in closer detail. The minor ambivalence over 'planned development of classrooms' was traced to small number of district high school teachers and secondary school principals. While the language was unfamiliar, and had been rejected, the intrinsic merit of the idea remained attractive to stakeholder executives. Ambivalence over feedback and appraisal from parents was felt most acutely by secondary and district high school teachers, district DEA personnel, and primary and secondary school principals. These findings came as no surprize; district high schools have special features and operate in a unique context, as noted above, while high schools tend to be less responsive structurally, and parents' complaints not resolved by schools are referred to district DEA personnel.

What was unanticipated was the degree to which teachers were generally reluctant to support links between the quality of teaching and promotion or school development, feedback and appraisal by more senior colleagues, parents having opportunities to co-plan or to develop as co-teachers, or for parents to be involved in planning improvements to teaching. Teachers, principals and DEA personnel were markedly more reluctant to accept independent or parental expertise than were parents. It was widely acknowledged by all groups surveyed and consulted that there is a worrying and learned incapacity about how to collect data, provide and accept feedback, and improve the quality of teaching in Tasmanian public schools.

In sum, Table 12.3 suggests that parents' desire to provide feedback and participate in the development of teaching services is unlikely to be satisfied until DEA personnel, principals and teachers become more responsive and broaden the strategic base of school improvement beyond (what is seen by parents to be) an over-exclusive reliance on teacher development. When the implications of Tables 12.1–12.3 are taken together, it suggests that parents would prefer that the improvement of learning and teaching be attempted through an integrated approach to classroom development and school improvement. Instead of more teacher development for individual teachers or groups of teachers, parents in policy forums asked for teaching accountability processes that will affirm and improve professionalism in a school community context.

This impression is confirmed in Table 12.4 where the responses to criteria proposed by all stakeholders to evaluate the quality of teaching are compared.

The presence of extensive touchstone criteria is strongly evident in Table 12.4, despite further evidence that educators are sometimes reluctant to accept school community and systemic perspectives. This could be explained by professionalism being biased by careerism to the stage where accountability is seen as 'politically incorrect' (Macpherson 1996d).

Compared to the learning criteria in Table 12.2, there is also less acknowledgement in Table 12.4 of the value of external profiles of competencies, research or extra-professional interests. Only three of the 15 criteria required systemic involvement; those concerning the effective implementation of school and curriculum policies, the use of K-12 achievement levels and teacher competencies in DEA job descriptions. All stakeholders assumed that 15 out of 18 criteria could be defined and operationalized by using or developing the expertise of each school community. Similarly, it was believed that the measurement and improvement of teaching should be a normal part of school life and that indicators of professionalism should relate teachers' knowledge, attitudes and skills to outcomes in the classroom.

In sum, it appears that all stakeholders expected accountability obligations concerned with the quality of teaching to be defined and discharged in each school community with supportive interaction between these learning organizations. While parents were asking educators to be more responsive to broader school community values, and join all other groups seeking better school pedagogical policies and practices, all stakeholders (including DEA personnel) appeared to limit the satisfaction of systemic priorities to the implementation of curriculum policies. Put another way, preferred accountability policies and practices related to the quality of teaching implied consistently assumed high levels of subsidiarity, pluriformity and complementarity in the public school system of Tasmania.

Parents' Policy Preferences concerning Leaders' Leadership

The third part of the APQ measured support for accountability processes and criteria concerned with the quality of leadership. The instrument defined leaders as those who provide leadership services in school communities. Table 12.5 summarizes the levels of support for all processes suggested by stakeholders to collect data, report on and improve leaders' leadership services.

Five striking features of Table 12.5 are:

(a) the area of policy touchstone available is comparatively more compact than in Tables 12.1 and 12.3,
(b) the extent to which preferred accountability processes assume that leadership is a responsive service to colleagues and school community, not to classroom or system,
(c) the broad yet integrated range of philosophical, strategic, political, cultural, managerial and evaluation capacities implied by the leadership duties supported by all stakeholders,
(d) the comparatively high number of leadership accountability processes favoured by some stakeholders that are in dispute, and
(e) proposed leadership appraisal processes are the most controversial issue.

Table 12.4: Support for criteria for evaluating teachers' teaching

Processes proposed by all stakeholders	Parents' views Mean	%	All groups' views Mean	%	Differences of means t	p <
Supported						
Classroom environment	1.68	93.2	1.60	95.3	1.18	ns
Organizational skills	1.78	93.2	1.79	92.0	0.15	ns
How well work is set, monitored and marked	1.79	91.1	2.00	83.5	2.87	0.01
Interpersonal communications within the classroom	1.84	90.3	1.70	94.7	2.38	0.02
Behaviour management skills	1.80	92.5	1.79	91.5	0.16	ns
Teachers' attitudes to students, parents and colleagues	1.86	88.2	1.97	85.1	1.56	ns
Student progress	1.82	86.9	1.99	81.0	2.03	0.05
Teachers' knowledge of subject and child/adolescent development	1.88	89.0	1.86	90.1	0.30	ns
The attitude of children (e.g. enthusiasm)	1.93	82.6	2.10	76.5	2.04	0.05
Instructional expertise	2.01	83.4	2.03	82.3	0.33	ns
Effective implementation of school and curriculum policies	2.12	82.1	2.02	85.7	1.71	ns
Willingness to engage in continuing professional development	2.13	77.1	2.08	77.9	0.51	ns
Support ambivalent						
Effectiveness of teachers' written records and plans	2.25	74.1	2.46	65.8	2.38	0.02
Students' achievement levels in K-12 Framework	2.25	62.3	2.39	62.0	1.81	ns
Teachers competencies in DEA job descriptions	2.42	56.5	2.40	63.6	0.25	ns
Communication skills with stakeholders	2.43	53.9	2.23	67.0	2.85	0.01
Teachers' participation in school and community activities	2.66	53.9	2.83	52.0	1.60	ns
Leadership services given by teachers in school	2.69	50.7	2.66	52.0	0.35	ns

The significant differences of means suggest that parents are yet to be as convinced as other groups are of the value of self-appraisal, peer appraisal, appraisal by colleagues and mentoring. The proposed involvement of parental, professional and DEA personnel in the selection of leaders evokes ambivalence between and within stakeholder groups. While district and high school principals (84% SA+A), district DEA personnel (80%), parents (67.4%), district high teachers (59.7%) and primary principals (58.6%) tend to favour cross level involvement, secondary teachers (32%), central

DEA personnel (43.8), primary teachers (48.1) do not. This result remains unexplained. On the other hand, the localization of leader selection is unsupported by all groups.

Table 12.5 also suggests that while all stakeholders are relatively clear about what they want leaders to be held accountable for, the yet-to-be-articulated leadership accountability processes will probably need to have reliable instrumentation, triangulated data that is handled sensitively, high responsiveness to classroom, school community and systemic perspectives and have explicit links to leader and institutional development. This summation is supported by the data presented in Table 12.6.

The impression created by Table 12.5, specifically that all stakeholders are relatively clear and in agreement about the purposes of holding leaders accountable while being far less sure about appropriate processes, is borne out by Table 12.6. Table 12.6 also confirms that ambivalence begins to arise when criteria are drawn from external research, the DEA, or the community. The general legitimacy of leadership accountability criteria fell with all stakeholders when associated with recommendations from school reviews, the physical environment, leaders' qualifications or national managerial competencies.

Six Sets of Accountability Policy Preferences

It was noted in Chapters 9 and 10 that when the 53 proposals supported by all stakeholder groups in Tables 12.1–12.6 were supplemented by the 27 items that attracted support with only minor levels of ambivalence in only one or two other groups, the 80 proposals were regarded by stakeholder executives as touchstone for site and system policy reviews. This treatment of data, and its outcome, were regarded by parents' representatives as 'common sense'.

The language of analysis changed depending on the stakeholder group. In consultations with parents' and teachers' representatives, the 80 items were reclassified into 'six sets of accountability processes and criteria', and are summarized in Table 12.7. Items were clustered into subgroups and given a descriptive title for convenience (in bold italics in Table 12.7). In Chapter 14, where principal's preferences and interpretations are discussed, the items are regarded as 'best practice', the cluster descriptors are retitled 'performance indicators' and the activity labels of the six sets are redesignated as 'key competencies of leadership service'. The differences in language were taken to indicate persistent assumptions about self, role and structure.

Implications of Preferred Forms of Accountability Activities

The first set the parents discussed focused on school communities setting directions and deciding 'what's best for the school'. It was more concerned with rightness and significance that techniques of decision making. Parents, like all stakeholders, want accountability processes and criteria that help with the clarification of purposes in each school community. They want them to provide a bridge between the evaluation of

228   EDUCATIVE ACCOUNTABILITY

Table 12.5: Support for processes to collect data, report on and improve leaders' leadership

Processes proposed by all stakeholders	Parents' views Mean	Parents' views %	All groups' views Mean	All groups' views %	Differences of means t	Differences of means p <
Supported						
Appraisal of support and feedback given to staff	1.84	90.4	1.78	91.8	1.02	ns
Provision and generation of a school vision	1.98	86.2	1.89	87.7	1.46	ns
Skill development programs, e.g. in governance and management	1.99	81.5	1.90	87.1	1.23	ns
Improved by using feedback from staff	2.01	87.5	2.06	84.3	0.82	ns
Quality of reporting to parents and community	2.01	87.5	2.11	82.1	1.80	ns
Evaluate the coherence between vision, plans and outcomes	2.06	82.6	1.84	90.3	3.86	0.001
Feedback from parents and students	2.08	80.6	2.26	71.7	2.48	0.02
Peer networks reflect on challenges of practice	2.16	72.6	1.97	79.7	2.95	0.01
Support ambivalent						
Parents, teachers, DEA collaborate in principal selection	2.24	67.4	2.42	60.3	1.87	ns
Survey of the school climate	2.28	67.6	2.17	74.4	1.53	ns
An appraisal of policy-making strategies used	2.30	69.6	2.18	74.7	1.86	ns
Self-appraisal	2.36	68.1	1.96	83.3	4.41	0.001
Appraisals should be reported to the DEA	2.43	58.5	2.76	43.3	4.09	0.001
An appraisal by school colleagues	2.43	64.1	2.24	74.0	2.25	0.05
An appraisal of the quality of external liaison	2.44	54.8	2.40	59.8	0.51	ns
An appraisal by the P and F/School Council	2.52	55.1	2.76	47.2	2.21	0.05
Peer appraisal	2.55	57.0	2.25	70.7	3.73	0.001
Improved by using a mentoring process	2.58	45.9	2.28	60.5	3.88	0.001
An appraisal by the DEA	2.65	51.7	2.78	45.5	1.44	ns
Appraisals reported to individuals and colleagues as part of the professional development program	2.75	43.3	2.99	34.2	2.76	0.01
Appraisals reported to parents as part of school development program	2.94	37.8	3.37	21.3	4.23	0.001
Fixed term and negotiated performance contracts	2.94	38.7	3.33	27.4	3.43	0.001
P and F/School Council set leadership service policies	2.97	32.7	3.35	21.9	3.88	0.001
An appraisal by the community	3.08	30.6	3.13	25.7	0.53	ns
Unsupported						
More localized selection of leaders	3.06	28.7	3.35	18.1	3.16	0.01
Overseas exchanges	3.30	20.0	3.02	28.8	3.14	0.01

learning, teaching and leadership and planning for improvement. They also want everyone involved to get better at doing this; each school to develop its capacities as a learning organization.

Such an approach is inconsistent with a neo-centralist prescription of what schools are for, a unitary concept of 'system', a uniform concept of 'school', a 'top-down' implementation of systemic policies, or accounting for the performances of schools, teachers or learners in comparative terms. Instead, the parental respondents to the APQ shared an accountability theory that emphasized a philosophical commitment to communitarian and problem-solving purposes, a democratic accommodation of pluralism and supportive interdependence.

A second set of valued activities in Table 12.7 is about the strategic role to be played by accountability processes and criteria. There is regular evidence that all stakeholders, including parents, value accountability for its provision of collaborative strategic analyses of the situation school communities 'find themselves in', the opportunity it gives participants to negotiate appropriate indicators of performance, and the imperatives it creates for classroom, professional and school development programmes. This theme in the evidence suggests that the current emphasis on the more technical aspects of 'self-managing' school finance, along with systemic attempts to standardize curriculum and assessment, fails to do justice to the desire among all stakeholders to participate in strategic analysis and direction setting. There is an expressed need for a holistic and inclusionary approach to accountability that integrates philosophical and planning activities.

The third set of activities parents and other stakeholders valued is a responsible and co-operative form of accountability politics. There are regular indications in the data that parents resent exclusionary forms of professionalism, that teachers are troubled by impersonal aspects of administration and that administrators are discomforted at being marginalized from education policy making. Hierarchy and social distance, and the use of arbitrary, coercive or manipulative power in education are anathema. Most respondents appear to believe that positional authority should grace and develop the moral economy of organizational micropolitics, that trust and support should preface regulation and sanctions, and that power relationships between stakeholders should be cast in a context of plural legitimate stakeholders, active citizenship and educational partnerships. The generally expressed preference for responsive and responsible accounting between stakeholders with mutual obligations runs counter to the contractual and technical forms of accountability promoted by the political, market and managerial mechanisms more typical of corporate 'self-management', LMS and SBM.

A fourth activity of importance is the development of supportive classroom and staffroom environments. Strongly supported references to positive attitudes in staff and students, quality communications, caring behaviours, open and participative decision processes, and development programmes in governance and management indicated that accountability was held to be central to the development and improvement of educational cultures. These group, institutional and systemic cultures were clearly assumed to be interactive and complementary, rather than independent and

Table 12.6: Support for criteria for evaluating leaders' leadership services

Processes proposed by all stakeholders	Parents' views Mean	Parents' views %	All groups' views Mean	All groups' views %	Differences of means t	Differences of means p <
Supported						
Capacity to hear and care for others	1.60	98.0	1.56	96.8	0.78	ns
Student and teacher morale and motivation	1.74	89.3	1.77	88.8	0.47	ns
Ability to plan outcomes and achieve priorities	1.84	93.7	1.78	94.2	1.15	ns
The extent to which staff support their leaders	1.87	89.3	1.93	84.5	0.88	ns
The openness and climate/tone of the school	1.91	81.5	1.76	88.8	2.07	0.05
Capacity to make and implement policy	1.93	87.8	1.86	90.9	1.20	ns
Management and organizational skills (evaluation, budgeting and governance)	1.93	87.1	1.89	88.3	0.58	ns
Valuing of creativity and productivity in school	1.99	82.5	2.04	79.5	0.76	ns
Evidence of the quality of teaching by the staff	2.09	79.0	2.21	74.6	1.56	ns
Extent of collaborative decision making	2.11	74.0	1.93	83.4	2.47	0.02
Evidence of learning by staff and students	2.13	77.3	2.16	73.6	0.37	ns
Quality of internal and external communications	2.17	77.4	2.10	79.9	1.01	ns
Support ambivalent						
Indicators from research literature used in plans to improve leadership	2.31	62.2	2.32	61.1	0.15	ns
Capacities as learners and researchers	2.32	76.0	2.29	68.6	0.41	ns
Extent of professional development within the school	2.32	69.0	2.16	74.9	1.92	ns
Performance indicators in guidelines provided by the DEA	2.34	59.0	2.36	62.9	0.30	ns
The expectations of the community	2.46	64.0	2.46	61.0	0.00	ns
Extent to which parents support school leaders	2.49	57.9	2.54	54.4	0.43	ns
Recommendations from school reviews	2.41	58.3	2.46	55.9	0.70	ns
Leaders' relevant qualifications	2.51	63.7	2.75	51.7	2.39	0.02
The quality of the physical environment	2.54	57.6	2.53	58.8	0.11	ns
National competency indicators for managers	2.54	48.6	2.69	40.6	2.04	0.05

competitive, and embedded in fiduciary rather than in market or power networks. Education is principally defined as a cultural activity in classrooms or work groups. Most organizational concepts evident in the items identified relationships in the classroom as the basic educational structure, with classroom relationships intermittently linked as 'school' or socially embedded in 'school community', while 'the system' was even less frequently evoked as a collective noun for 'schools'.

The fifth set of valued activities are about effective and efficient management. The evidence suggests that while all stakeholders expect position holders to discharge their management duties, they define these duties using a complex range of technical, professional and school community perspectives. Management duties were defined in three realms that outstrip the limits of economic rationalism. The first realm of duty implies accounting for the acquisition, management and development of resources. Examples might include collaborative programme budgeting and human resource development. The second set of duties mean accounting for the quality of support structures and processes in schooling, such as quality teams and co-operative programme evaluation. The third form of duty means accounting for the quality of the processes used to make and implement policies, such as the quality of governance, action research and information and decision systems. Accountability by these lights defines effective and efficient management in both immediate educational and broader communitarian terms in a context of multi-level democratic structures.

The sixth set of valued activities preferred by all stakeholders cluster around the idea that accountability practices will serve both summative and formative evaluation purposes. Demand for summative evaluation was seen in strongly supported proposals for the monitoring of outcomes and attitudes, the measurement of students' progress, self-esteem and life skills, and the surveying of school climate and stakeholders' policy preferences. Demand for formative evaluation was evident in proposals concerned with the provision of feedback and appraisal systems linked into classroom and school development programmes. While there was less surety over the most appropriate teaching and leadership accountability processes, as compared to those related to accounting for learning, the support for accountability criteria across all three activities was even and high. This suggests that the quality of learning, teaching and leadership require different accountability policy debates to determine best processes, the principles underpinning preferred evaluative criteria are available and relatively coherent.

Summary

There is little support in the Australian home of the 'self-managing school' for contractual accountability to employers, parents or designated corporate managers. Parents, teachers, principals and government education officials at district and state level also share a view that moral accountability to clients and professional accountability to peers should be set aside. They favour responsive and mutual accountability between stakeholders.

The 12 principles identified by ACCSO and APC were strongly supported. This

Table 12.7: Six sets of educative accountability criteria and processes

Evaluative activity	Philosophical activity
Monitors outcomes and attitudes	**Clarifies organizational purposes**
Samples student work (e.g. folios)	Provides and generates a school vision
Measures progress of individual students	Develops coherence between the school vision, the development plan and outcomes
Ensures that teachers use written checklists and running records to monitor student progress	**Links evaluation to planning**
	Helps teachers evaluate and plan their programs systematically
Uses reports from other support staff (e.g. guidance, welfare, speech and health specialists)	Helps teachers develop performance indicators at subject moderation meetings
Uses performance indicators in state and national policies (e.g. Frameworks, Profiles)	Helps teachers develop performance indicators in school
	Helps teachers evaluate and plan lessons thoroughly
Measures students' self-esteem and life skills	**Creates a learning organization**
Measures attitudes of students to school, teachers, peers, learning and homework	Shares reflections in peer networks on the challenges of practice
Measures teachers' attitude towards students, parents and colleagues,	Develop others' capacities as learners and researchers
Surveys school climate, uses feedback from staff	Negotiates new goals for professional development
Provides feedback and appraisal systems	Encourages teachers to read and do research
Uses feedback on teaching and leadership from parents and students	Develops the judgements made by teachers
Uses feedback and appraisal by peers	Enhances teachers' knowledge (of subject matter and child/adolescent development)
Uses appraisal of leadership services by school colleagues	Values creativity and productivity in the school
Encourages self-evaluation, self-appraisal	Reviews the quality of planning
Reports teacher appraisals to the individual teacher	Reviews the quality of policy-making strategies used
Develops formative evaluation systems	Uses a mentoring process
Uses objective assessment to plan improvements to students' learning	
Develops formative evaluation related to teaching objectives	

(continued opposite)

Management activity	Strategic planning activity
Manages the making and implementing of policies Supports participative governance processes Implements school and curriculum policies effectively **Organizes support structures and processes** Organizes the support and feedback given to staff Organizes the evaluation of teaching and learning Organizes and develops instructional expertise Organizes training and support to identify and cope with students 'at risk' Organizes how students' work is set, monitored and marked **Manages the use of resources** Organizes the use of time and other resources Organizes budgeting processes	**Provides collaborative strategic analysis** Negotiates priorities and outcomes Helps others make and plan the implementation of policy Provides conferencing between teachers and Helps parents/teachers/students jointly develop learning indicators Provides parents with curriculum goals, outcomes and individual expectations at the beginning of the year **Negotiates performance indicators** Helps teachers identify intended outcomes for each student Includes indicators from research as criteria when planning improvements to students' learning Negotiates indicators of learning by staff Negotiates indicators of quality teaching with staff **Prepares a professional development strategy** Plans the professional development of teachers Raises the willingness of staff to engage in continuing professional development

Cultural activity	Political activity
Develops supportive classroom environments Encourages collaborative classroom Develops supportive behaviour management skills Promotes positive attitudes in children (e.g. enthusiasm) Improves interpersonal communications in the classroom **Develops a supportive school environment** Leaders hear and care for others Develops an open and participative culture Promotes parent/teacher/student discussions (e.g. by mentoring and discussion between colleagues) Provides skill development programmes, especially in governance and management Provides a transition program for newly appointed teachers	**Raises commitment** Encourages collaborative decision making environments Develops student and teacher morale and motivation Encourages parent/ teacher interviews Develops staff support for their leaders **Communicates priorities and values** Provides parents with opportunities to give their views and to discuss teaching and curriculum policies Encourages co-operative learning between colleagues Reports to parents with clear and accurate descriptions of learning Reports student's learning through newsletters, school magazines, displays and positive public relations Reports achievements to parents and the community Documents best practices Values internal and external two-way communications

research added even more specific operational principles and organizational preferences. For example, the eighty proposals in the APQ that were supported or nearly supported by all stakeholders were regarded as touchstone for site and system policy reviews. They were taken to mean that all stakeholders value accountability processes and criteria that:

- help clarify purposes in each school community.
- provide a bridge between the evaluation of learning, teaching and leadership and planning for improvement,
- help each school develops its capacities as a learning organization,
- provide collaborative strategic analyses of the situation school communities find themselves in,
- give participants the opportunity to negotiate appropriate indicators of performance,
- create moral and practical imperatives for classroom, professional and school development programmes,
- generate a responsible and co-operative form of community accountability politics,
- develop supportive classroom and staffroom environments,
- help position holders discharge their management duties defined using a complex range of technical, professional and school community perspectives, and
- serve both summative and formative evaluation purposes.

More specifically, with regard to accounting for learning, the processes and criteria favoured by all stakeholders focus on improving relationships and collaborative action research in classrooms, while affirming the legitimacy and value of national and state profiles and performance indicators, research findings and teacher moderation.

Accounting for the quality of teaching was held by all stakeholders to be a school level responsibility with a broad consensus over appropriate processes and criteria. Special note needed to be taken of parents' disquiet over (a) exclusionary professionalism and (b) how reliance on professional development was displacing classroom development in a school community context.

Stakeholders were relatively clear and in agreement about the purposes of holding leaders accountable. On the other hand they were less sure about appropriate processes and many expressed concern about the reliability of intra-professional processes.

More broadly, with regard to organizational preferences, the policy touchstone identified by all stakeholders provided little support for:

- a neo-centralist and unitary concept of 'system',
- planning, co-ordination and policy implementation by corporate managers,
- comparative assessment of learning, teaching or leadership,
- partitioned curriculum and standardized resource management,
- communications within and between stakeholders being mediated by positional authority, or
- incentives being based on political or market devices.

Conversely there was strong support among all stakeholders for:

- a liberal, communitarian, pragmatic and pluralist philosophy of administration,
- an inclusionary, simultaneous and holistic approach to policy making, planning and implementation,
- a trustful, supportive and group-based approach to change management,
- classrooms relationships seen as primary educational structure, and
- improvement, accountability and legitimation seen as school community projects.

Until more targeted research can map actual practices, it was considered reasonable to assume that parents prefer greater subsidiarity, pluriformity and complimentarity in their schools and education system, only slightly more so that other stakeholders, and rather than neo-centralist and 'self-managing' corporate managerialism, uniformity and comparability. In the next chapter I explore teachers' interpretations of the findings.

13

Tasmanian Teachers' Policy Preferences Concerning Accountability: Individualistic or Communitarian Professionalism?

Introduction

This chapter is devoted to reporting and discussing the policy preferences of teachers when compared to other stakeholders' views. Primacy is given to teachers' interpretations expressed at the April 1995 forum and during subsequent consultations. From the outset, interpretations were strongly affected by the content of touchstone. Touchstone was used to generate a moral, political and practical mandate termed 'professionalism'. As the chapter proceeds it also becomes clear that while this term had a common basis in a constructivist and communitarian theory of effective teaching, learning and leadership, it was being modified in some circumstances by an individualistic philosophy of teaching.

Two major policy implications are drawn. First is that accountability policies will need to cohere with a communitarian model of accountability or run the gauntlet of the feral norms of uncontestable 'professionalism'. Second, the residual presence of individualistic professionalism will need to be diluted by professional development that provides teachers with regular and authentic participation in the co-leadership and co-governance of school communities. This strategy would enable teachers to simultaneously reconstruct their theories of accountability and professionalism in a supportive community context. The discussion begins by comparing the views of teachers with other stakeholders concerned with student's learning.

Accountability Processes related to Students' Learning

The processes that at least one stakeholder group believed should be used to collect data on, report and improve students' learning are summarized in Table 13.1. The support give by teachers and other stakeholders are compared. The percent symbol refers to the percentages that 'strongly agreed' plus 'agreed' to each proposal.

Attention focused on the items that had attracted support from all stakeholders. The statistically significant differences between supported items were set aside as politically and educationally irrelevant. The general view was that the 16 items supported by all

Table 13.1: Support for proposed processes to collect data, report on and improve students' learning

Processes proposed by all stakeholders	Teachers' views Mean	%	All groups' views Mean	%	Differences of means t	p <
Supported						
Parent/teacher interviews	1.56	95.5	1.46	97.2	1.05	ns
Teachers evaluate and plan lessons thoroughly	1.73	93.1	1.58	96.4	2.49	0.02
Teachers' observations	1.75	95.5	1.71	97.9	0.76	ns
Conferencing between teacher and student	1.81	82.5	1.74	93.2	1.16	ns
Reporting through publications and public relations	1.82	88.7	1.84	87.3	0.27	ns
The sampling of student work (e.g. folios)	1.85	89.3	1.77	91.1	1.12	ns
Teachers evaluate and plan programs systematically	1.86	89.4	1.77	91.8	1.26	ns
Student's own self-assessment	1.87	85.6	1.92	82.5	0.62	ns
Teachers identify outcomes for each student	1.89	88.6	1.80	89.3	1.29	ns
Teacher-designed mastery and diagnostic tests	1.95	89.6	1.93	87.7	0.32	ns
Support-staff reports; guidance, welfare, speech, health	1.96	84.8	1.99	83.4	0.39	ns
Parent/teacher/student discussions	2.02	86.9	1.80	91.9	2.86	0.01
Parent input and policy explanations	2.04	81.0	1.82	90.0	3.22	0.01
Reports—clear and accurate descriptions of learning	2.06	78.8	1.89	86.8	2.24	0.05
Teachers written checklists and running records	2.08	81.8	1.93	82.7	1.77	ns
Formative evaluation related to teaching objectives	2.24	70.4	2.17	73.6	0.94	ns
Ambivalent support						
Parents given goals, expected outcomes, and individual expectations at the beginning of each year	2.35	60.6	2.12	73.5	2.33	0.02
Peer appraisal	2.86	44.7	3.05	35.1	1.75	ns
State-wide, norm-referenced, standardized tests of literacy and numeracy	3.04	39.7	2.70	54.2	2.97	0.01
Reports with marks or grades	3.12	38.6	2.89	47.6	1.78	ns
P and Fs/Schools Councils review, discuss and report learning	3.25	22.7	3.09	33.4	1.57	ns
Unsupported						
Reports allow parents to compare child with others	4.20	7.6	3.89	17.0	3.10	0.01

stakeholders were consistent with 'high quality professionalism'. There was, nevertheless, some surprize expressed by teachers that a minority of colleagues apparently neither supported 'lesson evaluation' nor 'reporting in accurate language' as highly as they did or as highly as did other stakeholders. It was also regarded as 'odd' that there were also some who felt little obligation to relate openly with parents on plans or learning outcomes. Since the general levels of support were so high to these proposals, it was agreed that the minorities involved needed to made more aware of their colleagues' norms of professionalism and their leaders needed to help them come to terms with the implied responsibilities. It was concluded that the term 'professionalism' was giving carriage to some worrying ideas.

Discussion then considered teachers' ambivalence about 'parents being given goals, expected outcomes, and individual expectations at the beginning of each year'. In Table 13.2 all stakeholders' means are compared to the mean of all responses. The source of ambivalence to this proposal was solely in primary schools. Teachers involved in interpreting these findings came to the view that some of the minority of primary teachers involved were probably those teaching infants. They could, it was felt, quite reasonably feel unable to predict learning readiness with much certainty.

There was also some speculative comment concerning the situation in district high school communities. The difference between district high teachers' (SA+A = 60.7%) and district high principals' (85.7%) means was not statistically significant ($t = 1.38$, $p <$ ns). On the other hand, the political significance of the differences between district high teachers (60.7%) and parents (88.9%) was considered an issue worthy of attention in district high school communities, even though the difference was only approaching statistical significance ($t = 1.65, p < 0.10$). Teachers consulted were very aware of the special conditions that their colleagues face in district high school communities and noted how crucial it was for staff to participate in leadership and governance so they could understand, help develop and then explain 'the big picture' to colleagues.

There was also some surprize expressed by teachers to find that stakeholders had similar levels of general ambivalence to their own concerning the use of 'reports with marks or grades', indicating the death of myths to the contrary. Similarly they noted the general ambivalence regarding the proposal that 'reports should allow parents to

Table 13.2: Differences between stakeholders regarding the proposal that, in order to improve learning, parents should be given goals, expected outcomes, and individual expectations at the beginning of each year.

Respondents	n	Mean	SD	%SA+A	t	$p <$
Primary teachers	79	2.45	1.07	57.0	2.56	0.02
District HS teachers	28	2.22	1.05	60.7	0.49	ns
Secondary teachers	25	2.20	0.71	72.0	0.54	ns
All principals	125	2.19	0.96	71.6	0.72	ns
All parents	148	1.90	0.94	84.0	2.44	0.02
All DEA	26	1.84	0.69	84.0	1.96	0.05
All respondents	431	2.12	0.97	73.5	–	–

Table 13.3: Differences between stakeholders regarding the proposal that, in order to improve learning, reports should allow parents to compare their child with others

Respondents	n	Mean	SD	%SA+A	t	p <
Primary teachers	79	4.16	0.96	8.90	2.23	0.05
District HS teachers	28	4.29	0.90	7.10	2.23	0.05
Secondary teachers	25	4.24	1.01	4.00	1.67	0.10
Primary principals	89	4.10	1.15	13.5	1.56	ns
District HS principals	15	3.93	1.16	20.0	0.13	ns
Secondary principals	21	3.62	1.24	23.8	0.98	ns
Primary parents	96	3.80	1.18	19.0	0.68	ns
District HS parents	28	3.32	1.36	35.7	2.17	0.05
Secondary parents	24	3.50	1.25	25.0	1.49	ns
District DEA	10	3.80	0.92	10.0	0.30	ns
Central DEA	16	2.81	1.22	50.0	3.48	0.001
All respondents	431	3.89	1.17	17.0	–	–

compare child with others'. Again they were reassured to find that all educators were indifferent or antagonistic to the proposal but puzzled as to why central DEA personnel and district high school parents were virtually alone in providing support. Table 13.3 compares stakeholders' means with the means of all respondents.

The view reached by teachers was that district high parents and DEA central personnel were in special need of assurance over quality learning, the former because of the relative inexperience of staff, and the latter due to constant political pressure over the use of scarce resources. Again they suggested the value of opportunities to participate in leadership and governance in district high school communities. With regard to DEA central personnel, once anti-centralist views had been balanced by an acknowledgement of the Westminster system of government, it was suggested that touchstone items provided many and better forms of data for this purpose, providing they were aggregated *from* rather than *to* the classroom.

In sum, teachers interpreted Tables 13.1–13.3 as professionally affirming. They were, nevertheless, intrigued to find a minority of colleagues who were apparently less willing to support lesson evaluation, to report in precise language or to share plans or outcomes to the extent that they and other stakeholders regarded as appropriate. They saw the touchstone as a strong mandate for professionals to both facilitate the learning process and to question the use of negative practices in learning communities.

Five forms of 'best practice' relating to the quality of learning were reinforced. One- or two-way school-home links between adults were seen as less effective than three-way parent/teacher/student communications. Parental involvement was set aside in favour of parent and student participation in policy making and in setting both goals and individualized expectations. Thorough and systematic evaluation and planning of lessons and programmes was strongly affirmed, especially by using formative evaluation related to objectives. Data collection, it was felt, should include teachers' observations, samples of student work and self-assessment, teacher-designed mastery and diagnostic tests, support-staff reports and teachers written checklists and running records. Reporting, they said, should use precise descriptors of learning and be delivered to all stakeholders by sophisticated publications and public relations.

Negative practices, as defined by the touchstone items, included comparative reporting, the use of 'state-wide, norm-referenced, standardized tests of literacy and numeracy' and 'P and Fs/Schools Councils reviewing, discussing and reporting learning'. Despite the legitimation of 'best practices' afforded by the substantial touchstone available, there was unease concerning the most appropriate accountability relationship between teachers and parents with regard to students learning. Even when the special nature of infant school learning was set aside, there seemed to be some colleagues who defined their 'professionalism' in idiosyncratic and uncontestable ways. It was, therefore, considered to be vital that teachers had opportunities to participate in leadership and governance in school communities so they could help create a shared theory of professionalism that cohered with 'the big picture' of 'school policy'. This matter came to the fore again when analysing accountability criteria concerned with the quality of learning.

Accountability Criteria related to Students' Learning

Table 13.4 compares the support that teachers and all other groups gave to the student learning criteria suggested by all stakeholders.

Table 13.4: Support for proposed criteria for evaluating students' learning

Criteria proposed by all stakeholders	Teachers views Mean	%	All groups' views Mean	%	Differences of means t	$p <$
Supported						
Measures of individual progress	1.86	91.0	1.77	93.2	1.44	ns
Student attitudes to school, teachers, peers, learning, homework	1.91	87.2	1.90	89.4	0.15	ns
Results of objective assessment used in plans	1.94	90.9	1.85	92.7	1.65	ns
Measures of students' self-esteem and life skills	2.02	79.5	1.94	84.4	1.12	ns
Performance indicators developed by teachers	2.02	79.8	1.95	85.4	0.91	ns
Performance indicators in state and national policy documents (e.g. Frameworks, Profiles)	2.06	84.9	1.99	84.2	0.99	ns
Judgements by teachers	2.10	78.8	1.99	84.2	1.42	ns
Criteria developed by research in the classroom	2.34	70.2	2.24	71.4	0.00	ns
Ambivalent support						
Performance indicators developed by teachers through subject moderation	2.29	69.0	2.23	72.0	0.68	ns
Indicators developed jointly by parent, teacher, student	2.37	66.7	2.13	77.8	2.08	0.05
Indicators from research literature used in planning	2.39	62.1	2.17	72.7	2.47	0.02
Student participation rates (attendance, retention)	2.72	49.2	2.24	66.4	1.98	0.05
Parental expectations	3.14	28.0	2.96	37.6	1.88	ns

Teachers consulted were quick to define the touchstone items as 'best practice', and that, when used together, would provide a methodologically adequate approach to indicating quality learning. Some criteria would rely on quantitative data, such as measures of individual progress, student attitudes, students' self-esteem and life skills and objective assessment. Other criteria would be more qualitative and more customized; performance indicators developed by teachers, indicators drawn from state and national policy documents, with judgements by teachers added, perhaps developed by research in the classroom. Teachers involved in interpretation affirmed that a form of moderation should be designed for primary teachers as soon as possible; it would help build more sophisticated and more coherent theories of accountability.

They found it much less easy to interpret the views of teachers concerning the use of 'indicators developed jointly by parent, teacher, student' or that 'indicators from research literature should be used in planning'. Table 13.5 traces the source of ambivalence on the joint development of indicators to some secondary teachers and a few district high teachers, with all primary school stakeholders (especially parents) indicating high levels of readiness. This was regarded as an ideal opportunity for the collaborative development of school policy with a high likelihood of success and capacity building.

Table 13.6 suggests that primary school teachers and secondary principals are ambivalent about learning criteria developed by research in the classroom. For example, 52.4% of secondary principals indicated that they were 'not sure' about this proposal. These findings were interpreted as indicating limited research capacities rather than unwillingness to use research methods. Some of the teachers involved in interpretation recalled learning the value of action research when it had been set as an assignment during their masters degree but that it was yet to become common practice.

Table 13.5: Differences between stakeholders concerning the proposal that the criteria used to evaluate students learning should include indicators developed jointly by parents, teachers and students

Respondents	n	Mean	SD	%SA+A	t	$p <$
Primary teachers	79	2.20	0.91	74.7	0.27	ns
District HS teachers	28	2.50	1.00	64.3	1.70	0.10
Secondary teachers	25	2.79	1.02	44.0	2.96	0.01
Primary principals	89	2.23	1.02	67.8	0.51	ns
District HS principals	15	2.13	1.19	73.3	0.13	ns
Secondary principals	21	2.30	0.98	65.0	0.59	ns
Primary parents	96	1.83	0.77	84.1	3.73	0.001
District HS parents	28	2.43	1.10	67.3	1.22	ns
Secondary parents	24	1.96	0.96	83.3	1.04	ns
District DEA	10	1.90	0.74	80.0	1.13	ns
Central DEA	16	1.94	0.68	81.3	1.31	ns
All respondents	431	2.17	0.96	72.7	–	–

Table 13.6: Differences between stakeholders concerning the proposal that the criteria used to evaluate students learning should include indicators developed by research in the classroom

Respondents	n	Mean	SD	%SA+A	t	p <
District HS teachers	28	2.21	0.79	64.3	0.19	ns
Secondary teachers	25	2.16	0.55	74.0	0.69	ns
Primary principals	89	2.29	0.83	58.4	0.53	ns
District HS principals	15	2.14	0.54	78.5	0.26	ns
Secondary principals	21	2.43	0.87	42.8	1.16	ns
Primary parents	96	2.16	0.64	72.6	0.96	ns
District HS parents	28	2.15	0.66	85.2	0.63	ns
Secondary parents	24	2.09	0.67	73.9	0.97	ns
District DEA	10	2.00	0.94	80.0	1.02	ns
Central DEA	16	2.00	0.63	81.3	1.28	ns
All respondents	431	2.24	0.74	66.4	–	–

These findings about preferred accountability criteria related to students learning were then reviewed *en masse* by the teachers consulted. They saw first the need for professional development that was 'part and parcel' of the 'daily life of the school'. It was more than an article of faith, they explained, that professional development was the best and probably only route to improving the quality of learning in the classroom. Teachers held an informal veto on change. Professional development was widely regarded, they reported, to be a prior condition of school development and professional accountability, with 'school' defined first as a professional community of educators before being defined as a school community to include other stakeholders. How could teachers trust others if they could not trust themselves, it was often asked, rhetorically.

The most challenging implication was, therefore, to facilitate professional development in ways that took teachers from idiosyncratic notions of professionalism to collegial theories, and then from collegial to community theories. While professional educators should report using quantitative data, such as measures of individual progress, student attitudes, students' self-esteem and life skills and objective assessment, and other more qualitative performance indicators developed by teachers and drawn from policy documents, the reports would inevitably be mediated by personal professional judgements. It was recalled that many teachers' view of the ideal reporting process to parents (concerning students' learning) was still modelled on the confidential doctor–client consultation. Relatively few teachers had reportedly enjoyed the more holistic engagement of genuine action research or regular, undistorted and three-way discussions of achievements, limitations, plans and expectations.

While the introduction of touchstone mechanisms and criteria would be helpful, it was also concluded that the most appropriate target of professional development concerning the quality of learning should be the sophistication and coherence of professional judgement making. Such sophistication was to be defined in terms of its formative effect. Coherence was to be defined initially by a community of supportive colleagues, although increasingly by a broader community of stakeholders. The ideal described by most teachers, with regard to the quality of learning, was that a communitarian theory of mutual accountability should displace idiosyncratic and collegial theories of accountability.

Accountability Processes related to Teachers' Teaching

Table 13.7 compares the level of support given by teachers and all other groups to proposed processes related to the quality of teachers' teaching.

Teachers involved in interpreting this table noted that a touchstone of professional accountability was carrying both community and individualistic variants. Exceptions were then evaluated in terms of this belief system.

The variance in responses to the proposal to 'encourage teachers to read and do research' was traced to primary teachers. The mean of their responses was the only one that was significantly different to all stakeholders' mean response ($t = 4.20$, $p < 0.001$). The most favoured interpretation was that many primary teachers have a limited education in mathematics, and tending to be 'maths phobics', they would fear any need to conduct quantitative research.

The minor ambivalence over the proposal to use 'an appraisal of student outcomes' to evaluate teaching was traced to district high teachers ($t = 2.09$, $p < 0.05$) and secondary teachers ($t = 2.04$, $p < 0.05$) whose views were just significantly different than all responses. The advice here from teachers was that, as students get older, they take greater responsibility for learning outcomes, increasingly irrespective of the efforts and expertise of teachers.

The uncertainty over 'an appraisal of planning' being used to evaluate teaching was traced solely to primary teachers ($t = 3.46, p < 0.001$). An important feature of primary teaching, it was explained, was the capacity to depart from the planned programme to take advantage of incidental learning opportunities. This was also claimed to be a luxury that was largely denied to secondary colleagues who must teach an increasingly prescribed and examined curriculum.

Attention then focused on the proposals that had just failed to be included in the touchstone. Oddly, the 'planned development of classrooms' had failed to gain support from surveyed teachers, as low as 53.5% from district high teachers and 57.9% from secondary, yet was quickly accepted as 'an excellent idea' when discussed. The ambivalence was attributed to unfamiliarity. One teacher spoke for many when she said 'we have suffered badly from innovation overload; reaction is now automatic'. Two of the innovations referred to with some scepticism were the 'teacher appraisal scheme' and the 'school review scheme'. It was explained that any proposals in the survey that used these terms would be biased by negative connotations, confirming the influence of the historical context provided above in Chapter 9.

Another feature of the responses that interested the teachers who provided interpretations was that a significant minority of their colleagues seemed to reject any involvement or participation by parents. They reported being aware of colleagues that

Table 13.7: Support for proposed processes for collecting data, reporting on and improving teachers' teaching

Processes proposed by all stakeholders	Teachers' views Mean	%	All groups' views Mean	%	Differences of means t	$p <$
Supported						
Discussion between colleagues	1.28	100	1.38	98.4	2.10	0.05
Planned development of teachers	1.47	85.4	1.49	95.8	0.28	ns
Training and support to identify and cope with at-risk students	1.56	97.0	1.54	97.7	0.37	ns
Teacher appraisals reported to the individual teacher	1.70	93.2	1.64	92.7	0.95	ns
Co-operative learning between colleagues (e.g. mentoring)	1.82	90.9	1.80	92.0	0.30	ns
Self-evaluation	1.83	90.2	1.82	89.7	0.13	ns
Negotiating new goals for professional development	1.87	91.7	1.92	88.3	0.70	ns
Transition program for newly appointed teachers	1.93	77.3	1.86	80.4	0.83	ns
Feedback and appraisal by peers	2.02	83.3	2.00	81.5	0.26	ns
Encourage teachers to read and do research	2.06	81.8	1.86	88.0	2.40	0.02
Documentation of best practices	2.20	72.0	2.06	78.1	1.65	ns
Appraisal of student outcomes	2.31	71.9	2.09	80.3	2.72	0.01
An appraisal of planning	2.34	73.5	2.16	81.6	2.25	0.05
Support ambivalent						
The planned development of classrooms	2.36	59.8	2.28	65.3	1.06	ns
Individual and senior staff discuss appraisals	2.39	68.2	2.28	70.4	1.15	ns
The school review process	2.40	59.8	2.22	66.4	2.17	0.05
Standardized test results go to the individual teacher	2.54	57.6	2.47	63.4	0.76	ns
Opportunity for parents to develop as co-teachers	2.65	51.5	2.58	55.2	0.75	ns
Report teaching quality to the DEA for promotion and school development purposes	2.65	52.3	2.55	55.4	0.77	ns
Feedback and appraisal by senior school colleagues	2.66	53.8	2.44	65.7	2.50	0.02
Feedback and appraisal by students	2.70	51.5	2.59	65.9	1.10	ns
Network more effectively with the teachers' union	2.78	33.2	2.96	28.8	1.96	0.05
Assess teacher's contribution to school planning	2.87	46.4	2.72	52.3	1.54	ns
Feedback and appraisal by parents	2.89	41.6	2.60	52.8	2.86	0.01
Feedback and appraisal by an independent expert	3.12	27.3	2.89	36.0	2.13	0.05
Opportunities for parents to consult and co-plan teaching programs	3.21	28.1	2.88	42.3	3.27	0.01
Feedback/appraisal by the P and F/School Council	3.35	18.9	3.11	32.0	2.51	0.02
Unsupported						
The selection of teachers should be more localized	3.42	18.9	3.35	23.9	0.73	ns
Reporting of teacher appraisals to colleagues as part of professional development	3.53	19.7	3.20	30.6	3.34	0.001
P and F/School Council discuss teacher and classroom development	3.57	15.9	3.23	29.0	3.38	0.001
General reporting of teacher appraisals to parents as part of school planning and development	4.10	7.6	3.69	19.8	4.42	0.001

hold all parents in poor regard, attribute many of the problems being faced by students to the poor quality of parenting, and that the accountability of a teacher was limited to serving the 'best interests' of the student.

Similarly, another minority of teachers were reportedly unable or unwilling to accept feedback or appraisal from senior school colleagues, DEA personnel or independent experts. The reasons suggested included anti-authoritarian, anti-centralism and egoism. Some teachers were reputedly unable or unwilling to accept feedback or appraisal from students or parents on the grounds that their contributions lacked professional expertise. The apparent inability or unwillingness of teachers to accept feedback or participation by P and Fs or School Councils was attributed to the ambiguous legislation and limited expertise of the parents involved. It was noted again by teachers involved in interpretation that these attitudes are unlikely to change while relatively few teachers in Tasmania have experience as co-leaders and as co-governors of schools.

In sum, widely supported forms of accountability that would gather data on and report the quality of teaching tended to be limited to discussions between colleagues, co-operative learning, self-evaluation and feedback and appraisal by peers. Supported accountability processes that would improve teaching included self-planned professional and classroom development, specialist training to identify and cope with at-risk students, transition programmes for newly appointed teachers, teachers reading and doing research and documenting best practices, and learning from student outcomes and appraising how teachers plan.

Overall, the processes supported or nearly supported by all teachers and stakeholders were interpreted by teachers as involving more individualistic forms of professionalism than the ideal. Giving such priority to each teacher's autonomy was likely to impede the development of mutual accountability in a community of stakeholders. Idiosyncratic professionalism defined all teachers' views as incontestable, and yet, in absence of feedback or critique, made little provision for the growth of professional knowledge. Individualistic professionalism tended to define the views of other colleagues and stakeholders as inexpert or authoritarian, too easily set aside the need to raise the expertise of all stakeholders regarding the formative evaluation of teaching, and gave low priority to broader and collective social and cultural interests.

Accountability Criteria related to Teachers' Teaching

Table 13.8 compares levels of support expressed by teachers to the views expressed by all groups surveyed.

There was early agreement, again, that the theme behind the criteria was professionalism. Teachers affirmed that a positive classroom environment was characterized by supportive relationships and student enthusiasm, attitude, application and progress. This they attributed to each teacher's professional understandings and skills in organization, behaviour management, knowledge of subject and student development, and instructional expertise. This was explained, they reasoned, in terms of teachers' attitudes to students, parents and colleagues, their willingness to engage in

Table 13.8: Support for criteria for evaluating teachers' teaching

Criteria proposed by all stakeholders	Teachers views Mean	%	All groups' views Mean	%	Differences of means t	p <
Supported						
Classroom environment	1.66	94.7	1.60	95.3	0.94	ns
Interpersonal communications within the classroom	1.77	96.2	1.70	94.7	1.18	ns
Organizational skills	1.91	87.9	1.79	92.0	1.53	ns
Behaviour management skills	1.92	85.6	1.79	91.5	1.79	ns
Teachers' knowledge of subject and child/adolescent development	2.02	87.2	1.86	90.1	2.31	0.05
Effective implementation of school and curriculum policies	2.09	83.3	2.02	85.7	1.14	ns
Teachers' attitudes to students, parents and colleagues	2.16	78.1	1.97	85.1	2.52	0.02
Instructional expertise	2.19	77.2	2.03	82.3	2.41	0.02
Willingness to engage in continuing professional development	2.21	75.7	2.08	77.9	1.22	ns
How well work is set, monitored and marked	2.23	77.3	2.00	83.5	2.42	0.02
The attitude of children (e.g. enthusiasm)	2.28	70.5	2.10	76.5	1.94	ns
Support ambivalent						
Student progress	2.34	68.8	1.99	81.0	3.44	0.001
Communication skills with stakeholders	2.39	56.8	2.23	67.0	2.20	0.05
Teachers competencies in DEA job descriptions	2.47	63.7	2.40	63.6	0.79	ns
Students' achievement levels in K-12 Framework	2.52	60.6	2.83	45.6	1.48	ns
Leadership services given by teachers in school	2.85	44.7	2.66	52.0	1.98	ns
Effectiveness of teachers' written records and plans	2.88	48.5	2.46	65.8	4.19	0.001
Teachers' participation in school and community activities	3.01	39.4	2.83	52.0	1.69	ns

continuing professional development, and the degree of supportive leadership available concerning school and curriculum policies.

This causal story was then used to compare differences with all stakeholders' responses. Using teachers' 'knowledge (of subject matter and child/adolescent development)' was supported by all stakeholders, and thus the only significantly different mean (primary teachers, $t = 3.14$, $p < 0.01$) was disregarded. Similarly, statistically significant differences concerning 'teachers' attitudes to students, parents and colleagues', 'instructional expertise' and 'how well work is set, monitored and marked' were added to the touchstone criteria on political grounds.

The proposal that 'one of the criteria that should be used to evaluate the quality of teaching is student progress' generated considerable ambivalence among teachers, and

248   EDUCATIVE ACCOUNTABILITY

as Table 13.9 shows, was revealed to be a particularly controversial issue in district high settings.

When the mean responses of district high parents and teachers were compared, the differences were found to be very significant ($t = 5.04$, $p < 0.001$). District DEA personnel largely shared district high parents' views, although all principals, including district high principals, took a less extreme view that cohered more closely with all stakeholders.

Teachers interpreted these differences to mean that some teachers took the view that (a) student progress was a crude proxy of the quality of teaching, (b) the views of district high parents were inexpert and more of an indication of the special features of district high schools rather than the actual quality of teaching, and (c) DEA district personnel were driven by limited resources into blaming teachers rather than to providing supportive and developmental assistance. They also pointed out that some teachers asserted that the quality of teaching and the quality of learning had a limited correlation and that this relationship fell further as students took more and more responsibility for their learning as they progressed through schooling. They also recalled the warnings given by some colleagues that making 'student progress' an indicator of the quality of teaching would legitimate introducing 'the destructive effects of US-style testing'. I will come back to this issue when another 'hobgoblin' interpretation has been presented.

The next issue considered was the difference between stakeholders concerning the proposal that 'one of the criteria that should be used to evaluate teaching is the effectiveness of teachers' written records and plans'. Table 13.10 provides the details by comparing means to the mean of all stakeholders' responses.

The teachers' interpretation of Table 13.10 used much the same logic that was employed to explain Table 13.9. They reported that a significant minority of colleagues hold that (a) teachers' written records and plans are a crude proxy of the quality of teaching, and (b) the views of primary and secondary parents are inexpert. They remembered teachers asserting that the quality of teaching and the quality of teachers'

Table 13.9: Differences between stakeholders concerning the proposal that one of the criteria that should be used to evaluate the quality of teaching is student progress

Respondents	n	Mean	SD	%SA+A	t	$p <$
Primary teachers	79	2.21	0.95	73.4	2.33	0.02
District HS teachers	28	2.68	1.22	57.2	4.73	0.001
Secondary teachers	25	2.40	1.12	68.0	2.67	0.01
All principals	125	1.92	0.81	83.8	0.88	ns
Primary parents	96	1.85	0.92	85.2	1.60	ns
District HS parents	28	1.70	0.72	92.6	1.99	0.05
Secondary parents	24	1.79	0.78	87.5	1.28	ns
District DEA	10	1.44	0.53	90.0	2.32	0.05
Central DEA	16	1.63	0.72	87.5	1.90	ns
All respondents	431	1.99	0.93	81.0	–	–

Table 13.10: Differences between stakeholders concerning the proposal that one of the criteria that should be used to evaluate teaching is the effectiveness of teachers written records and plans.

Respondents	n	Mean	SD	%SA+A	t	p <
Primary teachers	79	2.85	0.98	49.4	4.11	0.001
District HS teachers	28	3.00	1.12	42.8	3.69	0.001
Secondary teachers	25	2.84	1.07	52.0	2.46	0.02
Primary principals	89	2.20	0.77	77.5	2.86	0.01
District HS principals	15	2.67	0.09	46.7	1.07	ns
Secondary principals	21	2.48	1.12	71.4	0.11	ns
Primary parents	96	2.27	0.94	72.6	2.15	0.05
District HS parents	28	2.32	0.91	75.1	0.96	ns
Secondary parents	24	2.08	0.83	79.2	2.41	0.02
All DEA	26	2.19	0.80	73.1	1.78	ns
All respondents	431	2.46	0.97	65.8	–	–

written records and plans had a limited causal relationship, especially in the early years of schooling. They also recalled warnings by some colleagues that making 'teachers' written records and plans' an indicator of the quality of teaching would legitimate introducing 'the pressure of Britain's OFSTED-style school inspections'.

The reported presence and reputed effectiveness of these two 'hobgoblin' arguments apparently relied on a belief held by a significant minority of teachers that demonized independent and external evaluation of teaching. This belief equated professionalism with absolute autonomy. A condition of this belief was that being held to account for (say) 'student progress' or the 'quality of plans' would place unwarranted limits on that autonomy. Other stakeholders' views were, therefore, defined (and could be dismissed) as inexpert or authoritarian, or as raising non-professional and political issues. As noted above, teachers interpreted these views as attempts to set aside (a) capacity building as a professional responsibility, and (b) having to respond to the broader stakeholder interests they represent. To summarize this section, teachers' interpretation of touchstone processes and criteria relating to the quality of teaching was sometimes constrained by the presence of a sizeable minority's belief system that favoured 'non-political', idiosyncratic and individualistic forms of professional accountability. This interpretation was supported by teachers' policy preferences concerning leadership criteria and processes.

Accountability Processes relating to Leaders' Leadership

Twenty-six accountability processes were suggested by all groups to collect and report data, and improve leaders' leadership. Teachers' responses were ranked by means and are presented in Table 13.11 alongside comparative data from all other groups. Interpretation gave primacy to teachers' perspectives.

The touchstone was, once again, the focus of attention and was used to generate rules for evaluating less well supported proposals. Teachers, like all stakeholders, expected leaders to provide professional leadership. They wanted leaders to be evaluated using

Table 13.11: Support for processes to collect data, report on and improve leaders' leadership

Processes proposed by all stakeholders	Teachers' views Mean	Teachers' views %	All groups' views Mean	All groups' views %	Differences of means t	Differences of means $p <$
Supported						
Appraisal of support and feedback given to staff	1.86	90.9	1.78	91.8	1.27	ns
Evaluate the coherence between vision, plans and outcomes	1.90	91.7	1.84	90.3	1.12	ns
Self-appraisal	1.93	87.0	1.96	83.3	0.38	ns
Skill development programs, e.g. in governance and management	1.95	84.1	1.90	87.1	0.73	ns
Provision and generation of a school vision	2.04	82.4	1.89	87.7	2.18	0.05
Peer networks reflect on challenges of practice	2.13	71.9	1.97	79.7	2.12	0.05
Improved by using feedback from staff	2.14	80.3	2.06	84.3	1.19	ns
An appraisal by school colleagues	2.18	76.5	2.24	74.0	0.74	ns
An appraisal of policy-making strategies used	2.20	72.7	2.18	74.7	0.30	ns
Peer appraisal	2.24	72.7	2.25	70.7	0.12	ns
Support ambivalent						
Quality of reporting to parents and community	2.30	69.7	2.11	82.1	2.87	0.01
Survey of the school climate	2.23	69.7	2.17	74.4	0.82	ns
Improved by using a mentoring process	2.37	56.8	2.28	60.5	1.23	ns
An appraisal of the quality of external liaison	2.51	55.3	2.40	59.8	1.44	ns
Feedback from parents and students	2.55	55.3	2.26	71.7	3.54	0.001
Parents, teachers, DEA collaborate in principal selection	2.72	47.7	2.42	60.3	2.77	0.01
Appraisals should be reported to the DEA	2.85	37.9	2.76	43.3	0.96	ns
An appraisal by the DEA	3.02	32.5	2.78	45.5	2.44	0.02
An appraisal by the P and F/School Council	3.13	34.8	2.76	47.2	4.41	0.001
Appraisals reported to individuals and colleagues as part of the professional development programme	3.14	29.5	2.99	34.2	1.58	ns
Unsupported						
Overseas exchanges	3.22	18.2	3.02	28.8	2.33	0.02
An appraisal by the community	3.30	18.2	3.13	25.7	1.99	0.05
More localized selection of leaders	3.39	12.1	3.35	18.1	0.46	ns
P and F/School Council set leadership service policies	3.49	15.2	3.35	21.9	1.48	ns
Appraisals reported to parents as part of school development programme	3.67	9.9	3.37	21.3	3.31	0.001
Fixed term and negotiated performance contracts	3.40	24.3	3.33	27.4	0.61	ns

processes that collected and reported evidence of how well they gave support to colleagues, and both gave feedback to and received feedback from staff. Professional leadership was also seen in demands for skill development programmes, especially in governance and management. The second general expectation was that data would be collected and reported concerning philosophical and strategic leadership; the provision and generation of a school vision, policy-making strategies used, and the level of coherence between vision, plans and outcomes. Methods widely supported included self-appraisal, peer appraisal and peer networks that reflected on challenges of practice, and the surveying of school climates.

The significant difference ($t = 2.87$, $p < 0.01$) between teachers and other stakeholders about evaluating leadership in terms of the 'quality of reporting to parents and community' was reviewed. The sole source of ambivalence was found to be primary teachers. It was reported that some were reluctant to have this process defined and centralized as a leadership responsibility, preferring instead to retain direct and frequent contact with parents and other community members.

Evaluating leadership by collecting 'feedback from parents and students' was then examined. Table 13.12 indicates that primary teachers were, once again the sole source of ambivalence.

Again the interpretation provided by teachers was based on 'professional' perspectives. First it was noted that about half of primary teachers surveyed apparently preferred to 'run their own show', and if they wanted feedback, they would want it 'first hand'. The most likely group to demand this, it was reported, were teachers of infant classes. The alternative claim, that primary teachers doubted the expertise of parents and students to make judgements about leadership, was soon discarded as it did not explain why district high and secondary teachers supported the proposal.

The next contested item concerning 'parents, teachers, DEA collaborating in principal selection'. The differences traced to the bimodal responses of primary and secondary teachers. About 30% of these two groups agreed and about 30% disagreed with this proposal. Given the relative inexperience of district high staffs, it was cautiously supposed in most forums that individualistic, careerist and exclusionary forms of professionalism could have been at play among some older staff. This was

Table 13.12: Differences between stakeholders concerning the proposal that the quality of leadership should be improved using feedback from parents and students

Respondents	n	Mean	SD	%SA+A	t	p <
Primary teachers	79	2.60	0.87	53.2	6.58	0.001
District HS teachers	28	2.46	0.96	64.3	1.38	ns
Secondary teachers	25	2.48	0.51	52.0	1.48	ns
All principals	125	2.22	0.77	75.0	0.00	ns
All parents	148	2.08	0.75	80.6	0.00	ns
All DEA	26	2.04	0.61	88.0	0.01	ns
All respondents	431	2.26	0.80	71.7	–	–

somewhat confirmed when the significant differences ($t = 4.41$, $p < 0.001$) between teachers and stakeholders views concerning an 'appraisal by the P and F/School Council' were examined (in Table 13.13).

Teachers involved in interpreting why some items failed to attract support from any stakeholders again evoked the presence of individualistic professionalism to explain outcomes. The proposals to use 'an appraisal by the community', 'P and F/School Councils setting leadership service policies', 'appraisals reported to parents as part of school development programme' and 'more localized selection of leaders' all reputedly failed the professional expertise test. On the other hand, it was agreed that the proposal to have 'fixed term and negotiated performance contracts' was probably seen by most respondents as an inappropriate method for determining leadership responsibilities in education as such contracts are reputedly too often narrow and inflexible and make use of simplistic and easily measurable indicators.

To summarize this section, teachers used touchstone to build expectations that leaders would provide 'professionally enhancing' leadership. They emphasized supportive rather than directive forms of accountability processes that doubled as skill development programmes, especially in governance and management. They also expected philosophical and strategic leadership to be a by-product of accountability methods and judgements. The participation of parents, students and PandF/School Councils was a source of concern to a sizeable minority of secondary teachers, and up to about 50% of primary teachers who apparently prefer to be self-led. They wanted to account personally through direct dealings with parents and students. In essence, professional norms concerning 'autonomy' from leadership were given priority over client accountability by about half primary and about one third of all secondary teachers.

This confirmed the interpretation developed above that suggested the presence of an individualistic professionalism in addition to the broad touchstone of mutual stakeholder accountability. District high teachers and all DEA personnel seemed to find the idea of leaders accounting to client groups for the quality of leadership more congenial. When proposals suggested wider involvement in career decisions, understandably, the bimodality of teachers' views became more sharply pronounced

Table 13.13: Differences between stakeholders concerning the proposal that the evaluation of leaders leadership should include an appraisal by the P and F/School Council

Respondents	n	Mean	SD	%SA+A	t	p <
Primary teachers	79	3.17	1.17	31.7	4.14	0.001
District HS teachers	28	2.89	0.99	46.4	0.89	ns
Secondary teachers	25	3.28	1.06	32.0	3.31	0.001
All principals	125	2.71	1.09	48.8	0.59	ns
All parents	148	2.52	1.14	55.1	2.95	0.01
All DEA	26	1.68	0.95	57.7	1.68	ns
All respondents	431	2.76	1.13	47.2	–	–

and indicated an increasing tension between careerism and professionalism. In general, however, teachers' preferences concerning processes used to evaluate the quality of leadership services cohered with other stakeholders, and despite a sizeable minority that emphasized autonomy and other self-referential values, the central theme was concerned with mutual accounting with reference to the needs and interests of a school community of stakeholders.

Accountability Criteria relating to Leaders' Leadership

It was noted in Chapter 12 that 22 accountability criteria were suggested to evaluate leaders' leadership services. They are ranked in Table 13.14 by the means of teachers' responses and then compared to all other's means.

Table 13.14: Support for criteria for evaluating leaders' leadership services

Processes proposed by all stakeholders	Teachers' views Mean	Teachers' views %	All groups' views Mean	All groups' views %	Differences of means t	Differences of means $p <$
Supported						
Capacity to hear and care for others	1.63	98.5	1.56	96.8	1.19	ns
Ability to plan outcomes and achieve priorities	1.81	93.7	1.78	94.2	0.58	ns
Student and teacher morale and motivation	1.83	87.2	1.77	88.8	0.84	ns
The openness and climate/tone of the school	1.84	87.9	1.76	88.8	1.09	ns
Capacity to make and implement policy	1.89	89.4	1.86	90.9	0.47	ns
Extent of collaborative decision making	1.92	87.1	1.93	83.4	0.15	ns
The extent to which staff support their leaders	1.96	82.6	1.93	84.5	0.42	ns
Management and organizational skills (evaluation, budgeting and governance)	1.97	85.6	1.89	88.3	1.24	ns
Quality of internal and external communications	2.18	74.3	2.10	79.9	1.03	ns
Ambivalent support						
Valuing of creativity and productivity in school	2.24	69.7	2.04	79.5	2.94	0.01
Extent of professional development within the school	2.34	66.6	2.16	74.9	1.97	0.05
Evidence of learning by staff and students	2.35	63.6	2.16	73.6	2.25	0.05
Performance indicators in guidelines provided by the DEA	2.42	60.2	2.36	62.9	0.92	ns
Indicators from research literature used in plans	2.43	53.1	2.32	61.1	1.61	ns
Evidence of the quality of teaching by the staff	2.47	65.1	2.21	74.6	3.09	0.01
Capacities as learners and researchers	2.51	55.3	2.29	68.6	2.70	0.01
The expectations of the community	2.53	58.3	2.46	61.0	0.84	ns
The quality of the physical environment	2.54	59.0	2.53	58.8	0.11	ns
Recommendations from school reviews	2.55	46.2	2.46	55.9	1.24	ns
National competency indicators for managers	2.67	38.7	2.69	40.6	0.29	ns
Extent to which parents support school leaders	2.70	45.5	2.54	54.4	1.89	ns
Leaders' relevant qualifications	2.79	50.0	2.75	51.7	0.39	ns

Two features and two implications of Table 13.14 stood out to teachers. They noted the extent of common ground between all stakeholder groups. They took the view that the agreed set of items meant that the evaluation of leadership services in schools should aim at enhancing the professionalism of leaders. Such educative leadership implied, they suggested, two complimentary aspects. First it was socially and communally rational. It emphasized hearing and caring for others, collaborative decision making, undistorted internal and external communications, student and teacher morale and motivation, an open climate or tone and reciprocal relationships. Second it achieved coherence between philosophy and action. It valued the capacity to make and implement policy, the ability to plan outcomes and achieve priorities and management and organizational skills to do with evaluation, budgeting and governance.

Attention focused on sources of ambivalence. Minor ambivalence relating to the criterion of 'valuing creativity and productivity' was traced solely to primary teachers ($t = 2.46, p < 0.02$), about one-quarter of whom had responded that they were 'not sure' or 'disagreed' with the proposal. It was explained that many primary teachers do not like the term 'productivity' being used in relation to leadership in education.

The proposal that 'the criteria used to evaluate the quality of leadership should include the extent of professional development within a school' drew diverse responses as summarized in Table 13.15 and compared to all stakeholders' views.

Further analysis showed that primary teachers' and primary parents' views have similar ($t = 0.38, p < $ ns) and somewhat ambivalent views on this proposal. In sharp contrast, both primary principals' and district DEA personnel have similar ($t = 1.41, p < $ ns) and highly supportive views. Not only were the differences between primary school principals and teachers significant ($t = 4.03, p < 0.001$), they were even greater between district DEA personnel and primary school teachers ($t = 4.30, p < 0.001$). The same patterns were found in district high schools but were apparently diluted by the presence of post-primary teachers.

It was decided by teachers that this situation revealed yet again the extent to which some primary teachers, most notably early childhood educators, believed that they should self-direct their professional development in consultation with their parents and that this function should not be a measure of effective school leadership. It was also felt that most teachers were equating 'leadership' with 'principalship' and that this in turn implied that most teachers had limited experience of leadership, that relatively few considered themselves to be providing 'leadership' and that 'leadership' was often considered antithetical to collaborative and individualistic professionalism.

This logic was then rehearsed to explain the mild ambivalence among all teachers to another proposal, that 'the criteria used to evaluate leadership should include evidence of learning by staff and students', while all other stakeholders supported it, with the added rider that what counted as 'evidence' would have to be negotiated carefully. It was used again to suggest why the minor ambivalence concerning using the 'capacities of leaders as researchers and learners' to evaluate leaders was traced only to district high ($t = 2.43, p < 0.02$) and primary teachers ($t = 2.19, p < 0.05$). It was explained that many teachers have no experience of collaborative forms of research or post-graduate learning and would set little store on lone activity that could

Table 13.15: Differences between stakeholders concerning the proposal that the criteria used to evaluate the quality of leadership should include the extent of professional development within a school

Respondents	n	Mean	SD	%SA+A	t	p <
Primary teachers	79	2.38	0.94	64.5	2.36	0.02
District HS teachers	28	2.29	0.94	71.4	0.89	ns
Secondary teachers	25	2.29	0.91	68.0	0.85	ns
Primary principals	89	1.85	0.78	88.7	3.49	0.001
District high principals	15	1.93	0.46	93.3	1.18	ns
Secondary teachers	21	1.86	1.06	85.8	1.81	ns
Primary parents	96	2.33	0.89	67.4	1.97	0.05
District high parents	28	2.48	0.94	66.7	2.20	0.05
Secondary parents	24	2.13	0.69	78.2	0.19	ns
District DEA	10	1.50	0.53	100.0	2.79	0.01
Central DEA	26	2.00	0.89	75.2	0.85	ns
All respondents	431	2.16	0.89	74.9	–	–

too easily be associated with personal career advancement. This reasoning was also believed to explain why only about half of all respondents supported the proposal that leaders' relevant qualifications should be used as a criterion to evaluate their leadership.

The interpretation of proposals by teachers sometimes considered the broad implications, usually in terms of professional development. Some had 'horror' stories to tell about a 'incompetent colleague' who drained the energy and enthusiasm out of peers but who had to be supported until they 'moved on'. Some were worried by the apparent incapacity of 'the system' to act faster in such cases yet recognized the need to protect the industrial rights and career of such colleagues. Nevertheless, most teachers tended to focus on the understandings and skills that school communities would need to develop in their leaders if accountability policies and practices were to be educative in intent and outcome.

The general view was that leadership needed to be seen as a school community capacity, not as the privilege of a position holder. While some clearly have been selected to provide leadership services, it seemed just as clear to teachers that any social distance such as hierarchy would undermine capacity in others. Accountability for leadership needed to be conceived in terms of how well the overall capacity building in the school community was proceeding. Again, the touchstone theme of professionalism was used to interpret how this should occur in practice. By bundling the supported and nearly supported processes and criteria listed in Tables 11.13–11.15 into clusters, they searched for a sequence that approximated to the policy cycle. They identified in the clusters a pragmatic sequence of testing ideas with trials, evaluating outcomes, reviewing purposes, reviewing options, planning, mustering support, modifying norms and further trials. This action research process also appeared to cohere with a constructivist theory of learning, a facilitative theory of teaching and an educative

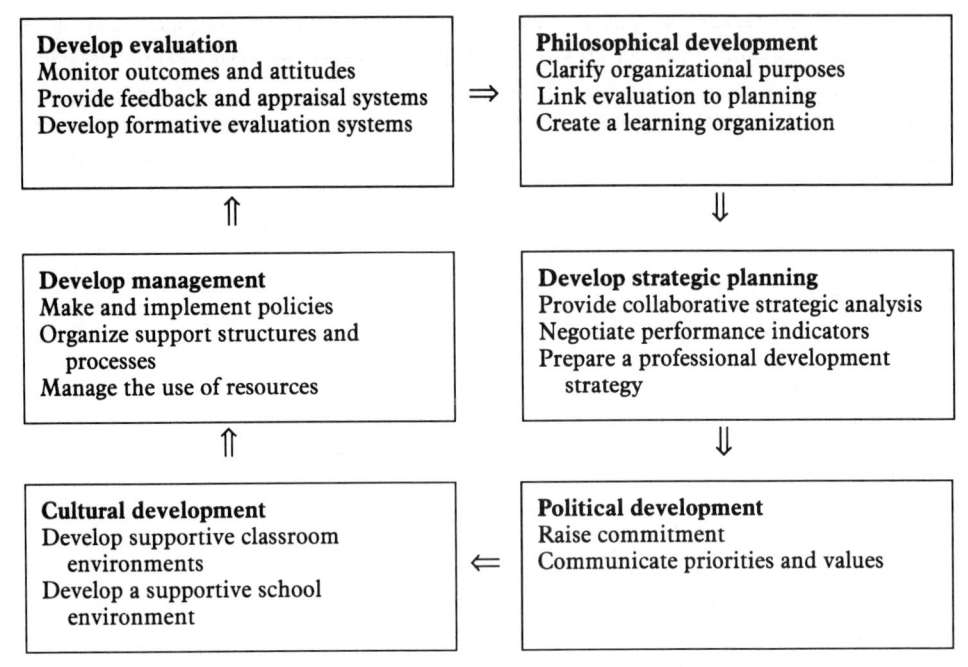

Develop evaluation Monitor outcomes and attitudes Provide feedback and appraisal systems Develop formative evaluation systems	⇒	**Philosophical development** Clarify organizational purposes Link evaluation to planning Create a learning organization
⇑		⇓
Develop management Make and implement policies Organize support structures and processes Manage the use of resources		**Develop strategic planning** Provide collaborative strategic analysis Negotiate performance indicators Prepare a professional development strategy
⇑		⇓
Cultural development Develop supportive classroom environments Develop a supportive school environment	⇐	**Political development** Raise commitment Communicate priorities and values

Figure 13.1: A school community development model of educative accountability.

theory of leadership. It is summarized in Figure 13.1 as a school community development model of educative accountability. The directional symbol ⇒ is used to indicate 'implies' or 'leads to'.

Summary

In this chapter teachers policy preferences were compared with other immediate stakeholders. The analysis permitted teachers' perspectives to identify implications. The main conclusion was that the accountability criteria and processes related to the quality of learning, teaching and leadership valued by all stakeholder groups, including teachers, provided a compelling case for a school community development model of educative accountability.

It is important to note that the interpretations above were provided on a conditional basis. A number of teachers insisted that their views would have to be revised in each staffroom before application. Others indicated that each staffroom was very different, and local students and parents would also need to be consulted. While they were pleased to draw on their experience and professional judgement to help interpret findings, and saw their own views as idealistic yet coherent, they struggled not to generalize while being willing to speculate.

Despite the conditional nature of their interpretations, the teachers and other stakeholder executives agreed that the policy touchstone was clear enough for 'the system' and schools to act on. Teachers affirmed the policy touchstone summarized in Tables 11.13–11.15 and clearly believed it to be positively and causally related to the quality of learning, teaching and leadership. They saw many the proposals in the touchstone area as highly consistent with 'professionalism', 'best practice' and 'union policy'. It was also evident at the April 1995 policy forum that they expected the DEA 'to implement' the supported proposals as 'policy'.

The basic reason for this degree of support was that teachers saw Tables 13.1–13.3 as professionally affirming. They accepted that a minority of colleagues were not as enthusiastic about lesson evaluation, having to report in accurate language or sharing their plans as much as other stakeholders expected. They, nevertheless, used the touchstone as a mandate to define 'best practice' relating to the quality of learning; three-way parent/teacher/student communications, parent and student participation in policy making and in setting both goals and individualized expectations, and sophisticated evaluation and planning of lessons and programmes. They also used it to identify ideal data collection and reporting methods and negative practices. They recommended opportunities for colleagues to participate in leadership and governance so that they could help build a shared theory of professionalism that cohered with over arching 'school policy'.

The touchstone accountability criteria related to students' learning were used to explain that professional development was a prior condition of school development and mutual accountability since 'school' had to regarded as a professional community of educators before it could become a school community of stakeholders. Professional development was required to help teachers build idiosyncratic views of professionalism into shared theories, and then into shared community theories. Until that happened, it was reasoned, many teachers' view of the ideal reporting process to parents concerning students' learning would continue to reflect the confidential doctor–client consultation. It could begin to happen, they also believed, if more teachers experienced action research and undistorted three-way discussions with students and parents. They predicted that this would improve the sophistication and coherence of professional judgement making. Ideally, a communitarian theory of mutual accountability would displace idiosyncratic and collegial theories of accountability.

Teachers were more cautious about involving non-professional stakeholders in accountability methods and criteria related to the quality of teaching. Sizeable minorities of teachers used more idiosyncratic and individualistic forms of professionalism, believing that 'autonomy' was to be defended. Giving high priority to teacher autonomy was found by teachers involved in the interpretation of findings to be at cost to the development of mutual accountability in a community of stakeholders. Idiosyncratic professionalism was seen as inappropriately incontestable and antithetical to the feedback and critique required for the growth of professional knowledge. Individualistic professionalism was also seen as setting aside the views of other colleagues and stakeholders as inexpert or authoritarian. Similarly, defining stakeholders questions as 'non-professional' or 'political' by the lights of individualistic professionalism gave low priority to (a) raising the expertise of all stakeholders regarding the formative evaluation of teaching or (b) respecting the expression of wider and collective social and cultural interests.

Teachers also used touchstone to suggest that leaders should provide 'professionally enhancing' leadership. They emphasized supportive accountability processes that also served as skill development programmes, especially in governance and management since they were believed to boost philosophical and strategic leadership capacities. The participation of parents, students and P and F/School Councils in leadership and governance was a source of concern to a sizeable minority of secondary teachers, and up to about half of all primary teachers, essentially on the basis that they have few such opportunities themselves.

In the interim, the pattern of preferences for leadership services indicated that individualistic professionalism persisted in addition to the broad touchstone of mutual stakeholder accountability. Understandably, teachers' views become polarized when career interests were evoked by leadership accountability criteria. A sizeable minority did emphasize autonomy and other self-referential values, and yet the central theme was concerned with mutual accounting with reference to the needs and interests of a school community of stakeholders.

The general view was also that leadership evaluation should consider capacity building in the school community. Processes and criteria listed in Tables 11.13–11.15 were bundled into clusters and arranged into a sequence that approximated to the annual school policy cycle. The pragmatic sequence of knowledge refinement steps involved testing ideas with trials, evaluating outcomes, reviewing purposes, reviewing options, planning, mustering support, modifying norms and further trials. This idealized school community action research process also seemed to cohere with a constructivist theory of learning, a facilitative theory of teaching and an educative theory of leadership. The next chapter compares principals' views with all other stakeholders.

14

Tasmanian Principals' Policy Preferences and Implications

Introduction

In this chapter I present and examine the policy preferences of Tasmania's principals concerning accountability criteria and processes, compare their views to other stakeholder groups and identify issues that they believe warrant attention. With some exceptions, there were many criteria and processes related to the quality of learning, teaching and leadership that were valued by all stakeholder groups, including principals. This touchstone was taken by them to provide a coherent mandate and set of moral imperatives for the provision of particular leadership services. The touchstone was also taken to imply that (a) Tasmanian state schools need to review and develop their accountability policies and (b) the professional development (PD) of principals needs to have specific content if it is going to help develop more educative forms of accountability practices.

The discussion begins by comparing the views of principals with other stakeholders concerned with student's learning. In each table, the percent symbol refers to the total percentage of respondents that 'strongly agreed' or 'agreed' to each policy proposal. It is important to note that the analysis and interpretations reported in this chapter were largely mediated by the perspectives of principals.

Accountability Processes related to Students' Learning

Table 14.1 compares the level of support that principals and all groups gave to the processes that at least one stakeholder group believed should be used to collect data on, report and improve students' learning.

The most immediately apparent feature noticed by principals was that their peers had given greater emphasis to the value of teachers evaluating and planning their programs than had other stakeholders. It was also realized that this emphasis could easily become offensive if overstated. Since all stakeholders supported the proposal, it was felt that principals might be well advised to proceed in a low key manner, assume

Table 14.1: Support for proposed processes to collect data, report on and improve students' learning

Processes proposed by all stakeholders	Principals' views Mean	%	All groups' views Mean	%	Differences of means t	p <
Supported						
Parent/teacher interviews	1.41	98.4	1.46	97.2	0.82	ns
Teachers evaluate and plan lessons thoroughly	1.41	98.4	1.58	96.4	2.20	0.05
Teachers evaluate and plan programs systematically	1.54	98.4	1.77	91.8	3.41	0.001
Teachers' observations	1.60	99.2	1.71	97.9	2.09	0.05
Conferencing between teacher and student	1.63	94.3	1.74	93.2	1.69	0.05
Parent/teacher/student discussions	1.63	96.6	1.80	91.9	2.38	0.01
Parent input and policy explanations	1.70	93.3	1.82	90.0	1.76	0.05
The sampling of student work (e.g. folios)	1.74	91.7	1.77	91.1	0.42	ns
Reporting through publications and public relations	1.78	86.6	1.84	87.3	0.67	ns
Student's own self-assessment	1.78	86.8	1.92	82.5	1.77	0.05
Teacher-designed mastery and diagnostic tests	1.83	94.5	1.93	87.7	1.50	ns
Teachers identify outcomes for each student	1.84	85.1	1.80	89.3	0.51	ns
Reports—clear and accurate descriptions of learning	1.93	86.8	1.89	86.8	0.56	ns
Support-staff reports; guidance, welfare, speech, health	1.99	84.3	1.99	83.4	0.13	ns
Formative evaluation related to teaching objectives	2.03	86.0	2.17	73.6	1.91	0.05
Teachers written checklists and running records	2.13	76.4	1.93	82.7	2.15	0.05
Parents given goals, expected outcomes, and individual expectations at the beginning of each year	2.19	71.4	2.12	73.5	0.71	ns
Ambivalent support						
State-wide, norm-referenced, standardized tests of literacy and numeracy	2.89	45.8	2.70	54.2	1.67	0.05
Peer appraisal	2.83	43.3	3.05	35.1	2.12	0.05
P and Fs/Schools Council review, discuss and report learning	3.25	30.6	3.09	33.4	1.46	ns
Reports with marks or grades	3.50	27.3	2.89	47.6	4.43	0.001
Unsupported						
Reports allow parents to compare child with others	4.02	15.7	3.89	17.0	0.92	ns

a broad base of support for systematic evaluation and planning, and focus on improving the routines in use. One such method would be to facilitate the inter-professional exchange of exemplary practices.

Another area that principals could have been a little ahead of their professional colleagues concerned the value of discussions between teachers, parents and students. Current practices, it was agreed, largely marginalized students and tended to reduce parent-teacher discussions to ritualistic exchanges at the end of school terms. Again, since support was widespread for such collaborative exchanges on students' learning, it was considered to be a wiser policy to improve the local processes incrementally, rather than campaigning publicly for the principle.

Another controversial area was the use of 'state-wide, norm-referenced, standardized tests of literacy and numeracy'. The rich mythology in the public domain concerning stakeholders' views was discussed in Chapters 10 and 12. It was noted that all parents (73.1% SA+A), particularly district high parents (84.6%), and district DEA personnel (80%), supported this proposal. It was unsupported by district high teachers (28.6%) with district high principals (33.4%) and secondary principals (33.4%) also tending to be unsupportive. Primary teachers (40.5%), secondary teachers (48.0%), primary principals (50.0%), central DEA personnel (56.3%) were found to be ambivalent. Principals concluded that it would be wise to be move cautiously on this matter, especially when district high teachers, principals and parents or when teachers' and parents' representatives were involved.

Table 14.1 was taken to suggest that while principals' views generally accorded with others concerning the level of parental involvement in reviews of student learning, parents sought significantly higher levels of participation. Principals also noted how this manifest desire was running against substantial professional conservatism; levels of support were as low as 20.0% SA+A among district DEA personnel, 22.7% among all teachers, with support among primary principals down at 23.5%. District DEA personnel and principals agreed that they needed to seek all means of generating more educative attitudes and capacities in themselves and in teachers.

Principals were the most sceptical of all groups regarding the proposal that 'reports with marks or grades should be provided'. It will be recalled that during the qualitative data collection phase of the project, most groups expressed what turned out to be myths about the preferences of other groups on this matter. The principals who discussed this matter came to the view that they would be well advised to be sceptical of all competing claims, including their own, and to research each controversial issue locally as part of devising a new school community policy.

Principals also felt that Table 14.1 gave leaders a mandate of some precision from stakeholders to encourage specific activities. They should promote discussions between teachers and parents and students, especially discussions that would provide a vehicle for (a) parent input, (b) policy explanations, (c) parents and students being given goals, expected outcomes and individual expectations at the beginning of each year, and (d) at the end of specific periods, teachers reporting to parents using clear and accurate descriptions of learning. Leaders should also encourage both lesson planning and evaluation and program planning and evaluation. They should encourage the use of particular data; student's self-assessment, teacher-designed mastery and diagnostic tests, teachers' observations, the sampling of student work, teachers' written checklists and running records, and information supplied by support staff. Finally, the principals involved in interpretation concluded that leaders should derive suitable material for public reporting *from* formative evaluation related to lesson and program objectives.

Accountability Criteria related to Students' Learning

Table 14.2 compares the support that principals and all other groups gave to the student learning criteria suggested by all stakeholders.

Principals, along with DEA personnel, were more willing than other groups to entrust to teachers the selection of criteria concerning students' learning. Parents sought greater trust being placed in the 'performance indicators outlined in state and

Table 14.2: Support for proposed criteria for evaluating students' learning

Criteria proposed by all stakeholders	Principals' views Mean	%	All groups' views Mean	%	Differences of means t	p <
Supported						
Judgements by teachers	1.66	95.1	1.99	84.2	4.41	0.001
Results of objective assessment used in plans	1.77	90.8	1.85	92.7	1.35	ns
Performance indicators developed by teachers	1.80	91.8	1.95	85.4	2.13	0.05
Measures of individual progress	1.81	89.2	1.77	93.2	0.62	ns
Student attitudes to school, teachers, peers, learning, homework	1.96	86.0	1.90	89.4	0.82	ns
Measures of students' self-esteem and life skills	1.99	83.1	1.94	84.4	0.68	ns
Performance indicators in state and national policy documents	1.99	83.1	1.99	84.2	0.79	ns
Indicators from research literature used in planning	2.20	73.8	2.17	72.7	0.47	ns
Ambivalent support						
Performance indicators developed by teachers through subject moderation	2.23	68.6	2.23	72.0	0	ns
Indicators developed jointly by parent, teacher, student	2.24	67.0	2.13	77.8	0.70	ns
Criteria developed by research in the classroom	2.28	59.2	2.24	71.4	0.78	ns
Student participation rates (attendance, retention)	2.65	48.4	2.24	66.4	1.17	ns
Parental expectations	2.97	38.0	2.96	37.6	0.09	ns

national policy documents (e.g. Frameworks, Profiles)'. Teachers suggested that more attention should be given to research literature when planning the use of criteria. It also appeared that teachers and principals had underestimated how much other stakeholders value the development of criteria through subject moderation meetings. Parents and DEA personnel appeared to value the collaborative development of learning criteria more than principals and teachers, to the extent that parental expectations were apparently given cursory consideration by principals and even less by teachers.

Implications for principals and district DEA personnel were easily identified by the principals consulted. First was the need to ensure that while teacher development remained integral to school development processes, internal professional perspectives must be far better balanced by national, state and client perspectives and by researched information during school policy making and planning. I will come back to this matter. For example, national curriculum profiles could be used to help review learning criteria in primary and secondary school communities. Another example suggested was that the successful secondary experience of moderation should be developed and extended to primary schooling.

The most challenging implication, however, was that principals would have to adopt, and help teachers develop, a more collaborative and inclusionary philosophy with regard to student and parental participation in determining learning criteria. The situation was seen as particularly acute in secondary schools where teachers (8% SA+A) apparently disregard 'parental expectations' as a criterion for the evaluation of students' learning. Other teachers and all principals are slightly more responsive (all >30% SA+A) to parental expectations concerning learning criteria than parents and DEA personnel (both >40% SA+A). Such moves towards more participative pedagogy and professional responsiveness would also cohere better with other touchstone views expressed by all groups, specifically that measures of individual progress, student attitudes to school, teachers, peers, learning and homework, and measures of students' self-esteem and life skills should be valued as key indicators of learning.

Accountability Processes related to Teachers' Teaching

Table 14.3 compares the level of support given by principals and all other groups to proposed processes related to the quality of teachers' teaching.

The most evident feature of the data summarized in Table 14.3 is the extensive degree of agreement over a number of processes that were believed to improve the quality of teaching. It was quickly interpreted as meaning that the array of collaborative strategies centred on teacher development with over 70% SA+A would attract general support relatively easily in most school communities.

It was also taken to imply that other proposals could be adopted with caution by using the 'general nature' of touchstone 'as a point of reference'. For example, it was noted that principals and DEA personnel have more faith than other groups in the efficacy of 'documenting best practice' with some ambivalence among primary parents (57.4% SA+A), primary teachers (63.3%) and secondary parents (65.2). The minor differences between principals and teachers over processes were all interpreted as issues

Table 14.3: Support for proposed processes for collecting data, reporting on and improving teachers' teaching

Processes proposed by all stakeholders	Principals' views Mean	%	All groups' views Mean	%	Differences of means t	$p <$
Supported						
Discussion between colleagues	1.23	99.2	1.38	98.4	2.75	0.005
Planned development of teachers	1.27	99.2	1.49	95.8	3.61	0.001
Training and support to identify and cope with at-risk students	1.48	98.4	1.54	97.7	1.34	ns
Report teacher appraisals to the individual teacher	1.52	94.2	1.64	92.7	1.89	0.05
Self-evaluation	1.54	96.7	1.82	89.7	3.90	0.001
Co-operative learning between colleagues (e.g. mentoring)	1.60	95.9	1.80	92.0	2.94	0.005
Negotiating new goals for professional development	1.65	95.9	1.92	88.3	3.79	0.005
Transition program for newly appointed teachers	1.76	84.3	1.86	80.4	1.05	ns
Feedback and appraisal by peers	1.78	88.3	2.00	81.5	2.56	0.01
Encourage teachers to read and do research	1.79	87.6	1.86	88.0	1.03	ns
Documentation of best practices	1.83	90.0	2.06	78.1	3.01	0.005
Appraisal of student outcomes	1.99	82.6	2.09	80.3	1.18	ns
An appraisal of planning	2.02	87.5	2.16	81.6	1.81	0.05
Support ambivalent						
The school review process	2.05	74.6	2.22	66.4	1.79	0.05
The planned development of classrooms	2.31	62.4	2.28	65.3	0.37	ns
Individual and senior staff discuss appraisals	2.32	67.3	2.28	70.4	0.29	ns
Feedback and appraisal by senior school colleagues	2.38	68.3	2.44	65.7	0.78	ns
Feedback and appraisal by students	2.53	60.0	2.59	65.9	0.60	ns
Standardized test results go to the individual teacher	2.57	60.0	2.47	63.4	1.09	ns
Assess teacher's contribution to school planning	2.59	57.1	2.72	52.3	1.21	ns
Opportunity for parents to develop as co-teachers	2.68	51.7	2.58	55.2	1.20	ns
Feedback and appraisal by parents	2.76	44.6	2.60	52.8	1.33	ns
Report teaching quality to the DEA for promotion and school development purposes	2.81	42.4	2.55	55.4	2.49	0.01
Opportunities for parents to consult and co-plan teaching programs	2.96	39.0	2.88	42.3	0.45	ns
Network more effectively with the teachers' union	3.02	23.2	2.96	28.8	0.74	ns
Feedback and appraisal by an independent expert	3.05	27.2	2.89	36.0	1.71	0.05
Reporting of teacher appraisals to colleagues as part of professional development	3.24	30.3	3.20	30.6	0.19	ns
Unsupported						
Feedback/appraisal by the P and F/School Council	3.35	22.4	3.11	32.0	2.17	0.05
P and F/School Council discuss teacher and classroom development	3.39	23.1	3.23	29.0	1.34	ns
The selection of teachers should be more localized	3.47	25.8	3.35	23.9	0.81	ns
General reporting of teacher appraisals to parents as part of school planning and development	3.95	12.4	3.69	19.8	2.27	0.05

that could be resolved locally and added to policy. Processes cited here were using empirical evidence of 'learning outcomes', the 'evaluation of planning', 'school review processes', 'feedback and appraisal by more senior colleagues', consulting or co-planning with parents, or the 'general outcomes of teacher appraisals being used for professional or school development purposes'.

Principals identified some issues as requiring careful attention. Their analysis traced each issue from the sources and levels of ambivalence in particular groups (defined as SA+A <70% >30%), compared them to 'the thrust of the agreed proposals' and then considered consequences in their setting. For example, secondary parents alone (65.2%) were concerned about 'negotiating new PD goals'. Primary parents (66.3%) alone were ambivalent about 'feedback and appraisal by peers'. Primary teachers (64.6%) were the only group harbouring minor concerns about the planning of teaching being evaluated. Minor doubts about using 'student learning outcomes' were located solely among the district high teachers (67.9%) and secondary teachers (68%).

It was noted that DEA personnel placed slightly more value than did principals on the use of particular criteria; 'student learning outcomes', the 'planning of teaching', the 'school review process', using students, outside experts and senior colleagues to give feedback and appraisals, parental consultation and co-planning, and providing greater involvement for the P and F/School Council in professional and school development. Principals interpreted these as differences in priority, and explained how the DEA's current policies had to be scheduled for implementation or teachers would suffer from further 'innovation overload'. On the other hand they could see that, by focusing negotiations on the ambivalent interest groups noted above, specified strategies should be developed quicker than generally anticipated. For example, the school review process was greeted with some ambivalence by all parents (61.9%) and all teachers (59.8%) yet supported by all DEA personnel (88.5%) and all principals (73.2%). It was felt that the ongoing state-wide evaluation of the school review process that was being co-ordinated by the ERU and the district superintendents would quite properly probably raise both the credibility and formative effect of the scheme. This matter is discussed in Chapter 15.

The principals also interpreted the apparent closedness of schools with great care. They took the view that principals should consult carefully to map parental demands for participation that are intended to improve teaching while also mapping the patterns of resistance. They noted that the proposal that 'parents should be given opportunities to be consulted and/or co-plan teaching programs' was regarded with ambivalence by all groups with only primary teachers (26.6%) and secondary teachers (28.0%) unsupportive. Similarly, all groups were also noted to be ambivalent about the suggestion that 'parents should to be given opportunities to develop as co-teachers' except district high parents (70.4%) and DEA personnel (70%). This was taken to mean that collaborative trials would be more likely to be successful in selected district high schools. On the other hand, 'feedback and appraisal of teaching by the P and F/School Council' was regarded as 'an idea well ahead of its time.' It was greeted with ambivalence by all parents (51.4%) and all DEA personnel (34.6%) or seen as insupportable by all teachers (18.9%) and all principals (22.4%).

The patterns of response to two proposals created puzzlement among principals. They noted that 'assessing each teacher's contribution to school planning as a means of improving teachers' teaching' was regarded with ambivalence by all groups, and yet, they also took the view that such participation was both crucial for team building and that it would improve the quality of teaching. They were also puzzled that 'the planned development of classrooms' was regarded with ambivalence by all groups yet greeted so favourably by secondary parents (83.4% SA+A), district high principals (73.4%) and secondary teachers (72%). Given the generally 'dull and boring' decorative state of many secondary classrooms, they came to the view that these two ideas were 'only just ahead of their time' and that broader consultations would probably lead to wider adoption.

Five proposals measured support for feedback and appraisal processes related to the quality of teachers' teaching. Peer-based methods were supported by all groups except by primary parents (66.3% SA+A) who exhibited minor levels of ambivalence. 'Student feedback and appraisal' was greeted with ambivalence by all groups less DEA personnel (84.6%) and district high school principals (85.7%) who were supportive. 'Parental feedback and appraisal' was also regarded with ambivalence except for primary parents (82.7%) who supported this proposal. Feedback and appraisal by 'an independent expert' was greeted with ambivalence by all groups except for primary teachers (22.8%), secondary principals (23.8%), primary principals (25.8%) and district high teachers (28.6%) who were all unsupportive. 'Feedback and appraisal by more senior school colleagues' was supported by secondary principals (85.7), district high school parents (85.2%), central DEA personnel (81.3%), secondary parents (79.2%), district high teachers (71.4%) and marginally by primary principals (69.7%), with all other groups ambivalent. Principals interpreted these patterns as meaning that each school policy community really needed to evaluate how teaching itself was being evaluated and developed, to consider using extra processes and to attend to these matters as a matter of strategic urgency. It was generally accepted that the quality of teaching was a school community responsibility that too was often 'slipping through the cracks'.

Accountability Criteria related to Teachers' Teaching

Table 14.4 summarizes levels of support expressed by principals and by all groups surveyed.

It was quickly agreed by principals that the first 13 proposals in Table 14.4 could be adopted in most schools with relatively little prospect of controversy. On the other hand, principals also noted that although using 'communication skills with stakeholders' as a criterion for evaluating the quality of teaching was supported by principals (87.2%) and DEA personnel (80.7%), teachers (56.8%) and parents (54.2%) were ambivalent. Explanations varied. 'Using student progress as a criterion to evaluate teachers' teaching' was supported by all groups, with only district high school teachers (57.3% SA+A) ambivalent. This was accepted as reflecting the special nature of district high schools.

Table 14.4: Support for criteria for evaluating teachers' teaching

Criteria proposed by all stakeholders	Principals' views Mean	%	All groups' views Mean	%	Differences of means t	p <
Supported						
Classroom environment	1.43	98.3	1.60	95.3	2.77	0.005
Interpersonal communications within the classroom	1.51	96.7	1.70	94.7	3.28	0.005
Behaviour management skills	1.66	96.7	1.79	91.5	1.93	0.05
Teachers' knowledge of subject and child/adolescent development	1.66	95.9	1.86	90.1	2.68	0.005
Organizational skills	1.71	94.9	1.79	92.0	1.14	ns
Effective implementation of school and curriculum policies	1.86	89.1	2.02	85.7	2.31	0.05
Teachers' attitudes to students, parents and colleagues	1.87	88.4	1.97	85.1	1.05	ns
Communication skills with stakeholders	1.88	87.6	2.23	67.0	4.60	0.005
Instructional expertise	1.92	83.4	2.03	82.3	1.77	0.05
Student progress	1.92	83.4	1.99	81.0	0.76	ns
Willingness to engage in continuing professional development	1.98	78.5	2.08	77.9	1.06	ns
How well work is set, monitored and marked	2.13	79.2	2.00	83.5	1.38	ns
The attitude of children (e.g. enthusiasm)	2.13	75.2	2.10	76.5	0.46	ns
Support ambivalent						
Effectiveness of teachers' written records and plans	2.31	72.7	2.46	65.8	1.66	0.05
Teachers competencies in DEA job descriptions	2.42	65.9	2.40	63.6	0.11	ns
Leadership services given by teachers in school	2.45	65.3	2.66	52.0	2.06	0.05
Students' achievement levels in K-12 Framework	2.48	59.7	2.83	45.6	1.04	ns
Teachers' participation in school and community	2.83	42.1	2.83	52.0	0	ns

The interpretations that principals made of ambivalence were specific to some items. Using 'the competencies embedded in departmental job descriptions' to evaluate teaching was supported by secondary teachers (80%), secondary principals (80%), district high parents (75.1%) and DEA personnel (88%) with all other groups ambivalent. This was interpreted as meaning that the evaluation of teaching was easier when it was subject related and mediated by moderation. On the other, 'using the effectiveness of teachers' written records and plans as a criterion to evaluate teachers' teaching' was supported by primary principals (77.5% SA+A), secondary principals (72.4%), parents (74.1%) and district DEA personnel (90%) while all others groups were ambivalent. This was taken to mean that teachers were ambivalent about their plans being used as a proxy for the quality of their teaching while tacitly acknowledging the value of planning lessons and programs.

Other patterns of responses were taken at face value. For example, it was accepted that all groups were ambivalent about using 'leadership services performed by teachers within the school' and 'teachers' participation in school and community activities' as teaching evaluation criteria. Since 'students' achievement levels in the K-12 Framework' were supported solely by secondary parents (79.2%) as a criterion for evaluating the quality of teaching, with all other groups ambivalent, this proposal was regarded as 'unworkable'.

In sum, it appeared that the principals recommended adoption of the first 13 criteria in Table 14.4, and to use the essence of touchstone past that point to guide cautious consultations with stakeholders expressing ambivalence in order to add other proposals to 'policy'.

Accountability Processes relating to Leaders' Leadership

It will be recalled that 26 accountability processes were suggested by all groups to collect and report data, and improve leaders' leadership. Principals' responses were ranked by means and are presented in Table 14.5 alongside comparative data from all other groups. As above, these data were then interpreted in policy forums and workshops.

Again, the appraisal processes proposed for collecting data, reporting on and improving leaders' leadership that attracted strong support across that board were seen as immediately useable and practical policy proposals. It was also agreed that principals could proceed with confidence with other ideas providing they consulted carefully, taking the sources of uncertainty and the nature of touchstone into account. 'Leaders' self-appraisal', for example, was supported by all groups except district high parents (60.7% SA+A) and secondary parents (50%) who were ambivalent. All groups supported 'an appraisal by school colleagues' except all parents (64.1%) who were marginally ambivalent. 'Peer appraisal' was also supported by all groups except primary parents (51%) and secondary parents (60.6%) who expressed some ambivalence. The proposed 'appraisal by DEA personnel' was regarded by all groups with ambivalence except secondary teachers (24%) who were unsupportive and district high principals (73.4%) who were, conversely, supportive. This pattern of support was, once again, explained in terms of the special nature of district high schools.

Principals were quick to note again that support for parental and community participation in leadership development was low. All groups were either ambivalent or unsupportive of 'appraisals by P and F/School Councils' and all stakeholders were unsupportive of an 'appraisal by the community' except primary parents (32.3%) and all DEA personnel (34.6%) who were ambivalent. The proposal to appraise 'the quality of external liaison' drew ambivalent support from all groups except district high parents (75%) and district DEA personnel (90%) who were supportive. These patterns were taken to mean that schools were not as open as they might be and that this was an issue for strategic attention.

The reporting of appraisals concerning leadership services was also seen as a potentially controversial area. It was noted that secondary parents (70.9% SA+A)

Table 14.5: Support for processes to collect data, report on and improve leaders' leadership

Processes proposed by all stakeholders	Principals' views Mean	%	All groups' views Mean	%	Differences of means t	$p <$
Supported						
Self-appraisal	1.57	95.1	1.96	83.3	4.65	0.005
Appraisal of support and feedback given to staff	1.60	95.0	1.78	91.8	2.73	0.005
Evaluate the coherence between vision, plans and outcomes	1.60	95.1	1.84	90.3	3.34	0.001
Peer networks reflect on challenges of practice	1.64	94.1	1.97	79.7	4.64	0.001
Provision and generation of a school vision	1.69	92.5	1.89	87.7	3.00	0.005
Skill development programmes, e.g. in governance and management	1.75	94.2	1.90	87.1	2.17	0.05
Improved by using a mentoring process	1.89	77.4	2.28	60.5	4.72	0.001
Peer appraisal	1.98	81.0	2.25	70.7	3.27	0.005
Improved by using feedback from staff	2.02	85.9	2.06	84.3	0.64	ns
Survey of the school climate	2.02	82.7	2.17	74.4	1.99	0.05
An appraisal of policy-making strategies used	2.06	80.2	2.18	74.7	1.90	0.05
Quality of reporting to parents and community	2.06	84.3	2.11	82.1	0.80	ns
An appraisal by school colleagues	2.14	79.3	2.24	74.0	1.23	ns
Feedback from parents and students	2.22	75.0	2.26	71.7	0.49	ns
Support ambivalent						
An appraisal of the quality of external liaison	2.29	45.8	2.40	59.8	1.49	ns
Parents, teachers, DEA collaborate in principal selection	2.34	65.9	2.42	60.3	0.75	ns
Overseas exchanges	2.50	53.7	3.02	28.8	5.65	0.005
An appraisal by the P and F/School Council	2.71	49.6	2.76	47.2	0.43	ns
An appraisal by the DEA	2.73	49.6	2.78	45.5	0.46	ns
Appraisals should be reported to the DEA	2.99	32.8	2.76	43.3	2.27	0.05
Appraisals reported to individuals and colleagues as part of the professional development programme	3.07	30.5	2.99	34.2	0.80	ns
Unsupported						
An appraisal by the community	3.10	26.7	3.13	25.7	0.32	ns
P and F/School Council set leadership service policies	3.58	15.7	3.35	21.9	2.22	0.05
More localized selection of leaders	3.58	15.7	3.35	18.1	2.29	0.005
Appraisals reported to parents as part of school development program	3.62	13.3	3.37	21.3	2.21	0.05
Fixed-term and negotiated performance contracts	3.81	15.1	3.33	27.4	3.84	0.005

believed that 'appraisals of leadership services should be reported to the DEA' while district high teachers (28.6%), district high principals (26.7%) and district DEA personnel (10%) took the converse view. All other groups were ambivalent. All groups were ambivalent about the suggestion that 'appraisals should be reported to the individual and to school colleagues as part of the professional development program' except primary teachers (27.9%), primary principals (27.6%) and central DEA personnel (20%) who were unsupportive. It was agreed that these ideas were 'probably unworkable' in most schools.

Principals also examined some generally supported improvement strategies that did not enjoy universal support. 'Peer networks to share reflections on the challenges of practice' was supported by all groups except by some primary teachers (67.1% SA+A). 'Surveying the school climate' was supported by all groups with the exception of district high teachers (53.5%), secondary teachers (64%), district high principals (60%) and primary parents (63.3%) who were, to various degrees, ambivalent. 'Using feedback from parents and students' was supported by all groups except all teachers (55.3%) and district high school principals (66.7%) who exhibited minor degrees of ambivalence. 'Improving leadership by using a mentoring process' was a proposal all groups supported except primary teachers (49.3%), district high teachers (60.8%) and all parents (45.9%) who were ambivalent. It was conceded, with some regret, that 'overseas exchanges' were unsupported by all groups except all principals (52.8%) and district DEA personnel (40%) who were ambivalent. In each of these cases, except the last, principals came to view that they could be modified and admitted to touchstone, especially when referred to school policy communities.

There were some issues that principals insisted on dealing with in a collective manner, rather than comparing the matter to the substance of touchstone, to infer consequences or refer it to school policy communities. An example was all proposals related to appointment processes and service conditions. They were very interested in the proposal that 'parents, professionals and the DEA should be involved in the selection of principals'. It was a current option for school communities although not always exercised. It generated complex but generally supportive or ambivalent responses. Teachers' views were bimodal and ambivalent; while 47.7% 'agreed' or 'strongly agreed', 22.7% were 'unsure' with 28.8% indicating that they 'disagreed' or 'strongly disagreed' with the proposal. Strongest support came from secondary principals (79% SA+A), district principals (86.7%), district DEA personnel (80%) and district parents (77.7%). Secondary teachers (32% SA+A), primary teachers (48.1%) and district teachers (60.7%) were all ambivalent. The detail is evident in Table 14.6.

Principals consulted were not able to come to one mind on how to interpret these patterns. The hardest issue to interpret was the bimodality of teachers' views. Two explanations were advanced tentatively in most forums; careerist and exclusionary forms of professionalism. Since 'the localization of leader selection' proposal had been greeted with ambivalence by parents (28.7% SA+A) and attracted no support in all other groups, it was agreed that decentralism was unlikely to be a viable explanation. Similarly, all groups were unsupportive of the idea that 'the P and F/School Council

Table 14.6: Distribution of responses to the proposal that, to improve the quality of leadership, there should be parental, professional and departmental involvement in the selection of principals

Respondents	SA(%)	A(%)	NS(%)	D(%)	SD(%)	Mean	Mode	SD
Sec principals	6(28.6)	11(52.4)	4(19.0)	0(0)	0(0)	1.91	2	0.70
Dist principals	4(26.7)	9(60)	1(6.7)	1(6.7)	0(0)	1.93	2	0.80
District DEA	3(30)	5(50)	1(10)	1(10)	0(0)	2.00	2	0.94
Dist parents	9(33.3)	12(44.4)	3(11.1)	3(11.1)	0(0)	2.00	2	0.96
Sec parents	7(29.2)	9(37.5)	7(29.2)	1(4.2)	0(0)	2.08	2	0.88
All parents	34(23.6)	63(43.8)	27(18.8)	18(12.5)	2(1.4)	2.24	2	1.00
All DEA	5(19.2)	10(38.5)	9(34.6)	2(7.7)	0(0)	2.31	2	0.88
All principals	23(18.7)	58(47.2)	23(18.7)	15(12.2)	4(3.3)	2.34	2	1.02
Prim parents	18(19.4)	42(45.2)	17(18.3)	14(15.1)	2(2.2)	2.36	2	1.03
Totals	79(18.6)	177(41.7)	89(21)	68(16)	11(2.6)	2.42	2	1.05
Central DEA	2(12.5)	5(31.3)	8(50)	1(6.3)	0(0)	2.50	3	0.82
Prim principals	13(14.9)	38(43.7)	18(20.7)	14(16.1)	4(4.6)	2.52	2	1.08
Dist teachers	6(21.4)	11(39.3)	4(14.3)	4(14.3)	3(10.7)	2.54	2	1.29
Prim teachers	9(11.4)	29(36.7)	17(21.5)	21(26.6)	2(2.5)	2.72	2	1.07
All teachers	17(12.9)	46(34.8)	30(22.7)	33(25)	5(3.8)	2.72	2	1.10
Sec teachers	2(8)	6(24)	9(36)	8(32)	0(0)	2.92	3	0.95

should set the policies about the preferred nature of leadership services' in schools. Teachers (24.3%) and principals (14.6%) were unsupportive while parents (38.7%) and DEA personnel (38.4%) were unsure about the suggestion that 'leadership would be improved by having fixed term and negotiated performance contracts'. It was eventually and tentatively agreed that perceptions of self-interest and ego were the most likely discriminating factors in responses.

To summarize this section, the implications for principals were clarified concerning processes that related to the quality of leadership services in schools. The first two ideas accepted were that there were many appraisal and reporting mechanisms available, and that there were many areas worthy of improvement. The third general interpretation made was that developing leadership capacities in schools was integral to improving the quality of teaching and learning. Fourth, it was concluded that principals and their executive teams would need to accept responsibility for the development of leadership services in their school community as a strategic and moral duty. Fifth, the differences in means in Table 14.6 were taken to indicate that leaders would probably have to perform slightly in advance of teachers' and parents' expectations to be successful. Sixth, and finally, it was accepted that leadership services in schools needed to improve, and in order to achieve this, most school communities would have to reclarify their values and create fresh policies, and then plan, develop and celebrate pro-active leadership.

Accountability Criteria relating to Leaders' Leadership

It was noted in Chapter 12 that 22 accountability criteria were suggested to evaluate leaders' leadership services. They are provided in Table 14.7, ranked by the means of principals' responses and compared to all other's means.

The agreed interpretation reached was that there was support across all stakeholder groups for judging leadership services in schools in quite specific terms; interpersonal style, school culture, planning capacities, achieving priorities, decision styles, morale and motivation, management skills (especially in evaluation, budgeting and governance), professional development, creativity, productivity, communications, staff support of leaders, learning by students and teachers, and the quality of teaching. Principals also noted that they tended to expect more of leadership services in schools

Table 14.7: Support for criteria for evaluating leaders' leadership services

Processes proposed by all stakeholders	Principals' views Mean	%	All groups' views Mean	%	Differences of means t	p <
Supported						
Capacity to hear and care for others	1.42	98.3	1.56	96.8	2.45	0.01
The openness and climate/tone of the school	1.50	96.7	1.76	88.8	3.64	0.001
Ability to plan outcomes and achieve priorities	1.69	94.4	1.78	94.2	1.32	ns
Extent of collaborative decision making	1.72	90.0	1.93	83.4	2.88	0.005
Student and teacher morale and motivation	1.73	89.1	1.77	88.8	0.53	ns
Capacity to make and implement policy	1.78	93.3	1.86	90.9	1.26	ns
Management and organizational skills (evaluation, budgeting and governance)	1.81	91.7	1.89	88.3	1.18	ns
Extent of professional development within the school	1.86	88.4	2.16	74.9	3.39	0.001
Valuing of creativity and productivity in school	1.92	84.9	2.04	79.5	1.39	ns
Quality of internal and external communications	1.94	78.4	2.10	79.9	2.15	0.05
The extent to which staff support their leaders	1.97	81.8	1.93	84.5	0.64	ns
Evidence of learning by staff and students	2.03	76.2	2.16	73.6	1.54	ns
Capacities as learners and researchers	2.10	73.5	2.29	68.6	2.26	0.05
Evidence of the quality of teaching by the staff	2.16	76.1	2.21	74.6	0.56	ns
Ambivalent support						
Indicators from research lit. used in plans	2.25	65.6	2.32	61.1	0.60	ns
Performance indicators in guidelines provided by the DEA	2.38	65.0	2.36	62.9	0.25	ns
The expectations of the community	2.41	63.3	2.46	61.0	0.57	ns
Extent to which parents support school leaders	2.41	60.0	2.54	54.4	1.42	ns
Recommendations from school reviews	2.46	61.9	2.46	55.9	0	ns
The quality of the physical environment	2.52	59.2	2.53	58.8	0.10	ns
Leaders' relevant qualifications	2.97	43.0	2.75	51.7	2.04	0.05
National competency indicators for managers	2.95	29.0	2.69	40.6	3.25	0.005

than did teachers, especially to do with the improvement of teaching and learning. They came to the view that it would be wise, nevertheless, to regard this list of expectations as 'system policy' and as most appropriately attended to by their school executive teams planning the development of leadership services in close consultation with their school's policy community.

The details of minor ambivalence were regarded as reinforcing this view. Using 'evidence of learning by staff and students' was supported by all groups except teachers (63.6% SA+A) who exhibited some minor ambivalence. All groups supported 'the extent of professional development' as an indicator of quality leadership with minor reservations; by teachers (66.6%), primary parents (67.4%) and district high parents (66.7%). Similarly, 'evidence of the quality of teaching by the staff' was supported as an evaluation criterion by all groups although district high teachers (53.6%), secondary teachers (56%) and district high principals (60%) apparently harbour some doubts. 'Leaders' capacities as learners and researchers' was also considered an appropriate criterion by all groups with minor ambivalence among teachers (55.3%), secondary principals (61.9%) and primary parents (69.9%). 'Using the recommendations of school reviews' was greeted with ambivalence by all groups except district DEA personnel (80%), who supported the option. Principals at all forums interpreted this group of proposals as warranting inclusion in touchstone policy, while remaining mindful of the minor doubts. The last item should be included, they argued, since school review processes were being upgraded each year as part of the interaction between themselves, district superintendents and the ERU.

A feature of Table 14.7 evident to principals was that criteria from ex-professional peer sources attracted markedly less support than those generated within. This inordinate trust in an immediate professional peer group was described as being 'typical' of professional conservatism in Tasmania. Indicators from 'research literature being used to plan improvements to leadership services', for example, was supported by secondary parents (75% SA+A) and district DEA personnel (70%) with all other groups ambivalent. Developing 'performance indicators from the guidelines provided by the DEA' was supported by DEA personnel (80%), district high school parents (81.5%) and district high school principals (73.3%) but all other groups were ambivalent. Adapting the 'indicators from national competency standards for managers' was an idea not supported by secondary teachers (24% SA+A), primary principals (28%) and secondary principals (23.8%) with district DEA personnel (70%) taking a supportive view and all other groups remaining ambivalent. Employing 'leaders' relevant qualifications' as a criterion drew ambivalent responses from all groups except secondary parents (75%) who supported this option.

This feature of professional conservatism was highlighted elsewhere by principals. Using 'parental and community participation' as a proxy measure of leadership service was, in general, not valued by all stakeholders, and taken as another illustration of the long standing conservative nature of Tasmanian educational culture. Using the 'expectations of the community' as a criterion was supported only by district high school parents (74.1% SA+A) with all other groups ambivalent. All groups were also ambivalent about evaluating leadership services in terms of 'the extent to which parents support their school leaders'. With some reluctance, principals involved in the interpretation of data came to two conclusions; degrees of authoritarianism were still endemic to leadership practice, and, principals and teachers were not as alert to or as open to advice or feedback on leadership services as stakeholders thought they should be.

Principals were, therefore, advised by their peers involved in interpretation to reflect on the limited or closed nature of participation in school leadership services that was implied by these findings. Parental, community and teachers' perspectives were found to be even more marginal than expected. It was noted with concern that many stakeholders regarded the current unsatisfactory situation as appropriate. Principals took the view that this was an area requiring determined leadership service.

There was another area characterized by paradox. General leadership services were regarded as vital to the quality of teaching and learning, implying that many more should share in providing leadership services. Simultaneously it appeared that stakeholders saw responsibility for the quality of leadership being located almost solely with those with positional authority, i.e. with principals. Paradoxically, all stakeholder groups gave high value not to authoritarian leadership but to collaborative and technically skilled forms of leadership, especially in the areas of evaluation, budgeting and governance.

This complex set of imperatives was taken to mean that principals would be well advised to advance the development of leadership services in a school community on as broad a base as possible, essentially by providing as wide a range of co-governance and co-management opportunities to parental, community and professional colleagues as they could tolerate. Another interpretation favoured by principals was that leadership needs to seen more as a shared responsibility in school communities, especially if it is to offer educative services that enhance the quality of learning and teaching. Primary school principals seemed to be in the vanguard in this matter. For example, they took the view that a relatively standard test of an organization's willingness to hear feedback, and to defend such processes, is that it has a negotiated formal complaints procedure that protects legitimate interests with due process and prevents victimization. Responses to this proposal are summarized in Table 14.8.

Principals were quick to note that all groups were either ambivalent or supportive. Teachers (59.8%) exhibited broadly similar degrees of support and ambivalence; secondary teachers (64%), primary teachers (60.8%) and district high teachers (53.6%). Principals (68%) exhibited a wider range of ambivalence; secondary principals (66.7%), primary principals (60.8%) and district high school principals (43.3%). Parents (80.6%) strongly supported the proposal; primary parents (79.6%), district high parents (82.4%) and secondary parents (82.3%). Oddly, while district DEA personnel (70%) also supported the proposal, DEA central personnel (50%) expressed considerable ambivalence. The special nature of district high schools was cited again to explain their relative defensiveness. There was some speculation but no reasons advanced that were regarded as 'sound' for the degree of resistance in DEA central personnel. Primary school principals concluded that they should accelerate their campaign by conferring with other principals' associations and teachers' and parents' organizations and the DEA.

Table 14.8: Distribution of responses to the proposal that, to improve the quality of teaching, learning and leadership, a formal complaints procedure should be negotiated by school communities within state guidelines that protect all legitimate interests with due process and prevent anyone from being victimized.

Respondents	SA(%)	A(%)	NS(%)	D(%)	SD(%)	Mean	Mode	SD
Dist parents	9(33.3)	13(48.1)	5(18.5)	0(0)	0(0)	1.85	2	0.72
All parents	41(28.5)	75(52.1)	22(15.3)	5(3.5)	1(0.7)	1.96	2	0.80
Prim parents	26(28.0)	48(51.6)	15(16.1)	3(3.2)	1(1.1)	1.98	2	0.82
District DEA	3(30)	4(40)	3(30)	0(0)	0(0)	2.00	2	0.82
Sec parents	6(25)	14(58.3)	2(8.3)	2(8.3)	0(0)	2.00	2	0.83
Totals	98(23.1)	196(46.2)	83(19.6)	38(9)	9(2.1)	2.21	2	0.97
Prim principals	22(24.7)	41(46.1)	13(14.6)	9(10.1)	4(4.5)	2.24	2	1.08
All principals	29(23.2)	56(44.8)	21(16.8)	14(11.2)	5(4.0)	2.28	2	1.07
Sec teachers	5(20)	11(44)	6(24)	2(8)	1(4)	2.32	2	1.03
All DEA	5(20.8)	9(37.5)	7(29.2)	3(12.5)	0(0)	2.33	2	0.96
Sec principals	5(23.8)	9(42.9)	3(14.3)	3(14.3)	1(4.8)	2.33	2	1.15
Prim teachers	13(16.5)	35(44.3)	21(26.6)	7(8.9)	2(2.5)	2.36	2	0.95
All teachers	23(17.4)	56(42.4)	33(25)	16(12.1)	3(2.3)	2.39	2	0.99
Dist principals	2(13.3)	6(40.0)	5(33.3)	2(13.3)	0(0)	2.47	2	0.92
Dist teachers	5(17.9)	10(35.7)	6(21.4)	7(25)	0(0)	2.54	2	1.07
Central DEA	2(14.3)	5(35.7)	4(28.6)	3(21.4)	0(0)	2.57	2	1.02

As noted in Chapter 11, senior DEA officials encouraged them in this line of action from the time they had the survey results.

The interpretation of proposals by principals occasionally moved back from the detail and addressed the broad implications. Many were mindful of the sometimes unflattering conclusions drawn concerning the quality of current leadership services in schools. Many were disappointed by the apparent incapacity of many school communities to build and implement developmental policies. Their attention focused on the understandings and skills that principals and others would need if they were to provide school communities with the leadership services that all stakeholders valued. This issue is now discussed.

Recommended Leadership Activities, Capacities and Professional Development

The interpretations recorded above were made by principals and others on a conditional basis. One such condition was that students' views and other interest groups in state education would have to have their ideas collected or measured as part of school community policy making. Another was that while the findings and comparisons did indicate the patterns of support, ambivalence or non-support, at the moment they were surveyed, their interpretations had drawn on experience and judgements to build a broad, idealistic and hopefully coherent explanation. The third condition, therefore, was that their explanations were only provisional, and that they were to be revised during application in each setting and the moment any part of the suggestions were shown to be inadequate. In sum, they saw themselves in ideal terms, and as leaders that were also learners about leadership service.

Given these conditions, the principals and other stakeholder executives came to the view that the broad patterns of support were clear enough to act on. Substantial common ground was considered to be available to help the Tasmanian state system of schools to develop a more integrated and comprehensive accountability policy, as well as providing the detail needed for better practices and policies at each school community site. The policy touchstone summarized in Tables 11.13–11.15 was believed to be positively and causally related to the quality of learning, teaching and leadership. The nature of the discussions at the April policy forum proceeded on the assumption that the DEA had accepted the touchstone ideas as adding to and extending 'current thinking'. Given the hierarchical norms of relationships in the DEA, this led inexorably to an expectation that district DEA personnel, principals and their executive teams would do what they could to embed them in school policies, although, as pointed out by a number of district superintendents, this would depend on the extent to which each school community engaged in policy-making activities and the influence that forum participants had in such processes. I come back to this issue in the next chapter.

The policy discussions at forums exhibited a number of interesting features, in addition to those discussed in Chapters 9 and 10. There was no one policy decision point involved. It appeared that 'policy', as a corporate view of the future, was changed in many tiny and subtle ways when the values of participants were affected by analysis, reflection and debate. The milestones in the fluid movement of commitment to the cultural artefact termed 'policy' was evident in language changes. A pertinent example is when 'preferred activities' to do with educative accountability processes and criteria (see Table 12.7) were somehow redefined as 'best practices' and as 'performance indicators' and then deemed to be suitable for 'bench marking' in schools. Another example occurred when these 'performance indicators' were gathered by a manual form of cluster analysis into 'sets of activities', and when given a collective label, they were defined as 'key competencies of leadership service'. I questioned a number of participants about these discourse strategies and compared their explanations to those offered in the appropriate literature.

Some saw this process was defensive since it co-opted the language of (and thus neutralized the agenda of) those trying to promote 'national management competencies' in Tasmania. This tactic also enabled the participants to display identity and their shared tacit moral theories as they resisted the ideology of an occasionally offensive national policy discourse (Gee 1990, p. 155). Others saw the process as progressive and noted that these linguistic tactics enabled familiar and comfortable metaphors to domesticate and legitimate the new ideas. I would suggest instead that the domestication and legitimation was more apparent than actual since the new 'common-sense' policy had actually been added to or reconstructed pre-existing webs of belief about professionalism. There had been a deep reverse take-over. A new moral order of 'educative accountability' had gradually colonized the prior components of collective webs of belief concerned with accountability. It seemed to me, much as suggested by Falk (1995, p. 77), that policy learning had actually been evident as a process of discursive colonization, and signalled a deeper reconciliation of the old and new sets of values to do with accountability.

Other external ideas informed this part of the policy process. When principals and other stakeholder leaders pressed hard for the sets of activities, the 'key competencies', to be 'organized in a coherent way', I eventually offered the six-part framework from a practical theory of educative leadership developed in Australian conditions for Australian practitioners (Duignan and Macpherson 1992, p. 183). It proved useful, and thus, was retained. On the other hand, this was hardly surprising since the framework had already proved useful in organizing the general skills and attributes that were felt by national bodies to be crucial in aspiring and current Australian primary and secondary principals for the next decade (Macpherson and Caldwell 1992).

As indicated in Chapter 10, this practical theory of educative leadership sees the quality of teaching, learning and leadership as causally interdependent. It rejects their disaggregation for developmental purposes on the grounds that there is no sound epistemological or practical justification for the partitioning of professional knowledge. It encourages balanced and critical attention to (a) educational ideas in context (i.e. purposes, strategic analysis, planning), (b) the inter-subjective realities of people involved (e.g. powers, structures, norms) and (c) the empirical reality of challenges in schools and systems (e.g. given objectives, resources, outcomes). In sum, this practical theory of educative leadership concludes that learning communities require six types of leadership service: philosophical, strategic, political, cultural, managerial and evaluative.

This framework was accepted provisionally and used to organize the 'key competencies'. By amalgamating stakeholders' policy preferences (Table 12.7), editing the touchstone items slightly into the managerial language of performance indicators, and reclustering them by type of leadership activity, it was possible to identify competencies embodying educative accountability criteria and processes broadly supported by all stakeholders. As an aside, given the known profile of Australian school principals (Grady *et al.* 1994) and how little Tasmanian principals differ from the national patterns, it was speculated by some principals that replicated inter-state research findings would not find significant differences. This idea awaits further research. In the interim, the outcomes of the analysis discussed above are displayed in Table 14.9 as 'key competencies' (in small capitals) and as 'performance indicators' (in italics) for those who are expected to provide educative accountability criteria and processes as part of their leadership services.

Summary

In this chapter I have compared principals' policy preferences with other immediate stakeholders. The analysis was constructed largely using the perspectives of principals to identify implications. The main conclusion was that the accountability criteria and processes related to the quality of learning, teaching and leadership valued by all

Table 14.9: Key competencies and performance indicators of educative accountability criteria and processes

EVALUATIVE LEADERSHIP	PHILOSOPHICAL LEADERSHIP
Monitors outcomes and attitudes	*Clarifies organizational purposes*
Samples student work (e.g. folios)	Provides and generates a school vision
Measures progress of individual students	Develops coherence between the school vision, the Development plan and outcomes
Ensures that teachers use written checklists and Running records to monitor student progress	
Uses reports from other support staff (e.g. guidance, welfare, speech and health specialists)	*Links evaluation to planning*
	Helps teachers evaluate and plan their programs systematically
Uses performance indicators in state and national policies (e.g. Frameworks, Profiles)	Helps teachers develop performance indicators at subject moderation meetings
Measures students' self-esteem and life skills	Helps teachers develop performance indicators in school
Measures attitudes of students to school, teachers, peers, learning and homework	Helps teachers evaluate and plan lessons thoroughly
Measures teachers' attitude towards students, parents and colleagues	
Surveys school climate, uses feedback from staff	*Creates a learning organization*
	Shares reflections in peer networks on the challenges of practice
Provides feedback and appraisal systems	
Uses feedback on teaching and leadership from parents and students	Develop others' capacities as learners and researchers
Uses feedback and appraisal by peers	Negotiates new goals for professional development
Uses appraisal of leadership services by school colleagues	Encourages teachers to read and do research
Encourages self-evaluation, self-appraisal	
Reports teacher appraisals to the individual teacher	Develops the judgements made by teachers
Develops formative evaluation systems	Enhances teachers' knowledge (of subject matter and child/adolescent development
Uses objective assessment to plan improvements to students' learning	
Develops formative evaluation related to teaching objectives	Values creativity and productivity in the school
	Reviews the quality of planning
	Reviews the quality of policy-making strategies used
	Uses a mentoring process

(continued opposite)

MANAGEMENT ACTIVITY	STRATEGIC PLANNING ACTIVITY
Manages the making and implementing of policies Supports participative governance processes Implements school and curriculum policies *Organizes support structures and processes* Organizes the support and feedback given to staff Organizes the evaluation of teaching and learning Organizes and develops instructional expertise Organizes training and support to identify and Cope with students 'at risk' Organizes how students' work is set, monitored and marked Helps teachers identify intended outcomes for each student *Manages the use of resources* Organizes the use of time and other resources Organizes budgeting processes	*Provides collaborative strategic analysis* Negotiates priorities and outcomes Helps others make and plan the implementation of policy Provides conferencing between teachers and students Helps parents/teachers/students jointly develop learning indicators Provides parents with curriculum goals, outcomes and individual expectations at the beginning of the year *Negotiates performance indicators* Includes indicators from research as criteria when planning improvements to students' learning Negotiates indicators of learning by staff Negotiates indicators of quality teaching with staff *Prepares a professional development strategy* Plans the professional development of teachers Raises the willingness of staff to engage in continuing professional development
Cultural activity	**Political activity**
Develops supportive classroom environments Encourages collaborative classroom environments Develops supportive behaviour management skills Promotes positive attitudes in children (e.g. enthusiasm) Improves interpersonal communications in the classroom *Develops a supportive school environment* Leaders hear and care for others Develops an open and participative culture Promotes parent/teacher/student discussions Encourages co-operative learning between (e.g. by mentoring and discussion between colleagues) Provides a transition program for newly appointed teachers Provides skill development programs, especially in governance and management	*Raises commitment* Encourages collaborative decision making Develops student and teacher morale and motivation Encourages parent/teacher interviews Develops staff support for their leaders *Communicates priorities and values* Provides parents with opportunities to give their views and to discuss teaching and curriculum policies Reports to parents with clear and accurate descriptions of learning Reports student's learning through newsletters, school magazines, displays and positive public relations Reports achievements to parents and the community Documents best practices Values internal and external two-way communications

stakeholder groups, including principals, provided both a political mandate and moral imperatives for particular leadership services. The principals interpreted this to mean that Tasmanian state school communities will need to review and develop their accountability policies and that the professional development of principals should co-opt the touchstone to specify content. There was some doubt about the extent to which schools actually perform as policy communities.

The mandate from all stakeholders regarding the quality of students' learning was interpreted in terms of encouraging specific activities. These activities were heavily interactional in nature; discussions between teachers, parents and students. These discussions were particularly valued when they enabled parent input, clarified policies, gave parents and students expected outcomes and individual expectations at the beginning of each year, and reported outcomes using clear and accurate descriptions of learning at the end of specific periods. They implied that leadership should be provided in lesson planning and evaluation as well as in program planning and evaluation. Particular data sources were recommended; student's self-assessment, teacher-designed mastery and diagnostic tests, teachers' observations, the sampling of student work, teachers written checklists and running records, and information supplied by support staff. There was also a mandate for leaders to take material for public reports from formative evaluation related to lesson and program objectives.

Principals' analyses of preferred criteria related to the quality of students' learning led them to conclude that many of their professional peers, especially secondary colleagues, would have to adopt a more collaborative and inclusionary philosophy concerning the quality of learning. Many teachers and principals were apparently setting parental expectations aside as a criterion for the evaluation of students' learning. Similarly, they came to the view that this appropriation of learning criteria was unwise and that it should be displaced by greater openness and responsiveness to other criteria. Other valued indicators of learning that should receive high priority, they argued, included measures of (a) individual progress, (b) student attitudes to school, teachers, peers, learning and homework, and (c) students' self-esteem and life skills.

Touchstone accountability processes intended to collect data, report on and improve the quality of teachers' teaching were also used as a mandate for leadership services. The facilitation of critical and professional interaction was heavily emphasized; transition programs for newly appointed teachers, self-evaluation, co-operative learning and research between colleagues, peer appraisals of planning and teaching discussed with the individual teacher, documentation of best practices and negotiating new goals for planned professional development. Data generally considered crucial were student outcomes and the needs of at-risk students. The line was drawn, however, on feedback and appraisal.

Many of the feedback and appraisal processes proposed that were related to the quality of teachers' teaching offended professional norms. Where peer-based processes were sanctioned by these norms, feedback from and appraisal by students, parents, independent experts and more senior school colleagues was not. This was taken to mean that each school policy community really needed to evaluate the development of pedagogy and to consider using extra-professional data sources and processes as soon as possible in order to develop those norms. It was concluded that accountability processes related to the quality of teaching were relatively rudimentary and had been rendered so by 'teacher development' processes.

This interpretation was reinforced when principals reviewed the touchstone criteria concerning the evaluation of teaching. The criteria included classroom environment and interpersonal communications, children's attitudes and their progress. They also included teachers' attitudes to students, parents and colleagues, knowledge of subject and child/adolescent development, willingness to engage in continuing professional development, behaviour management skills, organizational skills, effective implementation of school and curriculum policies, communication skills with stakeholders, instructional expertise and how well work is set, monitored and marked. Principals concluded that more extra-peer accountability processes to do with the quality of teaching were essential in the interests of validity, reliability and policy legitimation.

Concern was also expressed by principals about the methodological limits of leadership accountability processes. Apart from self-initiated peer processes, and the annual planning cycle, the school review process was 'virtually the only vehicle available' to help develop the quality of leadership services *in situ*. While some practices were said to be relatively common (self-appraisal, peer networks to reflect on challenges of practice and mentoring), there were fewer examples known of some widely valued preferences (peer appraisal, appraisals of support and feedback given to staff, skill development programs in governance and management, evaluation of reporting to parents and community, or generation of a school vision) and very few known examples of others (an evaluation of the coherence between vision, plans and outcomes, feedback on leadership services from staff, parents and students, school climate survey or an appraisal of policy-making strategies).

Five implications were drawn by those principals involved in interpretation. There were many appraisal and reporting mechanisms available and areas worthy of improvement concerning the quality of leadership services in schools. Boosting leadership capacities in schools was essential to improving the quality of teaching and learning. Executive teams had to accept responsibility for developing leadership services in their school community. From time to time they had to provide leadership in advance of teachers' and parents' expectations. School communities had to value, plan for, develop and celebrate pro-active leadership.

It will be recalled that touchstone criteria related to leadership services included interpersonal and decision styles and skills, school culture, planning capacities, achieving priorities, morale and motivation, creativity, productivity, staff support of leaders, learning by students and teachers, and the quality of teaching. Principals tended to expect more of leadership service than did other stakeholders, especially to do with teaching and learning. They were troubled by criticisms concerning leadership services and the implied incapacity of many school communities to generate improvements. Appropriate understandings and skills were drawn from the touchstone.

This process of interpretation simultaneously co-opted the managerialist language of competencies, displayed identity, shared tacit moral theories, and resisted the ideology of a national policy discourse. Familiar metaphors apparently domesticated and legitimated the new 'common-sense' policy that had been added to webs of belief. Under this discourse, a new moral order of 'educative accountability' had colonized components of collective webs of belief concerned with accountability. It was tentatively concluded that this discursive colonization indicated the reconciliation of different sets of policy values.

The most tangible evidence supporting this interpretation was that supported processes and criteria were activities that could be defined as 'best practice'. When these 'best practices' were grouped in sets, they were deemed to comprise 'performance indicators'. When these indicators were organized onto a framework of educative leadership, they were held to be the 'key competencies' of educative accountability leadership services. They are summarized in Figure 14.1.

In the next chapter I report and discuss the policy preferences of DEA administrators and central and district offices.

Evaluative leadership	Philosophical leadership
Monitors outcomes and attitudes	*Clarifies organizational purposes*
Provides feedback and appraisal systems	*Links evaluation to planning*
Develops formative evaluation systems	*Creates a learning organization*
Management activity	**Strategic planning activity**
Manages the making and implementing of policies	*Provides collaborative strategic analysis*
Organizes support structures and processes	*Negotiates performance indicators*
Manages the use of resources	*Prepares a professional development strategy*
Cultural activity	**Political activity**
Develops supportive classroom environments	*Raises commitment*
Develops a supportive school environment	*Communicates priorities and values*

Figure 14.1: Key competencies of educative accountability leadership services

15

System Administrators' Policy Preferences: Mediating Purposes and Politics while Supporting Community School Development

Introduction

The views of DEA central and district personnel are compared to all stakeholders' preferences in this chapter. Primacy is given to officials' interpretations as expressed at the presentation of preliminary findings in October 1994, at the April 1995 policy forum for stakeholders and during subsequent discussions in a range of central, district and informal settings. It extends the initial interpretation presented in Chapter 11. Both sets of interpretations were affected by the political and structural turbulence, policy paralysis and pressures on public expenditure that marked the period.

The Tasmanian Liberal Government faced serious budgetary problems in 1995 and went to the people early in early 1996, two weeks before a Commonwealth (federal) election. It lost its overall majority but stayed in office. The federal Labour Government, however, fell, and was replaced by a Liberal-National (conservative) coalition government. The incoming federal government soon announced an $8 billion projected deficit and began examining options for cutting public expenditure and grants to the states.

Since the opposition parties in Tasmania were not able to form a government, a new Premier and Deputy Premier were appointed by the Liberal parliamentarians to provide leadership and to manage the state's affairs. The Department of Education and the Arts was given additional duties and renamed the Department of Education, Culture and Community Development (DECCD). A new Minister (also Deputy Premier) was appointed; the Hon Sue Napier. Her CEO and Secretary of Education, Mr Bruce Davis, retired.

Unable to command a majority in the lower house, the Tasmanian government's legislative programme made slow headway. The selection of Mr Davis' replacement and the policy processes in the DECCD also moved slowly. All attention seemed to focus on the impecunious state of Tasmania's finances and the shaky state of its minority government. The Tasmanian community was then stunned in late April when a lone gunman massacred 35 tourists at Port Arthur. Apart from the appalling consequences for the families of victims and the community, state revenues from tourism halved overnight.

The new federal government foreshadowed cuts in public expenditure in the 22 August budget in the order of 10–15%. Salary negotiations in a number of portfolios gradually stalled. In late June, the New South Wales Department of School Education finally closed the Quality Assurance Review and Educational Audit Unit noted in Chapter 2. It had been 'black banned' in 1995 as part of industrial action over salaries. In July the Australian Education Union (Tasmania) starting rolling strikes to draw attention to its members' interests.

This context was not seen by stakeholders as conducive to the trustful negotiation of fresh accountability policies in public education. Nevertheless, it was a time of quiet achievement through consultations. The next section, therefore, attempts to summarize the effects of this context reported by central and district officials, especially on how they interpreted the views of stakeholders concerning accountability policy proposals in early 1996. It will be shown that this interpretation was central to understanding the nature and direction of the ongoing policy development process.

The Public Administration Context

Recurrent advice from officials in the period 1992–1996 emphasized that many of the core functional duties of the CEO and other central officials remained relatively unchanged while others were being reconstructed. The source of stability was clear. Ministers continued to need an executive arm to serve a number of generic functions; the management of the annual budget cycle, the co-ordination and monitoring of programme activity, the analysis and monitoring of major policy issues, planning and resource management in the context of portfolio-wide goals and priorities, the management of relationships within and external to the portfolio, and the establishment and maintenance of databases to support portfolio decision making and planning. Many senior officials' sense of public duty concerning these enduring responsibilities clearly informed their interpretation of accountability policy preferences.

On the other hand, they were equally clear about the sources and nature of changes to their responsibilities. The consolidation of portfolios and devolution of some responsibilities to agency and institutional heads in Tasmania over the same period, they claimed, blurred the relationship between government and executive. Ministers became more interventionist. They acquired more direct access to agency and institutional leadership. They steered system and institutional policies. They simplified administrative processes and cut costs by eliminating intermediate layers of managers. Two not uncommon interpretations from officials were that Ministers had tried to make education's change processes match the political cycles they lived by, and made structures more malleable to cope with financial emergencies.

One result reported by senior officials in the DEA 'Head Office' was that they had to co-ordinate policy making and implementation with flexible and temporary teams of functional directors and line managers, sometimes including principals. The dominant concern in the public service, almost irrespective of portfolio, was greater efficiency and effectiveness in 'service delivery'. Public administration had three main purposes; (a) to co-ordinate the business of government in a holistic manner, (b) to help government 'reposition' itself in the provision of services, and (c) to 'down-scale' in haste whenever budgets had to be cut. The press of economic rationalism, however, conflicted with the educational values of most DEA officials. They continued to question the government's values, whatever the general budgetary situation, and campaigned in a number of inter-departmental forums for changes in priorities in the lead up to the 22 August federal budget announcements.

Senior educational administrators at Head Office often mentioned the challenge of sustaining the professional commitment of colleagues. They reported using the language of 'human resource management and development' to raise the priority of professional development so that it could compete successfully against the three primary purposes of public management noted above. Another interesting change they reported was that central DEA officials were thinking about 'leadership competencies' less in terms of an individual's ability to implement given policies and far more in terms of building the capacities that would enable policy communities to become 'learning organizations'.

The district superintendents were acutely aware of these two trends and identified a third. They reported having to mediate the sometimes difficult exchanges between the organizational cultures of public administration and schools. While they had their differences within their own group, these district superintendents also regarded the 'core competencies' of leadership in education as actions that help all participants and stakeholders learn to take an active responsibility for efficiency, effectiveness *and* organizational learning about learning, teaching and leadership. The challenge, they regularly stressed, was not to redesign structures and announce functions, a common but educationally useless 'reform' strategy of the 1980s and early 1990s, but to build a critical culture of formative evaluation that balanced excellence and equity in each school community.

The district superintendents applied these views to community leadership services at both state and school level. They considered it important that central DEA officials and principals should aim to build learning organizations in a context of intensified competition between the principles of social justice, excellence, choice and competitive advantage in wider policy domains. Like most DEA central officials, they were all ex-teachers and usually ex-principals, and shared a profound concern for the disadvantaged. Despite fluctuations in support in the wider political context, they were keen to ensure that equity and excellence continued to be regarded as equally important policy principles in school communities.

Hence, while school councils were generally at an early stage of development, most DEA personnel valued the development of leadership services that would (a) acknowledge the plurality of Tasmanian society, (b) boost the philosophical and strategic planning capacities of policy communities, and (c) help school policy communities determine policy details within the state guidelines examined in Chapter 9.

How the district superintendents proposed to assist in this regard is clarified towards the end of this chapter when recent changes to the school review process are presented.

To conclude this section, many aspects of the complex context noted above and in Chapter 9 were reflected in the interpretations provided by DEA officials when discussing the results of this research. In the next section their views are compared with other stakeholders with regard to accountability processes relating to students' learning.

Accountability Processes related to Students' Learning

The processes that at least one stakeholder group believed should be used to collect data on, report and improve students' learning are provided in Table 15.1. The support given by DEA officials is compared to that given by all stakeholders, with '%' again referring to the percentages that 'strongly agreed' plus 'agreed' to each proposal.

The initial response by all DEA personnel to the data summarized in Table 15.1 was one of political relief. They were relieved to find that their views cohered with other stakeholders. They explained that they served in settings where their judgements were often contested at Ministerial level by assertions from interest group representatives and sometimes from other government departments. They noted that their own policy research processes were not as sophisticated as they might be.

They were also quick to compare how many touchstone proposals had been supported by each stakeholder group, and then to conclude that, as Table 15.2 shows, there was no significant difference between stakeholder groups' support for touchstone items ($\chi^2 = 2.93, p < $ ns). They came to the view that, despite a good deal of 'manufactured myth' to the contrary in 'the system', they were as much 'in touch' as any stakeholders were, and more inclusionary than most.

The second phase of response included setting aside statistical differences concerning all supported items as irrelevant in political and practical terms, regarding the touchstone as *de facto* policy and adopting 'its themes' in order to interpret other proposals regarded by some with ambivalence. These interpretations are discussed below.

The third response was to use the nature of accountability learning processes supported by all stakeholders to evaluate the school support services provided by DEA personnel. A logic of four components became evident; professional development, community school development, needs aggregation and the effectiveness of delivery systems. First they noted the need for teachers to be helped to evaluate and plan lessons thoroughly, to use formative evaluation related to their teaching objectives, to evaluate and plan their programmes systematically, to identify outcomes for each student, to observe systematically, to sample student work, to keep written checklists and running records, and to design and use mastery and diagnostic tests. They next saw that students needed to learn how to self-assess, to confer with teachers, and to participate in three-way parent/teacher/student discussions. They then accepted that parents would need far more educative opportunities and participative roles, such as more effective parent/teacher interviews, reports that gave clear and accurate descriptions

Table 15.1: Support for proposed processes to collect data, report on and improve students' learning

Processes proposed by all stakeholders	DEA views Mean	DEA views %	All groups' views Mean	All groups' views %	Differences of means t	Differences of means $p <$
Supported						
Teachers evaluate and plan lessons thoroughly	1.44	100	1.58	96.4	1.35	ns
Student's own self-assessment	1.50	96.1	1.92	82.5	3.46	0.001
Teachers evaluate and plan programs systematically	1.52	100	1.77	91.8	2.37	0.02
Parent/teacher interviews	1.54	100	1.46	97.2	0.63	ns
Teachers identify outcomes for each student	1.54	96.2	1.80	89.3	2.18	0.05
Teachers' observations	1.62	96.1	1.71	97.9	0.79	ns
Conferencing between teacher and student	1.65	88.5	1.74	93.2	0.65	ns
Parent/teacher/student discussions	1.72	96.0	1.80	91.9	0.72	ns
Reports—clear and accurate descriptions of learning	1.73	96.2	1.89	86.8	1.46	ns
Parent input and policy explanations	1.80	96.0	1.82	90.0	0.15	ns
Parents given goals, expected outcomes, and individual expectations at the beginning of each year	1.84	84.0	2.12	73.5	1.96	0.05
The sampling of student work (e.g. folios)	1.85	88.5	1.77	91.1	0.54	ns
Teachers written checklists and running records	1.88	80.0	1.93	82.7	0.34	ns
Formative evaluation related to teaching objectives	1.92	76.0	2.17	73.6	1.63	ns
Teacher-designed mastery and diagnostic tests	1.96	84.6	1.93	87.7	0.25	ns
Reporting through publications and public relations	2.19	73.1	1.84	87.3	1.46	ns
Ambivalent support						
State-wide, norm-referenced, standardized tests of literacy and numeracy	2.39	65.4	2.70	54.2	1.49	ns
Support-staff reports; guidance, welfare, speech, health	2.46	65.4	1.99	83.4	2.58	0.01
Peer appraisal	2.64	46.1	3.05	35.1	1.99	0.05
Reports with marks or grades	2.68	56.0	2.89	47.6	0.99	ns
P and Fs/Schools Councils review, discuss and report learning	3.19	30.8	3.09	33.4	0.41	ns
Reports allow parents to compare child with others	3.19	34.6	3.89	17.0	2.89	0.01

of learning, being given chances to provide policy input and to hear policy explanations, and being given classroom goals, expected outcomes, and individual expectations at the beginning of each year. Finally, they saw the need to improve how schools report through publications and public relations.

Table 15.2: Numbers of accountability touchstone items supported by each stakeholder group (SA + A > 70%)

	Learning processes	Learning criteria	Teaching processes	Teaching criteria	Leadership processes	Leadership criteria
Teachers	16	8	13	11	10	9
Parents	15	10	14	12	8	12
Principals	17	8	13	13	14	14
DEA	16	11	13	12	13	13

In sum, since many of these needs of teachers, students and parents were generic rather than site specific, and the resources available to serve them were limited and would probably to shrink still further, they concluded that support services had to be customized and delivered to groups of school communities *in situ*. There was much talk of how best to use school community-based evaluation, action research cycles and more inclusionary planning processes that both use internal and external data and build 'ownership'.

As reported in earlier chapters, a good deal of consideration was given by central DEA officials to the ambivalent support given to standardized tests. Their use was discontinued soon after the preliminary results provided in October 1994 and April 1995 helped bring matters to a head. The Australian Council of Educational Research was commissioned in 1995 to help design a teacher-based approach to the evaluation of learning. The 10R test's replacement, the 'Developmental Assessment Resource for Teachers' (DART), was designed to provide a broader evaluation of learning in a way that included teachers' judgements. Some items from the 10R were retained to sustain longitudinal monitoring for incidental political purposes. Trials began in 1996. The trialing of another DART to replace the 14R was scheduled for 1997. Decisions are yet to be made concerning replacements for the 10N and 14N tests.

Discussion then considered ambivalence in the DEA and other stakeholder groups to the proposal that 'the evaluation of students' learning should take reports from other support staff into account (such as guidance, welfare, speech and health specialists)'. The concern here was that school personnel might be far less sensitive to the needs of disadvantaged students than suggested by the concern for equity explicit in various systemic policies. In Table 15.3 stakeholders groups' means are compared to the mean of all responses.

While most of the ambivalence to this proposal resided with secondary principals and central DEA personnel, there was no interpretation advanced that drew broad support. Some suggested that special education and student-support services had been hampered by the cutbacks. Others claimed that systemic commitment to inclusionary principles was higher than ever. Some suggested that 'profiles of achievement' had standardized the evaluation of learning in secondary schools to the growing exclusion of personal factors. Others claimed that they actually encouraged teachers to take special circumstances into account. This uncertainty resonated with other concerns in the system about the extent which schools were attending to the principle of social

Table 15.3: Differences between stakeholders regarding the proposal that the evaluation of students' learning should take reports from other support staff into account (such as guidance, welfare, speech and health specialists)

Respondents	n	Mean	SD	%SA+A	t	p <
All teachers	132	1.96	1.96	84.8	0.39	ns
Secondary principals	21	2.57	0.75	57.1	3.45	0.001
All principals	125	2.00	0.72	84.0	0.13	ns
All parents	148	1.93	0.83	84.2	0.77	ns
District DEA	10	2.20	1.03	80.0	0.64	ns
Central DEA	16	2.63	0.81	56.3	3.11	0.01
All respondents	431	1.99	0.79	83.4	–	–

justice. It is notable that a new post of Senior Superintendent (Equity) was created in 1995 to co-ordinate a review of policies and programmes. It is fair to say that equity policy and programme evaluation persist as major challenges in Tasmania and in many other public education systems across Australia.

To summarize this section, DEA central and district personnel interpreted the learning processes touchstone in political and support service terms. They were relieved to find that they were as attuned as any stakeholder group to the political touchstone. They affirmed and adopted supported items as *de facto* policy and as representing 'best practice'. They used the touchstone to evaluate the school-support services provided by DEA personnel in terms of its capacity to (a) promote professional development, (b) facilitate community school development, (c) aggregate needs and (d) raise the effectiveness of delivery systems. Standardized tests were discontinued and a search began for a new and more teacher-based approach. The findings also added impetus to a review of the needs of disadvantaged students and systemic equity policies and programmes.

Accountability Criteria related to Students' Learning

Table 15.4 compares the support that DEA and all other groups gave to the student learning criteria suggested by all stakeholders.

DEA personnel again defined the touchstone items as 'best practice' indicators, and that, when used together, would provide a methodologically more adequate approach to indicating quality learning than in the past. By this they meant that some criteria would rely on quantitative data, such as measures of individual progress, student attitudes, students' self-esteem and life skills and objective assessment. Simultaneously, other criteria were more qualitative and more customized. Examples included performance indicators developed by teachers and indicators drawn from state and national policy documents, with judgements by teachers added that were perhaps developed by research in the classroom. The challenge noted by the ERU and some district superintendents was that the current one-year school review process did not

Table 15.4: Support for proposed criteria for evaluating students' learning

Processes proposed by all stakeholders	DEA views Mean	%	All groups' views Mean	%	Differences of means t	p <
Supported						
Measures of individual progress	1.39	100.0	1.77	93.2	3.70	0.001
Judgements by teachers	1.54	100.0	1.99	84.2	4.22	0.001
Results of objective assessment used in plans	1.65	100.0	1.85	92.7	2.00	0.05
Performance indicators developed by teachers	1.76	92.0	1.95	85.4	1.55	ns
Indicators developed jointly by parent, teacher, student	1.92	80.7	2.13	77.8	1.75	ns
Performance indicators in state and national policy documents (e.g. Frameworks, Profiles)	1.96	84.6	1.99	84.2	1.17	ns
Criteria developed by research in the classroom	2.00	80.8	2.24	71.4	1.59	ns
Indicators from research literature used in planning	2.04	76.0	2.17	72.7	1.39	ns
Measures of students' self-esteem and life skills	2.08	84.0	1.94	84.4	1.08	ns
Student attitudes to school, teachers, peers, learning, homework	2.15	80.7	1.90	89.4	1.70	ns
Performance indicators developed by teachers through subject moderation	2.24	76.0	2.23	72.0	0.00	ns
Ambivalent support						
Student participation rates (attendance, retention)	2.50	61.6	2.24	66.4	0.18	ns
Parental expectations	2.69	42.3	2.96	37.6	1.37	ns

insist on the use of ex-school or empirical data, confined itself to the planning and management of three issues each year, and that there was no mechanism to weld together the benefits of external reviews and school community action research. I come back to this set of issues below.

It was interesting to watch how DEA personnel interpreted the ambivalence of all groups to the proposals that student participation rates (attendance, retention) and parental expectations should be used to evaluate students' learning. Both were regarded as lacking construct validity. Nevertheless it was decided that participation rates had some practical and political validity and that they would be retained in practice. Parental expectations apparently had to be mediated by governance processes that gave play to professional expertise before they could serve students' best interests. While this allocation of political privileges can be contested, what it did indicate was that DEA personnel tended to interpret ambivalent responses by professionals in terms of what 'they can't' rather than 'they won't' do. They used educative rather than deficit analysis.

They tended to explain and almost ingratiate each professional groups' position on issues rather than evaluate the logic used rigorously or consider political motives. They shared a sociological imagination when interpreting plural perspectives rather than an epistemically critical approach when evaluating claims. They rarely used a political imagination that identified the extent to which stakeholders' interests were being served. This meant that they tended to propose normative and re-educative strategies, rather than the empirical and rational or power and coercive strategies reportedly preferred by public service and political élites, or considered the limitations of their own often subjectivist epistemology.

The touchstone of preferred accountability criteria related to students' learning were conceived in terms of four recurrent themes; professional development, school improvement, how needs might be aggregated and how to redirect support service delivery towards school community capacity building. DEA personnel took the view that professional development had to be embedded into each school community's culture and strategic action research. However, instead of being seen as a prior condition of school improvement, the position taken by teachers, professional development was regarded by DEA personnel as a school community development strategy. The key issue was not whether teachers should be trusted to be professional but that all stakeholders should be entrusted to build a learning community with mutual, rather than professional or client forms of accountability.

To summarize this section, it was acknowledged that accountability current practices concerning learning sometimes reflected idiosyncratic notions of professionalism among a minority of teachers rather than the touchstone of collegial or community theories of accountability. This situation was, however, not a matter of blame but a matter of learned capacity. Relatively few teachers had used action research involving students to evaluate learning or had experienced open three-way discussions of students' learning. Until more recently there had been limited opportunities to share and develop the capacity to make judgements concerning learning. The primary sector was yet to enjoy a form of moderation. Hence the importance, it was concluded, of targeted support services that moderated personal theories of how students learn best, the personal theories held in all stakeholder groups.

Accountability Processes related to Teachers' Teaching

Table 15.5 compares the level of support given by DEA personnel and all other groups to proposed processes related to the quality of teachers' teaching.

The initial response by DEA personnel to Table 15.5 was one of professional affirmation. They noted that touchstone items explicitly valued the current support services provided by DEA personnel.

The second response was of concern. The main systemic mechanism used to guarantee the improvement of schools was the school review process, yet the proposal that it 'should be used to improve teaching' encountered minor ambivalence. As Table 15.6 indicates, it was just traceable to all teachers by comparing stakeholders' mean responses to all responses.

Closer analysis showed that teachers and parents had similar views ($t = 1.18, p <$ ns). Parents and principals, however, had significant differences ($t = 2.14, p < 0.05$) as did principals and the DEA ($t = 2.14, p < 0.05$). The major differences found were, however, between the DEA personnel and parents ($t = 3.76, p < 0.001$), and even more so, between DEA and teachers ($t = 4.44, p < 0.001$).

Table 15.5: Support for proposed processes for collecting data, reporting on and improving teachers teaching

Processes proposed by all stakeholders	DEA views Mean	%	All groups' views Mean	%	Differences of means t	p <
Supported						
Planned development of teachers	1.39	100.0	1.49	95.8	0.97	ns
Discussion between colleagues	1.42	100.0	1.38	98.4	0.39	ns
Report teacher appraisals to the individual teacher	1.39	66.2	1.64	92.7	2.16	0.05
Co-operative learning between colleagues (e.g. mentoring)	1.50	100.0	1.80	92.0	2.87	0.01
Self-evaluation	1.52	96.0	1.82	89.7	2.46	0.02
Training and support to identify and cope with at-risk students	1.62	100.0	1.54	97.7	0.78	ns
Negotiating new goals for professional development	1.65	100.0	1.92	88.3	2.69	0.01
Encourage teachers to read and do research	1.69	96.2	1.86	88.0	1.23	ns
Appraisal of student outcomes	1.73	92.3	2.09	80.3	2.92	0.01
Feedback and appraisal by peers	1.73	88.5	2.00	81.5	1.97	0.05
Transition program for newly appointed teachers	1.76	80.0	1.86	80.4	0.63	ns
Documentation of best practices	1.77	92.3	2.06	78.1	2.39	0.02
An appraisal of planning	1.85	96.2	2.16	81.6	3.21	0.01
Support ambivalent						
The school review process	1.73	88.5	2.22	66.4	3.55	0.001
Feedback and appraisal by students	2.04	84.6	2.59	65.9	4.82	0.001
Feedback and appraisal by senior school colleagues	2.20	76.0	2.44	65.7	1.79	ns
The planned development of classrooms	2.40	52.0	2.28	65.3	0.90	ns
Opportunity for parents to develop as co-teachers	2.44	52.0	2.58	55.2	0.95	ns
Individual and senior staff discuss appraisals	2.46	57.7	2.28	70.4	0.82	ns
Assess teacher's contribution to school planning	2.54	61.5	2.72	52.3	1.03	ns
Report teaching quality to the DEA for promotion and school development purposes	2.68	48.0	2.55	55.4	0.64	ns
Standardized test results go to the individual teacher	2.76	52.0	2.47	63.4	1.54	ns
Feedback and appraisal by parents	2.62	46.1	2.60	52.8	0.01	ns
Feedback and appraisal by an independent expert	2.89	42.3	2.89	36.0	0.00	ns
Opportunities for parents to consult and co-plan teaching programs	2.58	53.8	2.88	42.3	1.79	ns
Reporting of teacher appraisals to colleagues as part of professional development	2.92	36.0	3.20	30.6	1.45	ns
Feedback/appraisal by the P and F/School Council	3.08	61.5	3.11	32.0	0.16	ns

(continued opposite)

Unsupported

P and F/School Council discuss teacher and classroom development	2.85	42.3	3.23	29.0	1.79	ns
The selection of teachers should be more localized	3.60	12.0	3.35	23.9	1.28	ns
Network more effectively with the teachers' union	3.60	4.00	2.96	28.8	4.74	0.001
General reporting of teacher appraisals to parents as part of school planning and development	3.67	20.8	3.69	19.8	0.09	ns

Table 15.6: Differences between stakeholders concerning the proposal that one of the strategies which should be used to improve teaching is the school review process

Respondents	n	Mean	SD	%SA + A	t	p <
All teachers	132	2.40	0.85	59.8	2.17	0.05
All parents	148	2.29	0.85	61.9	0.86	ns
All principals	125	1.75	0.91	73.8	1.75	ns
All DEA	26	1.73	0.67	88.5	3.55	0.001
All respondents	431	2.22	0.87	66.4	–	–

These patterns were interpreted by DEA personnel in a number of ways. First they noted that the school review process was not originally expected to improve the quality of teaching. It had focused on school management and planning. Teachers could assume, it was argued, that the review process had been peripheral to their performance in the classroom and that new applications were yet to be negotiated. Second, most parents had limited experience of school management and planning, and were therefore, it was proposed, unlikely to value its potential with regard to the quality of teaching. These two possibilities were judged by DEA personnel to indicate further significant limits of the current one-year model of school review. This matter is taken up in a later section in this chapter.

The next issue that attracted a good deal of discussion was the access that teachers wanted to feedback and appraisal. While they apparently valued feedback and appraisal from peers, they exhibited considerable ambivalence with regard to obtaining evaluations from students, senior school colleagues, parents or independent experts. Two general views were expressed by DEA personnel. The first was that the term 'appraisal' had acquired a range of meanings often prejudiced by the 'now largely symbolic' teacher appraisal policy and process. The second view was that many school communities had lost many of the arts of evaluating and developing pedagogy by focusing on curriculum development. The leadership capacities that promote instructional expertise, it was argued, had been overlooked in an era of administrative devolution that stressed budgetary expertise and personnel management. Some district superintendents felt that 'pedagogical leadership' should be identified as a priority area for school reviews, once the current one-year model's application had been explicitly broadened beyond curriculum policy implementation to include the quality of learning, teaching, leadership and governance.

Another general issue raised was the limited role given to parents in the development of classrooms and schools. Some DEA personnel were surprized to see how little support there was in general for parents to work with teachers, for parents to acquire supportive skills and understandings, for parents to be able to affirm teaching practices or for parents to have means of legitimating school policy decisions related to the quality of teaching. Others noted a destructive paradox involved. While these norms appeared to offer teachers 'professional autonomy', the quality of teachers' service could actually be judged by any stakeholder group using almost any criteria or process. In such circumstances, it was reasoned, teachers would tend to invest their trust in peer judgements and deny the validity of all others, despite the implausibly circular methodology involved. This was understandable, it was felt, since self-respect is a minimum condition of professionalism.

Hence, aware of the feral and sometimes unreasonable nature of parental expectations endured in some communities, a number of DEA personnel noted the importance of encouraging the formation of school councils, and as soon as capacities permitted, extending their role into supporting school and classroom improvement. This somewhat indirect strategy had, it was tentatively argued, the triple virtues of being an approach that was gradual rather than radical, likely to be professionally respectful and able to build co-governance capacities in school communities.

To summarize this section, DEA personnel interpreted the teaching touchstone items to do with accountability processes as affirming the array of support services provided at central and district levels. They also saw them as identifying major limits to the school review process. There was a need perceived to extend the focus on management and planning so that the quality of teaching, pedagogical leadership capacities and governance services could be better evaluated and developed. A new model also needed to value feedback and appraisal from *all* stakeholders, use action research techniques and to make use of comparative and empirical data. Third, governance opportunities for parents needed to be provided so that their expectations could be better moderated and contribute to school and classroom development. Exactly which criteria should be used to evaluate the quality of teaching is the subject of discussion in the next section.

Accountability Criteria related to Teachers' Teaching

Table 15.7 compares levels of support expressed by DEA personnel to the views expressed by all groups surveyed.

Again the themes explored were professional development and school improvement, needs aggregation and support services that would build school community capacities. District superintendents, for example, drew on their experience to reiterate that their more successful classrooms exhibited expert teaching, supportive relationships and student progress. Their theory of accountability related to quality teaching had three

Table 15.7: Support for criteria for evaluating teachers teaching

Processes proposed by all stakeholders	DEA views Mean	%	All groups' views Mean	%	Differences of means t	p <
Supported						
Student progress	1.56	92.0	1.99	81.0	3.18	0.01
Interpersonal communications within the classroom	1.58	96.2	1.70	94.7	1.02	ns
Willingness to engage in continuing professional development	1.65	92.3	2.08	77.9	3.23	0.01
Organizational skills	1.69	92.3	1.79	92.0	0.67	ns
Behaviour management skills	1.69	100.	1.79	91.5	1.02	ns
Classroom environment	1.76	92.0	1.60	95.3	1.31	ns
Effective implementation of school and curriculum policies	1.81	92.3	2.02	85.7	1.81	ns
Teachers' knowledge of subject and child/adolescent development	1.85	88.4	1.86	90.1	1.86	ns
Instructional expertise	1.89	88.5	2.03	82.3	1.17	ns
The attitude of children (e.g. enthusiasm)	1.96	77.0	2.10	76.5	0.84	ns
Teachers' attitudes to students, parents and colleagues	2.00	88.0	1.97	85.1	0.23	ns
How well work is set, monitored and marked	2.04	83.4	2.00	83.5	0.28	ns
Support ambivalent						
Teachers competencies in DEA job descriptions	1.84	88.0	2.40	63.6	3.66	0.001
Students achievement levels in K-12 Framework	2.00	79.1	2.83	45.6	2.87	0.01
Communication skills with stakeholders	2.04	80.7	2.23	67.0	2.04	0.05
Effectiveness of teachers' written records and plans	2.19	73.1	2.46	65.8	1.65	ns
Leadership services given by teachers in school	2.50	57.7	2.66	52.0	0.83	ns
Teachers' participation in school and community	2.85	46.2	2.83	52.0	0.11	ns

major components. The first concerned teachers' capacities in the classroom; being able to organize learning, manage a micro-culture and encourage student development with high expectations. The second component concerned teachers' school capacities; their attitudes to students, parents and colleagues and their willingness to engage in professional development. The third was the need for an appropriate mix of policies, culture and facilities sustained by supportive leadership. This theory for successful accountability of teaching implied collective and reciprocal relationships between stakeholders. It was then used to explore differences with all stakeholders' responses.

The minor ambivalence among teachers over using 'teachers competencies in DEA job descriptions' as a criterion for evaluating the quality of teaching was probably due, it was explained, to two possible reasons. Generic competencies appeared to run counter to the need for customize capacities according to the needs of each school community.

Table 15.8: Differences between stakeholders concerning the proposal that one of the criteria which should be used to evaluate teaching is students achievement levels as listed in the K-12 Framework

Respondents	n	Mean	SD	%SA+A	t	p <
District HS teachers	28	2.96	1.14	42.8	2.60	0.01
All teachers	132	2.52	0.89	60.6	1.48	ns
District HS principals	15	3.00	1.07	46.7	2.18	0.05
All principals	125	2.48	0.87	59.3	1.02	ns
All parents	148	2.25	0.80	62.3	1.81	ns
District DEA	10	1.80	0.79	80.0	2.33	0.02
Central DEA	16	2.14	0.54	78.5	1.77	ns
All respondents	431	2.39	0.85	62.0	–	–

Second, as an innovation, it would automatically be regarded with suspicion by some. This latter explanation provided tacit acknowledgement of some 'innovation overload' as referred to above.

The more substantial ambivalence concerning the use of 'students' achievement levels as listed in the K-12 Framework' to evaluate the quality of teaching was traced to reluctance by about one in three teachers and principals to adopt this new policy, with the figure rising close to 50% in district high schools. Table 15.8 summarizes the situation by comparing groups' means to all means. DEA personnel recognized again the special circumstances in district high schools.

The significant differences in views concerning the use of teachers' 'communication skills with stakeholders' to evaluate the quality of teaching were traced to two major sources. First was the general contrast between the high support given by district superintendents and primary school principals (90–100% SA + A) and the ambivalence exhibited by parents (53.3%) and teachers (56.8%). Second was the sharp differences in primary schools between principals and their colleagues ($t = 5.21, p < 0.001$) and between principals and parents ($t = 6.27, p < 0.001$). Some district superintendents ventured to suggest that these findings indicated that about half of all primary teachers and parents were not accustomed to seeing non-professionals as legitimate stakeholders, and that many primary principals realized that this had to change. This interpretation added support to the analysis in Chapter 13 suggesting that an individualistic professionalism persisted in primary schools in addition to the broader touchstone of mutual stakeholder accountability.

Similarly, the ambivalence over the use of 'the effectiveness of teachers' written records and plans' was traced, as illustrated in Table 13.9, to only about 50% of teachers and district high school principals agreeing or strongly agreeing when all other groups expressed over 70% support for the proposal. DEA personnel took this as indicating a propensity of some schools with high levels of individualistic professionalism to develop a culture that favoured professional accountability rather than supportive and mutual mechanisms. It was this not unreasonable possibility, they argued, why teachers and principals had to be guaranteed occasional access to external perspectives, in addition to internal planning and action research.

Finally, DEA personnel noted that no more than 50% of any group had seen 'leadership services', or 'teachers' participation in school and community activities' as appropriate criteria for evaluating teachers' teaching. This was, it was claimed, mainly due to a mix of four persistent beliefs; responsibility for providing leadership was vested in positions, teachers' leadership responsibilities were limited to the classroom, parents and students had no leadership responsibilities in schools, and 'the school' was a professional community requiring professional leadership. The presence of such beliefs, it was argued strongly by district superintendents, was another reason why the leadership expertise available to schools had to move away from the facilitation of curriculum development and towards building governance, strategic leadership, planning and change management capacities. Other reasons given included the two hobgoblin arguments identified in Chapter 13 concerning the dangers of US tests and OFSTED inspections, and the symbiotic myth of 'professional autonomy'.

In sum, justifications by DEA personnel for their interpretation of findings related to the quality of teaching consistently backed up into four integrated concerns; professional development, community school development, needs aggregation and the effectiveness of delivery systems. This touchstone was also used to interpret findings related to leadership services.

Accountability Processes relating to Leaders' Leadership

It might be recalled that 26 accountability processes were suggested by all groups to collect and report data, and improve leaders' leadership. The views of DEA personnel were ranked by means in Table 15.9. Comparative data from all other groups are also provided and tested for differences.

While the items were ranked in a slightly different order than seen in earlier chapters, and emphasized the need for philosophical and strategic leadership just ahead of the supportive aspects of cultural and political leadership, DEA personnel were again intrigued to find that their views were not particularly different than other stakeholders. Again they took this to mean that they 'had their finger on the pulse of the portfolio', and moreover, that governance and leadership development in school communities apparently warranted high priority.

The second realization was that the systemic policy encouraging supportive yet systematic leadership styles and planning approaches had not been effectively implemented and that both adapting to and developing local capacities were under emphasized. Some school communities were not receiving the leadership services expected by stakeholders. Tacit acknowledgement of 'innovation overload' was also evident in descriptions of intense and sometimes poorly planned or under-funded attempts to implement changes, occasionally without adequate consultation.

In such challenging circumstances, it was reiterated, those who provided leadership services needed to be aware of their impact, to encourage feedback so that learning about leadership might improve, and to take a structured approach to skill development, especially in governance, leadership and management. The immediate implication noted was that school review processes had to be far better at collecting and reporting

Table 15.9: Support for processes to collect data, report on and improve leaders leadership

Processes proposed by all stakeholders	Principals' views Mean	Principals' views %	All groups' views Mean	All groups' views %	Differences of means t	Differences of means p <
Supported						
Evaluate the coherence between vision, plans and outcomes	1.46	96.2	1.84	90.3	3.22	0.01
Provision and generation of a school vision	1.64	92.0	1.89	87.7	1.93	ns
Self-appraisal	1.69	88.5	1.96	83.3	1.93	ns
Peer networks reflect on challenges of practice	1.77	84.7	1.97	79.7	1.39	ns
Appraisal of support and feedback given to staff	1.85	88.4	1.78	91.8	0.57	ns
Skill development programs, e.g. in governance and management	1.92	96.0	1.90	87.1	1.05	ns
Quality of reporting to parents and community	1.92	96.0	2.11	82.1	2.26	0.02
Peer appraisal	1.96	80.7	2.25	70.7	1.86	ns
An appraisal by school colleagues	2.00	84.6	2.24	74.0	2.02	0.05
Survey of the school climate	2.04	84.0	2.17	74.4	1.16	ns
Feedback from parents and students	2.04	88.0	2.26	71.7	1.75	ns
An appraisal of policy-making strategies used	2.08	80.0	2.18	74.7	0.86	ns
Improved by using feedback from staff	2.12	76.9	2.06	84.3	0.42	ns
Support ambivalent						
Improved by using a mentoring process	1.96	76.0	2.28	60.5	2.12	0.05
An appraisal of the quality of external liaison	2.15	76.9	2.40	59.8	2.22	0.05
Parents, teachers, DEA collaborate in principal selection	2.31	57.7	2.42	60.3	0.61	ns
An appraisal by the DEA	2.50	50.0	2.78	45.5	1.59	ns
An appraisal by the P and F/School Council	2.50	57.7	2.76	47.2	1.74	ns
Appraisals should be reported to the DEA	3.08	26.9	2.76	43.3	1.62	ns
Appraisals reported to individuals and colleagues as part of the professional development programme	3.20	24.0	2.99	34.2	1.04	ns
An appraisal by the community	2.81	34.6	3.13	25.7	1.96	0.05
Fixed term and negotiated performance contracts	2.85	38.4	3.33	27.4	1.89	ns
Unsupported						
Overseas exchanges	2.92	26.9	3.02	28.8	0.66	ns
Appraisals reported to parents as part of school development programme	3.16	24.0	3.37	21.3	1.45	ns
P and F/School Council set leadership service policies	3.60	16.0	3.35	21.9	1.35	ns
More localized selection of leaders	3.69	7.60	3.35	18.1	1.90	ns

philosophical and strategic aspects of leadership and to help people take a more parenthetical view of the part they played in school community development. Activities highly valued in this regard were those that required collective reflection, such as creating a school vision, inclusionary and educative policy-making strategies, and searching for higher levels of coherence between vision, plans and outcomes. 'Self-appraisal', 'peer appraisal', 'peer networks that reflected on challenges of practice', and 'surveying the school climate' were all useful methods, it was argued, that had to be learned to be effective.

This touchstone was then used to evaluate the proposals that nearly attracted support from all stakeholders. An examination of responses to the 'use of a mentoring process' to improve the quality of leadership showed that disagreement was usually well under 10% in all stakeholders groups and that about 30–45% of teachers and parents were simply unsure about the practice. It was concluded that such uncertainty was probably due to some leaders not wishing others to know that they were being mentored, some individualistic professionalism ruling out co-operative learning about teaching, mentoring not yet being regarded as customary, and the benefits of mentoring yet to be widely shared with stakeholders.

'An appraisal of the quality of external liaison' was proposed as a method of evaluating leaders' leadership. It was supported by DEA personnel and district high parents while given ambivalent yet majority support by teachers, principals and other parents. This finding was interpreted as indicating the special nature of district high school communities as well as a broad expectation that leaders would provide expert boundary management.

The next proposal that met with significantly different responses was that 'there should be parental, professional and departmental involvement in the selection of principals'. About 60% of all stakeholders favoured this idea, with 20% unsure and 16% disagreeing. It was immediately noted that while primary principals shared the view of all stakeholders, district high principals ($t = 2.30, p < 0.05$) and secondary principals ($t = 3.17, p < 0.01$) were particularly supportive. Similarly, while parents took the view of all stakeholders, district high parents were especially supportive ($t = 2.23, p < 0.05$). In contrast, while district high teachers shared all stakeholders' views, primary teachers ($t = 2.30, p < 0.05$) and secondary teachers ($t = 2.54, p < 0.02$) were significantly more negative. DEA personnel saw this as indicating yet again the limited exposure that some primary and secondary teachers have had to stakeholders' views, and how such levels of exposure tended to be related to an exclusionary professionalism. Their view was that such attitudes would continue to change as more collaborative selection practices became more common and more stakeholders experienced the benefits.

The next bloc of proposals concerning appraisals 'by the DEA', 'the P and F/School Council' and 'the community', also suggested that such appraisals should be reported 'to the DEA' and 'to individuals and colleagues as part of a professional development programme'. All were regarded as 'worthy ideas but ahead of their time'. Some school communities were reportedly testing some of these ideas or were planning to trial them. In many settings, however, it was argued, they would still be regarded as 'an innovation too far'.

In sum, once the DEA personnel realized that findings about leadership accountability processes reaffirmed many of their own commitments, they tended to argue that such methods were needed to further advance governance and management in school communities. Shortfalls in service were often regarded as a matter of undeveloped leadership capacities. This was particularly so with regard to philosophical and strategic aspects of leadership intrinsic to creating a school vision, policy-making strategies that involved and educated others in the arts of governance, and to guaranteeing coherence between vision, plans and outcomes. They could see mentoring, the collaborative selection of staff and a wider suite of appraisal processes all being used in time. In the interim, they concluded, it was important to build on increasingly trusted and valued processes such as the school review process to obtain more educative forms of accountability concerning leadership.

Accountability Criteria relating to Leaders' Leadership

The 22 accountability criteria suggested to evaluate leaders' leadership services are ranked in Table 15.10 by the means of responses of the DEA personnel surveyed, and then compared to all other's means.

There was a reaffirmation of the need for proactive leadership in philosophical, strategic, political and cultural realms, especially to promote professional development, community school development, aggregate needs and raise the effectiveness of delivery systems. The DEA personnel revealed their professional roots as educators. They reiterated the logic identified in Chapter 13; the most appropriate leadership was socially and communally rational because it emphasized hearing and caring for others, collaborative policy and planning processes, undistorted communications, sustaining morale and motivation, and open and reciprocal relationships. Second, such leadership promoted coherence between philosophy and action with an emphasis on reflective action or *praxis*. It valued the ability to jointly make and reflect critically on policy, to plan outcomes and achieve priorities, and to manage and organize others, especially to evaluate, budget and govern. Then, using this theory of ideal leadership service, some items with indications of ambivalence were then 'factored into' the touchstone.

One example of quick inclusion was the proposal that one of the 'criteria that should be used to evaluate leaders' leadership should be their capacities as learners and researchers'. The ambivalence was traced solely to primary teachers ($t = 2.15, p < 0.05$) and to primary principals ($t = 2.16, p < 0.05$). One interpretation was that, since many primary school personnel were probably unable to conduct empirical research, they should be helped to develop skills as action researchers. Another was that a minority of primary school personnel would see this as a infringement of their 'professional autonomy', until they took clients' interests into account or experienced the benefits of action research. An associated proposal, that 'indicators from research literature should be included in the criteria used to plan improvements to leadership services', was found to attract agreement from between 60–70% from all stakeholders. The conclusion reached was that all educators, especially those providing leadership, had

Table 15.10: Support for criteria for evaluating leaders' leadership services

Processes proposed by all stakeholders	Principals' views Mean	%	All groups' views Mean	%	Differences of means t	$p <$
Supported						
Capacity to hear and care for others	1.65	88.5	1.56	96.8	0.65	ns
Ability to plan outcomes and achieve priorities	1.65	96.2	1.78	94.2	1.15	ns
Capacity to make and implement policy	1.72	100.	1.86	90.9	1.47	ns
Evidence of the quality of teaching by the staff	1.73	88.5	2.21	74.6	3.48	0.001
Management and organizational skills (evaluation, budgeting and governance)	1.80	92.0	1.89	88.3	0.63	ns
Student and teacher morale and motivation	1.81	84.6	1.77	88.8	0.20	ns
Extent of professional development within the school	1.81	84.7	2.16	74.9	2.15	0.05
The openness and climate/tone of the school	1.85	92.3	1.76	88.8	0.65	ns
The extent to which staff support their leaders	1.89	84.6	1.93	84.5	0.25	ns
Valuing of creativity and productivity in school	1.89	84.6	2.04	79.5	1.13	ns
Extent of collaborative decision making	1.96	80.0	1.93	83.4	0.17	ns
Evidence of learning by staff and students	2.00	84.6	2.16	73.6	1.35	ns
Quality of internal and external communications	2.12	80.7	2.10	79.9	1.09	ns
Ambivalent support						
Capacities as learners and researchers	1 92	88.4	2.29	68.6	3.17	0.01
Performance indicators in guidelines provided by the DEA	2.08	80.0	2.36	62.9	1.72	ns
Indicators from research lit used in plans	2.11	69.2	2.32	61.1	1.46	ns
Recommendations from school reviews	2.27	61.5	2.46	55.9	1.39	ns
The expectations of the community	2.42	65.4	2.46	61.0	0.23	ns
National competency indicators for managers	2.44	60.0	2.69	40.6	2.09	0.05
The quality of the physical environment	2.46	57.7	2.53	58.8	0.38	ns
Extent to which parents support school leaders	2.50	50.0	2.54	54.4	0.19	ns
Leaders' relevant qualifications	2.92	38.5	2.75	51.7	0.89	ns

to both demonstrate their commitment to professional development and learn how to manage school community action research.

Another item also quickly added was the proposal that 'the criteria used to report on leaders' leadership should include the performance indicators in the guidelines provided by the DEA'. The sole source of ambivalence was found to be district high school parents ($t = 2.43, p < 0.05$). Similarly, ambivalence about the idea that 'the expectations of the community' should be used to evaluate leaders' leadership was due to disagreement from 20% of district high principals and 17% of district high teachers. These findings were held to indicate the special need to accelerate co-governance in district high school communities in order to help mediate parents' expectations, to offer many more co-management opportunities to boost commitment and to offer more customized leadership services.

Ambivalence concerning the proposal that the 'criteria used to report on leaders' leadership should include recommendations from school reviews' was traced solely to between 40–50% of teachers being unsure and disagreement from 14% of primary principals. This finding was held to indicate, yet again, the need to revise the school review process.

To summarize this section, DEA personnel indicated that leadership should be valued across philosophical, strategic, political and cultural areas of activity with special reference to professional development, community school development, aggregating needs and the effectiveness of delivery systems. They used a theory of educative leadership that stressed a social and communal rationale, coherence between philosophy and action, and *praxis*. They saw the need to accelerate opportunities for co-governance and co-management and for more customized leadership services. DEA personnel focused on their responsibility to improve school review process, the topic addressed in the next and penultimate section.

Revising the School Review Process

By 1996, a one-year model of school review was in use; the 'School Plan Implementation Review'. The process was led by each 'District Superintendent working collaboratively with the school principal and the school community in reviewing the implementation of the school plan, with the aim of achieving continuous improvement of teaching and learning' (DEA 1996, p. 1). Focus questions were sent at the beginning of the year to schools selected for review so that they could develop brief written answers. Review teams were formed, often including another principal. They usually visited their school once or twice for a few days on each occasion to meet with 'as many members of the school community as is practicable. This could include school council/parents and friends members, staff, students, parents and may involve visits to classrooms in connection with the review areas' (p. 1). The visits began after Easter, most occurred in Term 2, and all were concluded early in Term 3 to allow findings to inform planning for the next year.

Each review team evaluated planning and management in three priority areas. In 1996 these areas were science, parent participation and schools councils, and a third area chosen by each school. Each superintendent's report of about two pages was expected to record achievements in the three priority areas, specify agreements reached and recommend future directions. It was understood that principals would share the report with their school community and that superintendents could share it with the ERU.

When the initial results of this research were presented to senior officials in October 1994, and more sophisticated analyses tabled at the policy forum mounted by the ERU and the University in April 1995, the ERU and the district superintendents were encouraged by all stakeholders 'to revisit' the design of the one-year model in use. The district superintendents consulted extensively in 1995 on the matter. In early 1996, a working party of three district superintendents, Rosemary Wardlaw, Elizabeth Daly and David Hanlon, were able to report that a three-year 'review and school performance' model had drawn general support.

The rationale of the new model noted that a one-year cycle was neither providing the forms of support needed by 'under-performing schools' nor satisfying wider school or systemic needs for improved accountability. The new three-year process proposed was designed to include the use of external and comparative data every third year and to encourage implementation and action research evaluation in the years between. To clarify, Year One activity was to comprise an 'assisted school self-review', Year Two was an 'implementation of the school improvement plan' using 'action research evaluation', Year Three was a repeat of Year Two, and Year Four was to repeat the Year One 'assisted school self-review' process.

It was proposed that the 'assisted school self-review' use three forms of data, (a) 'best practice indicators' covering teaching and learning, leadership and management, and professional and staff development, many to be drawn from this research, (b) perceptions collected by surveys and other methods from parents, students and staff, and (c) empirical data such as examination results, DART scores, key intended numeracy and literacy outcomes data, retention rates and student outcomes data.

The recommendations for action from the 'assisted school self-review' were to become the basis for a 'school improvement plan'. As in the past, such plans were to be approved by the district superintendent. In some schools, it was suggested, superintendents could 'require particular action to be included in the school improvement plan'. Examples given were professional development in budget and personnel management. In a few schools, presumably the seriously 'under-performing' schools, superintendents could specify a 'next-year audit' of particular issues and help such schools implement their plans.

The systemic implications of this quickly-adopted model were substantial. The newly appointed Assistant Secretary (Audit), Mr Ralph Spalding, immediately initiated a policy research process that had four general aims; design data collection instruments and methods, review and negotiate licenses to use and modify survey materials and software, create 'clearing house' infrastructure to share school self-review resources, and create a networked accountability information system that permitted site and systemic input and comparative analysis.

There were also four wider policy and practical implications to be considered. In the past the DEA had provided review priorities and a standard school planning process. The three-year model would instead expect school communities to develop their own planning and development strategies in order to (a) contribute to and draw data from the systemic accountability information system, (b) respond to external evaluations of learning, teaching and leadership, (c) use action research methods and cycles, and (d) use planning processes that attended to site-specific priorities. It was an open question among the district superintendents as to whether schools had or could develop the necessary capacities.

Among other practical challenges was that DEA and school personnel had to design and learn how to use an electronic accountability information system and a 'clearing house' of accountability resources. The hope expressed by some district superintendents was that all would see the benefits of sharing 'best practices' and an appropriate mix of data for local and state accountability purposes.

The third challenge involved changing the forms of support service personnel available to schools. While 'curriculum officers' had been appropriate in the past, there was now a clear need for school community development expertise, especially in the areas of governance, strategic leadership, planning and change management. The fourth challenge, as ever, was how to carry through these improvements in a contracting budgetary context.

Summary

The views of DEA central and district personnel were compared to all stakeholders' preferences in this chapter and used to interpret findings, about 18 months after the initial analysis had been presented. Interpretation was affected by the context of structural and political turbulence, periods of policy paralysis and tightening constraints on public expenditure. Although it was not a propitious time to be negotiating fresh accountability policies, the evidence is that the research project helped to sustain the ongoing policy review process.

Many of the core duties of central DEA officials had remained relatively unchanged. Ministers continued to require an executive arm to serve a number of generic functions, including the need to provide accurate information on what was being achieved in the portfolio. On the other hand, it was also shown that structural rationalization had made the contractual accountability relationship between government and executive much more immediate. All public administrators were expected to help co-ordinate the business of government in a holistic manner, help 'reposition' the government in the provision of services, and occasionally, to rationalize in haste to cope with financial crises.

Some special effects of this context were noted during the interpretation of findings. One was the challenge of sustaining the professional commitment of colleagues. Another was that appropriate 'leadership competencies' were being considered less in terms of an individual's ability to implement given policies and far more in terms of building the capacities that would enable policy communities to become 'learning organizations'. Central personnel and district superintendents reported having to mediate exchanges across the boundaries between the organizational cultures of public administration and education. For example, DEA personnel reported having to defend equity and excellence as equally important educational policy principles. Another was that DEA personnel valued leadership services that acknowledged the plurality of Tasmanian society, boosted the philosophical and strategic planning capacities of policy communities, and helped school communities determine policies within state guidelines, despite the more instrumental priorities of the wider public service.

DEA central and district personnel interpreted the supported accountability processes concerned with the quality of learning as politically and professionally affirming. They adopted supported items as *de facto* policy and regarded items as representing 'best practice'. They reviewed school support services in terms of their capacity to (a) promote professional development, (b) facilitate community school development, (c) aggregate needs and (d) raise the effectiveness of delivery systems. Standardized tests were discontinued in favour of a new and more teacher-based approach. Impetus was added to a review of equity policies, programmes and practices.

DEA personnel also accepted and applied the touchstone accountability criteria related to learning. They acknowledged that current practices sometimes reflected idiosyncratic notions of professionalism rather than collegial or community theories of accountability. They deemed this to be not a matter of blame but a matter of capacity, experience and leadership. They aggregated needs and identified support services that would moderate personal theories of how students learn best in all stakeholder groups.

DEA personnel also interpreted the touchstone items to do with teaching accountability processes as affirming the array of support services provided at central and district levels with one serious exception; the one-year school review cycle. The focus during school reviews on management and planning associated with curriculum development, they believed, had to be extended to boost the quality of teaching and pedagogical leadership capacities. They also came to the view that feedback and appraisal from stakeholders and the use of comparative empirical data had to be improved. More parents needed to participate in governing school policies so that their expectations could be moderated and be reflected in school and classroom development projects.

It was realized that four beliefs had to be challenged, specifically that responsibility for providing leadership was vested in positions, teachers' leadership responsibilities were limited to the classroom, parents and students had no leadership responsibilities in schools, and 'the school' was a professional community requiring professional leadership. Another challenge was to ensure that the leadership available to schools could build governance, strategic leadership, planning and change management capacities, provide plausible alternatives to US tests and OFSTED inspections and displace the myth of 'professional autonomy' with a policy of collective accountability.

The findings about leadership accountability processes were taken to reinforce many of the methods being used to advance the quality of governance and management in school communities. DEA personnel recognized that it was important to build on increasingly trusted processes to obtain more educative forms of accountability concerning leadership. They used a theory of leadership that stressed a social and communal rationale, coherence between philosophy and action, and reflective action or *praxis*. They saw the need to accelerate opportunities for co-governance and co-management and for more customized leadership services.

A new three-year school review cycle was designed to gain greater coherence between internal and external accountability criteria and processes, provide more appropriate forms of support to 'under-performing schools' and to satisfy school and systemic needs for better accountability processes and criteria. Year One activity will be an 'assisted school self-review', Year Two will include 'implementation of the school improvement plan' and 'action research evaluation', Year Three will repeat Year Two, and Year Four will repeat the 'assisted school self-review' process.

The adopted 'assisted school self-review' will use three forms of data. 'Best practice indicators' will relate to teaching and learning, leadership and management and professional and staff development. Perceptions will be collected by surveys and other methods from parents, students and staff. Empirical data such as examination results, DART scores, key intended numeracy and literacy outcomes data, retention rates and student outcomes data will be used. Each 'assisted school self-review' will then provide recommendations for action that are to be embedded in the 'school improvement plan'. Each plan will be approved by the district superintendent and 'under-performing' schools may be subjected to a 'next-year audit' of particular issues.

A systemic policy research process was initiated by the ERU to design data collection instruments and methods, review and negotiate licenses to use and modify survey materials and software, create 'clearing house' infrastructure to share school self-review resources, and create a networked accountability information system that will offer site and systemic comparative performance analysis.

Wider implications were also identified. The three-year review cycle will expect school communities to contribute to and draw data from the systemic accountability information system. School communities will be expected to help design, participate in and respond to internal-external evaluations of learning, teaching and leadership. School communities will need to learn and use action research evaluation methods and cycles. Each school community will need to employ planning processes that develop and attend to site-specific priorities. DEA personnel will have to design and learn how to use an electronic accountability information system supported by a 'clearing house' of accountability resources. Instead of curriculum officers, as in the past, appointees familiar with school community development will be valued, especially if they are expert in the areas of co-governance, strategic leadership, evaluation co-planning and change management, all in contracting budgetary contexts. In the next and final chapter I synthesize the implications of these and other perspectives for the Tasmanian system of public schools and for the field of Educational Administration.

16

Reflections on Educative Accountability Policy: Theory, Practice and Research

Introduction

The purpose of this final chapter is to offer reflections on the nature of an educative accountability policy and to suggest directions for further research. The current situation warranting attention in Tasmania is summarized. The policy research approach used and its context are presented in brief. General reasons for having an educative accountability policy in public education are rehearsed, along with lessons drawn from comparative settings that have experienced other approaches. The policy touchstone created by the research process is then reported along with the implications of interpretations by stakeholders.

The set of pragmatic purposes, processes, criteria and structures offered for consideration comprises a theory of mutual and educative accountability. This theory features democratic governance, educative leadership and communitarian parenting and professionalism. These three themes were found in Tasmanian stakeholders' interpretations of expressed and measured policy preferences. The potential relevance of the post-positivist policy research methodology used and the pertinence of the proposals in systems of public schools with high levels of devolution together suggest the possibility of applications elsewhere.

The general position presented is that an educative accountability policy in education is an expression of theories about (a) what constitutes valuable knowledge, (b) how learning, teaching, parenting, leadership and governance can be demonstrated and improved, and (c) how obligations should be discharged between stakeholders in a policy community. In practical terms, this means that accountability policies in educational systems and institutions need to define the processes (procedures, actions or methods) and criteria (standards, benchmarks or indicators) that are to be used to collect data, report on and improve learning, teaching, parenting, leadership and governance. At the very least, these processes and criteria need to be fair, effective, improvable and reasonably coherent. One consequence is that both the current content and the methods of making accountability policy require recurrent evaluation.

The Accountability Policy Challenge in Tasmania

Current accountability policies in Tasmania exhibit the influence of many factors. Historical events, political and economic contingencies and administrative drives for efficiency and effectiveness have all played their part. The ideological commitments of policy actors and the development of educational values have also affected how knowledge is valued and beliefs about how learning is to be demonstrated. There has been a parallel waxing and waning of interest and the evolution of plural beliefs concerning the quality of teaching, leadership and governance in wider contexts. Federal politics, ministerial style and national curriculum have all constrained the discharging of obligations, sometimes limiting practices to line management, at other times encouraging local and mutual forms of stakeholder accountability. Accountability policy has even been used on occasion to bridge the contradictions between the devolution of administrative powers to 'self-managing' schools and the use of centralized governance powers. A recent example has been the introduction of new national curriculum accountability and productivity mechanisms while 'devolving' responsibility for budgetary contraction in public education.

The accountability policy documents of the 1990s in Tasmanian embedded accountability in planning, resource management, monitoring, reviewing and reporting functions. This bureaucratic design was supported by a systems theory of organization and a corporate model of 'self-management'. Schools were expected to have an 'accountability cycle' that cohered with the DEA's strategic plans and evaluation activities. Senior school personnel were expected to implement policy statements in consultation with district superintendents. Schools councils were *not* empowered to provide governance or accountability. The stress was on 'proving' rather than 'improving' education, reporting outcomes using different forms of evidence and co-ordinating the information collected by schools through the system's line management hierarchy.

Preliminary data and documentary evidence collected in 1992 and 1993 showed that three perspectives dominated a largely paralysed policy discourse in order; technical, professional and client interests. On the other hand it also showed a growing awareness that accountability was an important means of reconstructing policy legitimacy, the rationality of public schooling and the commitment of participants and stakeholders. The establishment of the ERU was an expression of this awareness. It was to both demonstrate and develop the quality of schools and boost confidence in public schooling, essentially by co-ordinating reviews of systemic policy and institutions. While 'proving' the performance of schools apparently required no more than positivist techniques of quality assurance, it was increasingly realized that 'improvement' and public legitimacy would require complex forms of capacity building in school communities using more inclusionary methods.

While positivist research uses empirical facts to prove and improve the theories used in policies, post-positivist research uses many forms of information and a process of conjecture and refutation to advance the quality of theories being used by policy actors to justify policies. The practical and methodological implications are immense, especially if the pluralism of stakeholders in public education is acknowledged. An immediate implication is that policy research needs to focus not on the empirical evidence of practices but on the quality of theories used to provide practices with legitimacy and problem-solving potency.

There were also differences between national and state jurisdictions on whether accountability policy should both prove *and* improve public schools. Australian national policy communities have shown markedly less interest in school community capacity building in recent years. National policy discourse has featured commodification, vocationalism and quantification. The emphasis has been on empirical performance and outcomes indicators that correlate with, and thus justify, accounts of expenditure. The stress has been on productivity and political forms of accountability.

In Tasmania, successive governments and the DEA have sustained a strong additional interest in the accountability criteria and processes that enhance learning conditions, professional development and school improvement. The focus has been on professionally effective forms of accountability. The strategic challenge for the DEA has been to balance political and productivity accountability with more professional forms of accountability. The three-part policy challenge was to identify the forms of 'educative accountability' preferred by immediate stakeholders, to sustain a dualistic state education accountability policy that promises to both 'prove' and 'improve' public education, and to advance accountability policy-making capacity in schools. The post-positivist policy research process developed to assist with this strategy is summarized in the next section.

Epistemically Critical Policy Research

The DEA commissioned policy research to compliment the policy review processes managed by the ERU and the school review processes managed by the district superintendents. It began by mapping the theories of stakeholders in natural settings over two years. School and community workshops, group interviews and focus groups were used to gather theories about what 'educative accountability' might be. These theories were elicited as 'policy preferences'. Two general research questions were used. What processes (procedures, actions or methods) should be used to collect data, report on and improve students' learning, teachers' teaching and leaders' leadership? What criteria (standards, benchmarks or indicators) should be used to evaluate students' learning, teachers' teaching and leaders' leadership? Feedback processes were designed to verify the data and collect interpretations.

All policy preferences articulated by school community and stakeholder groups were accepted, refined and converted into Likert items. All policy proposals were included. The APQ then measured the extent to which each stakeholder group, including administrators at all levels, supported each policy proposal. At this point the theories in use were considered to be mapped in qualitative, conceptual, quantitative and political terms.

International resources and stakeholders' interpretations were then added to the process of systemic reflection. Policy seminars facilitated the collective interpretation and application of findings in central, district and school settings. A number of papers were drafted, presented, debated and published. Meetings with DEA and Treasury officials in 1995 and 1996 related the findings to the formative evaluation of schools and to the quality of systemic service delivery systems.

This approach trialed three post-positivist ideas concerning the growth of knowledge. The first idea is that policy knowledge grows not by testing hypotheses with empirical proof but by an interactive process of conjecture and refutation that develops people's theories; theories conceived as webs of belief. The second idea is that stakeholder-sensitive policy research can identify where theories in use already overlap, despite the existence of apparently mutually exclusive perspectives reflecting group interests. The third idea is that negotiations between stakeholders can develop the areas of overlap found, and through constant comparison with ideas from elsewhere, expand the policy touchstone available.

None of these three ideas about the growth of policy knowledge were refuted in practice. Although there were moments when ego, altruism and creativity where given epistemic privileges by policy actors, the endurance of these three ideas and the production of substantial policy touchstone suggests that a major part of the problem of accountability policy making in the past has been epistemological in nature.

The history of ideas in the field of educational administration supports this possibility. The field has experienced three major epistemological watersheds; positivism, subjectivism (humanist and socially critical), and non-foundationalism. The third watershed has also meant setting aside paradigm-based research, remaining sceptical of post-modernism/post-structuralism, and considering Evers and Lakomski's variant of scientific realism; naturalistic coherentism.

Despite these significant advances, positivism is still very influential. It is not uncommon for international accountability policy research to use a bureaucratic and systems view of organization, an objectivist view of social reality and logical empiricism to identify performance indicators and select quality assurance mechanisms. The purposes of public education and accountability are regarded as given. Learning, parenting, teaching, leadership and governance services are treated separately with regard to accountability, denying the holistic and interdependent nature of knowledge in daily use in school communities. Accountability policy is typically developed out of context. In many settings the stakeholders that desire or help create learning outcomes are not permitted a voice. A discourse of ingratiation is not unusual in such circumstances to explain the uneven distribution of power. Examples include 'lay persons' and 'provider capture'. Many approaches to policy research use only empirical data, an ideological discourse and exhibit outcomes that reflect a consensus of the personal preferences in a policy élite.

Post-positivist theory building focuses on the theories in use, evaluates them using ideas from research and elsewhere, and advances on the ground left unaffected by refutation. The evaluation processes inevitably highlights values and then provides a place for moral knowledge in policies. The acceptance of responsibility in public education, for example, is taken to oblige policy makers to provide forms of accountability that will cohere with formative evaluation, supportive reporting relationships, politically sensitive planning and strategic direction setting. To be self-referentially coherent, such accountability policies also need to be developed and justified in an demonstrably educative manner. Appropriate methods of policy development, therefore, include reflective action (or Aristotelian *praxis*) and reconciling stakeholders' perspectives through action research. Appropriate forms of justification also include appeals to practical consequences, the standards in touchstone, and the constant comparison and testing of theories that are used to justify public policies.

In sum, such epistemically critical research also offers the possibility of greater policy legitimacy, organizational rationality and commitment to 'best practices' in a pluralistic organization, be it a school, system, or society. This possibility is examined by considering the reasons for having an educative accountability policy in public school systems.

A Rationale for an Educative Accountability Policy in Public Education

A pragmatic reason for having an educative accountability policy in public education is to ensure that evaluation is done, seen to be done, seen to be reported and seen to be applied in a manner which is educational in intent and outcome. Since an accountability relationship between parties exists in the knowledge that outcomes have 'high stakes' consequences for all stakeholders, a public policy needs to clarify at least three matters. It needs to formalize collective and personal obligations between the stakeholders. These obligations need to give practical expression to the metavalues of the organization. The policy needs to make accountability obligatory in terms of criteria and processes.

The moral obligations involved trace from the unique distinction that educational systems and institutions have as organizations; they are expected to both reproduce and transform societies' values. Hence, a second reason for having an accountability policy in public education is ideological in nature. The nature of evolving society must remain a policy issue in the public domain. The specification of obligations and criteria for evaluation identifies the values being promoted so that they can be contested and affirmed. Various comparative sources have argued that accountability schemes in public education have to be fair, valid, intelligible, objective, verifiable, remedial, non-damaging, comprehensive, provide appropriate data and maintain confidence. Such values appreciate the sense of security, significance and solidarity intrinsic to a school community. Such an appreciation is not gratuitous. The Thatcherite version of economic privatism in the UK denied the existence of society, devalued the aesthetic and social purposes of publicly financed institutions, and saw public schools not as assets of the state but as unnatural liabilities.

A third major reason for having an accountability policy in education concerns the capacity to sustain the moral economy of public organizations. A public accountability policy gives expression to the reconciliation of plural principles, beliefs, values and interests in a policy community. It needs to do this in a way that enables professional and administrative activity to proceed confident of its significance and moral standing, and the appropriateness of its structures and technology. This also means that the structures of participation in educational governance are intimately related to the learned capacities of policy makers, as well as to the organizational structures that are used to convert complex and plural public values into effective policy settlements. They can be improved with use and with critical reflection.

It follows that a fourth reason for having an accountability policy in education is that it may provide the conditions for improving the moral order of human relationships and structures in social institutions. This moral order may improve further as stakeholders find that their views *are* recursively honoured, reconciled as reasonable obligations and discharged publicly. Such development requires three standards of public accountability to endure. Accountability processes must remain open and fair. The accountability criteria used must remain explicit, be used in decision making and be documented. Those who make decisions must be held accountable for their consequences.

A fifth and equally important reason for having an accountability policy in education is strategic in nature. Governments constantly review their roles as demographics and political priorities change. The move to SBM, LMS and 'self-management' structures has cut the capacity of systemic accountability policies to articulate the value to society of public education. Restricted views of 'the public interest' have redefined public accountability as client accountability, which in turn, reduces the moral obligations to maintaining market and or political relationships. Conversely, a public accountability policy may be the source of valuable social and political commentary and an effective marketing strategy for public education. It is likely to be especially potent if it reiterates the civic purposes of public education, valuable knowledge in a complex democracy, offers strategies for demonstrating educational productivity, evokes communitarian values in key constituencies, and further, takes out long-term insurance by specifying means of nurturing organizational learning capacities.

A sixth and related reason for having an accountability policy is to sustain the quality of public policy discourse concerning education. Although 'public accountability' means accounting publicly in terms of the 'public interest,' the term has been co-opted in some settings to justify subordinating the goals of education to the goals of the state, transferring power from professionals and school communities to the state, and shifting attention onto the means rather than the ends of accountability. Similarly, while 'contractual accountability' means answering in terms of an employment contract, it has been altered to mean demonstrating policy subordinacy. 'Contractual accountability' has moved in some jurisdictions to become 'political accountability' (accounting to political masters).

Some terms are euphemistic. 'Moral accountability' has been degraded to mean 'answerability to clients'. As noted above, while this defines 'answerability' as a moral obligation, it limits the obligation to consumerist or political processes and criteria and may exclude the public interest. Similarly, 'professional accountability' has been debased to mean 'answering to colleagues' and has been used to justify school image management. Other terms have become synonyms. For example, although they have been used interchangeably, 'responsive accountability' (accounting to stakeholders during policy making) may be distinguished from 'communitarian accountability' (accounting to a policy community in terms of the public interest). The point being stressed here is that an educative accountability policy should be conceptually rigorous, add to the quality of public policy discourse and display its educational and moral philosophy.

A seventh reason for having accountability policies in public education is that the values of the seemingly objective technologies involved may be contextualized and made contestable. For example, while the OFSTED model of school evaluation in England and Wales has improved in technical terms, it is yet to demonstrably discount SES effects, guarantee the reliability of inspectors, define 'value adding,' measurably 'add value' to school community development, acknowledge the pluralism of local (e.g. multi-cultural) communities, or cohere with other review and accountability mechanisms used in UK public education. Its monist and centralist tendencies contrast with those evident in the trends away from greater systemic coherence, towards greater local responsiveness and to differentiated zones of legitimacy. If school communities continue to assert the need for greater distributional equity, and increasingly enact their legislated role in accountability in England and Wales, then a communitarian ethic in associational life can be expected to challenge the values of OFSTED technology with pluralistic local and regional public policy projects.

An eighth reason for having an accountability policy in education concerns improving the structures and legitimacy of government. Comparative experience suggests that the contestability of an accountability policy helps to insulate it from the vagaries of, and cynicism attached to, political credibility. The existence of many constituencies in complex and plural social systems, many of which have the capacity to exercise various forms of power, implies an inevitable trade-off between fidelity and adoption during top-down implementation. Alternatively, an educative accountability policy committed to democratic pluralism and a post-positivist view of knowledge would respect current theories of 'the situation' in all constituencies with inclusionary and recurrent policy processes. It would also acknowledge the complexity of structures in public education with differentiated organizational purposes and tasks, respect the need for mediation and coherence between interdependent limited governments, and employ multiple assessments to establish multiple zones of legitimacy.

A ninth reason for having accountability policies in public education concerns the quality of society. Having an accountability policy in public education potentially freshens the democratic impulse in a civilized community. Both education and democratic society rely on accountability to recreate the legitimacy of structures, policies and practices, to sustain the rationality of organizations, and to retain the confidence of all participants and stakeholders.

Unlike other options, democratic accountability ensures that governors remain responsible to the governed for the quality of their stewardship. Three principles are involved. The delegation of the authority to govern proceeds only with the consent of the governed. This consent is conditional on the stewards of that delegated authority remaining accountable. Accountability processes and criteria must remain contestable. When these principles are in operation, and when democratic accountability invests political authority in 'the people,' four values are potentially celebrated; individualism (the achievement of potential), liberty (freedoms consistent with order), equality (equity of opportunity, outcomes and access to power), and fraternity (co-operation in building a wholesome society). The history of accountability in the UK and United States suggests that when these principles and values are put aside, public education is degraded.

Lessons Learned from Other Systems

The evolution of accountability policies in the UK has been a difficult process marked by the negotiation of pragmatic and contested policy settlements. Most stakeholders accept the general proposition that being accountable is to be responsible, to be explicit about obligations and to be answerable in a 'high stakes' process. On the other hand, resistance to political aspects of accountability has risen each time governance has exhibited incapacities. One example is structures that have limited participation and offended democratic norms. Another is when policy actors' ideological commitments impede the conversion of complicated public values into workable policy agreements. There are many examples of where new policies and practices have been resisted until processed by multiple constituencies and then added to their stocks of knowledge. In each case the consent of the governed has been withheld.

Accountability policies have also unwisely partitioned aspects of learning, parenting, teaching, leadership and governance. Partitions employed in the UK have included standards of achievement, curriculum content, participation by parents and managerial responsibilities. This ideological partitioning traded off the benefits of 'partnership' in favour of 'client accountability'. The partitions both valued local market and local political accountability and insulated centralist policy actors from public accountability. The upshot was a legacy of cynicism, simplistic policy dualism's and a partial withdrawal of consent by educators. There were other reasons for the legitimacy crisis, such as stakeholders holding confused priority sets in complex policy structures, that denied the possibility of multiple legitimacies. In sum, it can be speculated that the departitioning of policy knowledge to accommodate pluralism potentially encourages the reconstruction of sectional priorities in an ethos of fraternalism.

Such speculation appears to be warranted given the many converse cases. Policy research cycles, for example, have consistently exhibited positivist assumptions about the growth of policy knowledge in England and Wales. Legislation was employed to install selected values. Curriculum and assessment were nationalized on the theory that when parents exercised choice between schools, schools would be forced to improve. LMS localized accountability through financial, managerial, political and market mechanisms on the theory that this would improve resource efficiency. Unfortunately, neither the Acts nor the accountability mechanisms recognized schools as policy communities comprising well-intentioned people. Legislation did not change parents' partitioned capacity to exercise choice. The localization and commodification of resource accountability partitioned understandings and was found to be divisive, confusing, incoherent and unreliable as a methodology of improvement.

What is also striking is how these experiences of ideological politics and falling systemic coherence in England and Wales were associated with growing levels of responsive forms of accountability in differentiated zones of legitimacy. School communities started learning the arts of local government. They realized the need for greater distributional equity. An unintended consequence of the New Right intervention, therefore, appears to be a communitarian liberalism and growing expertise in local governance. This consequence is more typical of the experience of European social democracies based on stakeholder economies than the Thatcherite theory of economic privatism. It can also be argued that, in a modern and mixed economy, there needs to be a coherent philosophical relationship between the accountability of central government, and the purposes, rationality, legitimacy and accountability of public education. This argument for coherence and pluralism is supported by the evolution of accountability policy in US public education. Conversely, this evolution provided very little support for the proposition that such a coherent and pluralistic relationship could be generated recurrently by a positivist policy research process.

Four positivist assumptions have constrained accountability policy research in the United States for decades; the purposes of schooling imply *instrumental* ends and means, schooling *can* be expressed as an outcomes production function, mandated policies *will* govern behaviours, and policy implementation *requires* a behavioural science of management. The result is that accountability has become a largely bureaucratic response to a political impulse reflecting economic conditions.

The 'first wave' mandates provided conditions that impaired the improvement of teaching and learning. The 'second wave' reforms empowered schools and encouraged improvement but did not review the methods and criteria that should be used to define responsibilities and publicly discharge obligations. The 'third wave' reforms valued 'outcomes accountability' and comparative 'high stakes' mechanisms but largely prevented the evolution of responsive accountability to local stakeholders.

The learned incapacities of educational administrators traced to positivism meant that the epistemic base of the 'education indicators' policy research could not be challenged effectively until more recent times. Indicators research claimed to guarantee empirical facts that might be used to build functional, beneficial, behaviourist, pragmatic, objective, analytic, uncomplicated and responsive accountability subsystems. Simplistic production functions gained in political credibility and were installed as policy. Indicator systems were mandated, structured and legitimated. The actual outcomes included professional control within schools, political or systemic control of external accountability criteria and processes, and concerns about public legitimacy.

State interventions in the late 1980s stressed goals, tighter controls, targeted support and incentive schemes and sanctions. They exhibited positivism, mixed purposes and produced multiple effects and marginal gains. New curriculum and accountability policies raised expectations beyond their capacities to deliver improvement. The systems theories in use lacked the communicative prerequisites needed to boost the external legitimacy of schools. The resultant accountability systems discouraged risk taking, entrepreneurial problem solving and action research. Indicators made purposes, goals, and school missions uncontestable. Standardizing teaching and curriculum discouraged educators from researching children's learning. Political confidence cemented in the ubiquitous tests, indicator systems and school and report cards.

In the early 1990s the states became even more assertive and moved attention away from district test results to the broader quality of school performance. Schools and systems trialed a widening array of evaluation and accountability methodologies. Some were distinctively more educative in intent such as forms of 'authentic assessment' and mastery learning. Some models combined school self-review and external review with an aim of creating a culture of permanent and reflective self-appraisal. The OTL movement reflected a growing desire to make US public schooling more accountable for its degree of fairness. This was partially counter-balanced by the scrapping of compliance regulations (in order to enable school improvement, it was supposed) at immediate cost to equity and probity. Again it can be concluded that the politics of accountability can distort development when governance structures are unable to provide educative accountability policy settlements.

As in England and Wales, it is instructive that policy researchers in the US setting have continued to reiterate serious technical, conceptual and ethical problems with accountability policies and practices. When a major study used educational experts and post-positivist methods to link governance, leadership, teaching, learning and support services, the indicators created were found to have high construct relevance. Since most other indicator systems have been conceptually impoverished and educationally irrelevant, this supports the contention that a good deal of accountability policy research has been epistemically flawed.

Reviews explained that administrators tend to use the easiest strategies available to satisfy mandates and confuse the *presence* of indicators with the need for accountability systems that could *make use of* such information. School community governance, leadership and action research infrastructure, however, remain relatively undeveloped. The orthodox accountability policy of state-wide testing programmes is frozen in place by state political culture. And while accountability policy has a high political profile it has minimal influence. Political appeals to productivity are contested by calls for equity, economy and 'local control,' and soon portrayed by educators as unfair, non-remedial, too costly or authoritarian. Populist concerns target rising costs, 'big government' and 'outcomes-based education'. Professional resistance focuses on the technical-rational language of reform proposals and the controversial role and value of tests. In some states tests are being replaced by more authentic forms of assessment that are based on the taught curriculum and emphasize skill mastery, understanding and application. These examples are exceptions.

Recent changes appear to be intensifying past practices with marginal improvements. Two surges of national political interest were 'finessed' by presidents and state governors using a rich rhetoric of symbolic reform. Scrutiny and reporting intensified with little real change to purposes intended or served. The data and processes used to discharge obligations remain largely unchanged. There was no broad advance towards more moral, responsive and formative forms of accountability. The high mobility of students and teachers made calculations of value adding by schools implausible. The general concern for equity was directed into OTL policy making and more authentic assessment. And while school communities are regarded as part of the problem, they are yet to be widely called by policy actors on to be part of the solution.

It appears that school community governance and leadership infrastructure is needed to construct accountability policy settlements between educators, who prefer fair and reliable processes of showing what valued learning has occurred, and other stakeholders and policy actors, who seek robust criteria and inexpensive data. This need is also seen in the search for networking, reculturing and restructuring strategies, and in the generally poor alignment of accountability purposes, policies, theories and mechanisms. Accountability policy research in seriously disadvantaged communities has also shown that policy-making methodology needs to be far more sensitive to the political, economic and social development of district communities. Instead of methods that inevitably 'prove' 'failure', processes are needed to help school communities develop and redevelop locally shared goals and understandings, and then translate those moving expectations into accountability policies, criteria and practical processes.

It has been demonstrated by exception that positivist methodologies sustain rather than solve the policy problem of accountability. One unusual approach called for more appropriate mixes of accountability types determined by level and by policy issue; a practical example of creating multiple zones of legitimacy. Another suggested guaranteeing equitable access to resources in order to encourage local initiatives, fairer distributions of teaching staff and more customized curriculum and support options. This approach began by accepting pluralism and the need to develop local reconciliation's of the principles of excellence, equity, choice, community and so on. A third approach attempted to transcend the jurisdictions of limited governments in a coherent manner by identifying student-centred educational imperatives that paid balanced attention to the quality of learning, teaching, leadership and governance.

In sum, ten themes were found common in two international case studies of accountability policy making. They have immediate implications for systems of public schools;

- accountability policy clarifies responsibilities, obligations, evaluation and consequences,
- accountability is a major means of sustaining or changing the 'moral economy' of an organization,
- the accountability policy-settlement process is broadly cyclical, philosophical in significance, strategic in scope, political in process, and cultural and pragmatic in effect,
- the accountability policy-settlement process converts diverse public values into workable policy agreements by partially departitioning policy knowledge,
- the accountability policy-settlement process can be impeded when ideological commitments partition knowledge, participation from stakeholders' constituencies is constrained, or structures deny the possibility of multiple zones of legitimacy,

- low systemic accountability policy coherence can permit the development of local responsive accountability, permit differentiated zones of legitimacy, highlight distributional equity, evoke communitarian liberalism and encourage expertise in stakeholder governance,
- the philosophical coherence between the accountability of central government and the purposes, rationality, legitimacy and accountability of public education is moderated by the theory of knowledge used in policy making,
- positivism is associated with instrumentalism, economic rationalism, a behavioural science theory of motivation and educational administration, and government by mandate and bureaucracy,
- a systems theory of organization inhibits communication, the external legitimacy of schools, risk taking, entrepreneurial problem solving and action research, and
- political cultures can inhibit the evaluation of accountability programmes, the development of governance structures, ingratiate serious technical, conceptual and ethical problems, and consign accountability policy to a high profile role principally in symbolic politics.

The Pragmatics of Educative Accountability

The policy touchstone considered worthy of consideration in Tasmanian public schools and at state level was interpreted in various ways. The first response by all stakeholders was to consider the touchstone as comprising a check list that could be used to suggest or affirm 'best practices'. While the language of interpretation and the implications drawn depended on the stakeholder group, the 80 touchstone items were soon regarded as six 'sets of accountability processes and criteria' (see Table 12.7). The items were clustered manually into subgroups and given a descriptive title. Teachers referred to items as 'best practices', subgroups as 'areas of professionalism' and the entire set as 'professional competencies'. Principals spoke of 'best practices', 'performance indicators' and 'key competencies of leadership service'. The differences in language indicated key assumptions about self, role and structure. Parents regarded them as 'common sense' and 'overdue'.

The second general interpretation was that this touchstone should be made available as a resource to school policy communities and stakeholder executives as soon as possible, for their own interpretation and use. There was no discussion about the need to rewrite the current official policy. It was assumed that the ideas would only be used when they were added to the understandings in each community, and that in doing so, the process would help build their policy making and implementation capacities. While the ERU was not able to issue posters and summaries, as agreed, all stakeholder executives, professional associations and many schools were encouraged enough to organize briefings, conference presentations, workshops or focus groups. The DEA also recognized wider audiences and responsibilities when it arranged to present the findings in interstate settings and encouraged international dissemination.

The third general interpretation in each stakeholder executive or school policy committee was the 'good sense' of applying the standards explicit in the touchstone. It was used to evaluate items that had failed to attract support, to set aside policy myths and to affirm a rational and moral basis for practice. The 'common ground' was added to the normative base in many settings and to the standards used by district superintendents for reviewing practices and policies at each school community site. The construction of touchstone can therefore be seen as a process of departitioning policy knowledge about what 'educative accountability' might be. Conversely, the persistent partitions in policy knowledge were found to relate to the structures and priorities established by prior governance structures and practices. This drew attention to the governance values of stakeholders.

It is notable that the research-based proposals for multiple zones of legitimacy in England and Wales, and for issue-based and more authentic accountability policies in the United States, both began with an acknowledgement of pluralism and the need to review the structures of accountability policy governance. Similarly, Tasmanian stakeholders were found to prefer communitarian limited government over representative, centralist or hierarchical governance structures. With parents in the vanguard, they expressed a desire for co-operative politics and greater structural subsidiarity, pluriformity and complimentarity. This suggests that the current formal accountability policy, which displays neo-centralism, 'self-managing' corporate managerialism, uniformity and comparability, is probably becoming obsolete in political and governance terms. The practical implications are immense.

A rewriting of policy in the Australian home of the 'self-managing school' might begin with the following assumptions. Each school policy community is as unique as the people that constitute it. The boundaries of each school community may be defined by its immediate stakeholders; students, teachers, parents, the local community and the state of Tasmania. These stakeholders comprize a policy community with limited rights to self-government. This form of limited government shares sovereignty with state and federal forms of government. Each of these equally valued forms of government requires a democratic constitution that specifies limited and interdependent sovereignty. The limits of sovereignty prescribe a zone of legitimacy.

To avoid the extremes of isolationism and sectarianism, the constitutions of school communities will also need to specify to the nature of relationships with other jurisdictions. The vertical governance relationship between federal, state and school community zones of legitimacy should be marked by respectful subsidiarity rather than neo-centralist hierarchy. The horizontal governance relationship between them should be marked by co-operative complimentarity rather than cool or contrived comparability. Given the special responsibility of public schools to both reproduce and transform the values of a civil society, the political relationships within each school community should be structured to reflect the values of a liberal democracy (Dunleavy and O'Leary 1987).

The biggest shift proposed is away from monism, the belief that there is a unitary state in which one nation or people in an unbroken territory should have one set of laws, values and social relations, towards pluralism. Accepting the pluralistic views of Tasmanian stakeholders will mean guaranteeing pluriformity in organizational structures, celebrating minority rights and tolerance for degrees of separatism, and accepting diverse customs, religious and moral beliefs and habits of association.

An immediate implication in Tasmania is that boards of school community trustees need to replace school councils. School community governance and accountability powers need to be vested in these boards of trustees. While boards of trustees will need to be structured to guarantee a technical majority to students and parents, to acknowledge of their prior constitutional rights concerning education, membership should otherwise exhibit a balanced representation of stakeholders. The rights, responsibilities and obligations of all stakeholders will require constitutional articulation by school communities within general democratic principles provided above and operational guidelines suggested by the sponsoring jurisdiction, Tasmania, without any intimation of superordinacy.

Applying the principles of subsidiarity, complimentarity and pluriformity in such a form of limited government will require significant changes to current forms of accountability. Contractual accountability to employers, parents or designated corporate managers will have to be replaced by locally negotiated mutual accountability. Agreements will have to specify the responsibilities, obligations, evaluation criteria and processes that apply to each group of stakeholders. Similarly, the concept of 'moral' accountability will have to decoupled from clients and recognized as a basis for the justification for all bilateral and co-operative practices between stakeholders. The moral imperatives associated with obligation need to be spelled out by boards of school community trustees and regularly contested. Obligations will need to include codes of professional practice, classroom and staffroom relationships, parenting, leadership and governance. Some of the implications for each different stakeholder group are now clarified.

Educative Accountability for Parents and Students

It will be recalled that the principles in Chapter 12 developed by the ACCSO and APC were strongly supported by the touchstone. They can be adopted. The broader set of perspectives used to create the touchstone also identified additional responsibilities that cohered with these principles. Table 16.1 summarizes the expanded set of accountability roles for the parents and students associated with a school community. These roles would be shared with other stakeholders.

This formidable set of obligations would be best regarded as long term aims in most of the school communities that contributed to the findings presented in Chapter 11. A less daunting approach would involve regarding these obligations as learnable capacities, some of which are already available in school communities in a number of forms. The more appropriate forms will vary by community. Some of the more familiar roles might not be recognized immediately for the significant roles they play in generating accountability in a school community. There are, nevertheless, many examples that can be cited.

Governorship is provided by parents when elected to New Zealand's school boards of trustees alongside elected teachers, community representatives and the principal, *ex officio*. They help negotiate a 'school charter' with central government in order to acquire public funding. The charter is a statement of purposes and strategies. It is used every few years by a central agency as the basis for external evaluation.

REFLECTIONS ON EDUCATIVE ACCOUNTABILITY POLICY

Table 16.1: Educative accountability obligations for parents and students in a school community

Obligation	Specification
Governor	Clarifies and determines purposes in the school community.
Legitimator	Create moral imperatives for classroom, home, professional and leadership improvement.
Researcher	Assists with home, classroom and school action research.
Strategist	Participates in collaborative analyses of the situation and strategic decision making.
Politician	Participates in a responsible and co-operative form of community accountability politics.
Cultural agent	Helps improve the school community's capacities as a learning organization.
Motivator	Helps improve supportive classroom, staffroom and school environments.
Educator	Co-teacher, mentors others and helps peers to take responsibility for their own learning.
Negotiator	Negotiate appropriate indicators of service or performance with other stakeholders.
Manager	Helps acquire and use resources well to achieve school community aims.
Evaluator	Uses summative and formative evaluation re-learning, parenting, teaching and leadership.

A method of parents serving as school community researchers has been pioneered in Rokeby, Tasmania. Suburban parent networks arrange 'kitchen conferences' to discuss school community policy proposals. An example of parents engaging as strategists and politicians was evident at the Tasmanian Council of State Schools Parents and Friends Associations (TCSSP & F) 1996 Annual Conference. The conference theme was 'Blackboard to Keyboard.' The Minister of Education was also invited to launch a policy statement that the TCSSP & F had helped formulate; 'School Council Guidelines'.

Tasmanian students have participated as cultural agents whenever they have negotiated forms of accountability policy in the 'Supportive School Environments Program'. They have acted as educators in the 'Peer Mediation Program'. Parents in NSW have taken up active roles as pre-school educators and evaluators in the 'Parents as Teachers Program'.

The point here is that such mechanisms potentially ameliorate parents' concerns reported in Chapter 12 over exclusionary forms of professionalism, the doubtful reliability of intra-professional processes, and how 'professional development' displaces 'classroom development in a school community context'. Parents and students may also be assured that they share many views with other stakeholder groups and may campaign with some confidence of support. The political accountability touchstone for such campaigning is summarized in Table 16.2.

One notable omission to Table 16.2 is any reference to accountability criteria and processes that relate to being a parent or a student. One major reason is that, like

Table 16.2: Political accountability touchstone among Tasmania's public education stakeholders

Obsolete political touchstone	Current political touchstone
A neo-centralist and unitary concept of 'system'.	Classrooms relationships seen as primary educational structure.
Planning, co-ordination and policy implementation by 'self-managing' corporate managers.	An inclusionary, simultaneous and holistic approach to policy making, planning and implementation.
Comparative assessment of learning, teaching and leadership.	Co-operative mutual accountability featuring formative evaluation of learning, teaching and leadership.
Partitioned curriculum and standardized resource management.	A liberal, communitarian, pragmatic and pluralist philosophy of administration.
Communications within and between stakeholders being mediated by positional authority.	A trustful, supportive and group-based approach to change management.
Incentives being based on political or market devices.	Improvement, accountability and legitimation seen as school community projects.
Hierarchy, uniformity and comparability.	Subsidiarity, pluriformity and complementarity.

governance, they were not associated with 'educative accountability' at the outset of the policy research process. However, since any holistic and coherent school community accountability policy would have to integrate the rights and obligations that attend being a student or a parent, this oversight will need to be rectified in follow-up research.

Educative Accountability for Teachers

It was noted in Chapter 13 that the touchstone created by this research was used by teachers to generate a fresh moral, political and practical mandate they termed 'professionalism'. This interpretation was, in the main, drawn out of a humanist, constructivist and communitarian theory of effective teaching, learning and leadership. In some circumstances, a minority of teachers employed a more individualistic philosophy of teaching while others legitimated their practices using peer norms.

The first of two conclusions is that accountability policies are needed to ensure that teachers move from the feral norms of uncontestable and hyper individualistic 'professionalism' to more collegial norms, and then, to move from a closed peer culture of collegialism towards an open communitarian model that anchors the legitimacy of professional practice in school community development. The second conclusion reached was that the move from hyper individualistic professionalism might be facilitated by collaborative forms of professional activity, and then to more communitarian norms by authentic participation in the co-leadership and cogovernance of school communities. Such an enabling strategy would initially help committed teachers to reconstruct their theories of accountability and professionalism in a supportive peer group, and then in a supportive community context. The enablement strategy proposed is further clarified in Table 16.3 ($\Rightarrow$ means 'ideally develops into').

Another limitation of the research is apparent at this point. While the methodology drew out the ideal of communitarian professionalism, it was also intimated that actual practices were probably clustered around collegialism with a tail extending in both directions. Follow-up research might use semantic differential analysis to measure the extent to which practices differ from the ideal in each stakeholder group. This method would also signal the potential need for intervention.

Table 16.3: Hyper individualistic, collegial and communitarian forms of professional accountability

Professional accountability	Hyper individualistic $\Rightarrow$	Collegial $\Rightarrow$	Communitarian
Main purposes	Report student achievement and improve home support for the teaching program.	Report learning and improve teaching and home	Reports and improves learning, support for the teaching, school program, parenting and school–community leadership and governance.
Appropriate strategies	Professional evaluation and reporting of student's learning.	Peer professional formative evaluation of teaching and learning.	Formative evaluation and mutual accountability of learning, teaching, parenting, leadership and governance.
Source of evaluation criteria standards	An empirical project; students' comparative ability and application.	A collegial project; achievement profiles and negotiated peer	A community policy project; to balance excellence, equity, choice, community …
Appropriate accountability processes	Confidential consultation alone with parent.	Collective methods such as moderation and student/teacher /parent conferencing	Collective methods located within school–community action research.
Strengths	Focus on student progress. Provides implications for supportive parenting.	Focus on supportive classroom and home relationships and on professional development	Openness to feedback and critical evaluation. Community organizational learning.
Limits	Uncontestable norms. Decouples learning from teaching and governance. No feedback implies limited professional development.	Closed intra-collegial norms. Decouples learning and teaching from parenting leadership and governance.	Vulnerability to external and non-educational political agendas.
Appropriate professional development	Co-operative peer evaluation, joint planning and team teaching.	Co-operative internal–external consultancies. Participate in leadership and governance.	Participate in school–community action research and governance.

Table 16.4: Action research and school community development processes

Action research phases objectives, processes, key issues	Teachers' preferred school community development model
1. Define accountability together. Collective mapping of the situation. Plural stakeholder views, interests and values.	Clarify organizational purposes. Create a learning organization. Communicate priorities and values.
2. Describe the policy challenge of accountability in context. Joint specification of key aspects. Antecedents, causes, effects, terms.	Provide collaborative strategic analysis. Link evaluation to planning.
3. Review wider knowledge of policies and practices. Research and search teams commissioned and report. Alternative ideas, policies, trials and findings.	Provide collaborative strategic analysis. Link evaluation to planning.
4. Gather and evaluate consequences of options. Co-operative evaluation of the relative merits of options. Competing explanations and solutions.	Provide collaborative strategic analysis. Link evaluation to planning. Raise commitment
5. Deciding policies and planning practices. Select strategies and joint planning. Feasible? Beneficial? Plausible? Educational?	Make and implement policies. Manage the use of resources. Organize support structures and processes.
6. Action. Teams are organized to accomplish specific tasks. What? Why? Who? When? With? How? Evaluation?	Negotiate performance indicators. Prepare a professional development strategy. Develop supportive classroom environments. Develop a supportive school environment.
7. Evaluate policy outcomes. Collective review and report. History of ideas, intended and unintended effects.	Monitor outcomes and attitudes. Provide feedback and appraisal systems. Develop formative evaluation systems.
1a. Redefine accountability together ... Joint formative evaluation and mapping of the situation ... Newly revealed challenges, views, interests and values ...	Clarify organizational purposes. Create a learning organization. Communicate priorities and values.

It was also shown in Figure 13.1 that teachers conceived of professional development as a cyclical process that reviewed and advanced policy knowledge in an incremental manner. In essence, teachers re-interpreted the manual clustering of touchstone items (see Table 12.7) to illustrate how a school community development model of educative accountability could work sequentially in practice. The similarities between their model of school community development and an action research process in order to create an educative accountability policy (see Table 8.1) are clarified in Table 16.4.

The differences in emphasis concerning the clarification and communication of organizational purposes are again traceable to the extent to which teachers take a collegial or communitarian view of what constitutes 'the school' as a policy community. The similarities reflect the ubiquitous touchstone of educative accountability. A central theme of Table 16.4 is the close relationship between participative learning and the growth of holistic organizational knowledge. Another is that this close relationship coheres with a communitarian theory of mutual accountability.

Educative Accountability for Principals

In Chapter 14 it was shown that the touchstone was interpreted by principals as providing a coherent, moral and educational mandate for particular forms of leadership and governance services. These services were required, it was argued, so that Tasmanian state schools can better review and develop their accountability policies as policy communities. Hence the conclusion reached that the professional development of principals needs to help prepare practitioners in specific skills and understandings concerning leadership and governance. The skills and understandings were identified in Table 14.2 as 'key competencies'. Give the analysis above, they can also be seen to be 'key' in the sense that they are essential to the development of action research processes, communitarian professionalism and mutual accountability in a policy community of stakeholders.

The general response provided by principals was complex and compelling. It co-opted the managerialist language of competencies, reinforced the collective identity of principals and reiterated their moral stewardship as educators. It used the touchstone to supplement and interpret the ideology of national policy that stresses productivity and political accountability. It used familiar metaphors to present a new 'common-sense' policy when touchstone had been added to shared webs of belief. Diverse values on accountability were colonized by and thus reconciled with 'educative accountability'.

On the other hand the principals involved in interpretation were very aware of the massive pragmatic challenges implied by these conclusions. They reported that many of their peers would have to (a) adopt a more collaborative and inclusionary philosophy concerning parental expectations and participation, (b) offer greater openness and responsiveness to alternative evaluation criteria, (c) more actively facilitate critical dialogue between stakeholders, and (d) enhance the quality of appraisal mechanisms and feedback data in many aspects of school community policy making. Accountability processes related to the quality of teaching, leadership, governance and parenting were all regarded as being at a relatively early stage of development, well behind the sophistication of 'teacher development' processes.

Another major challenge identified was the quality of pedagogy. While principals found the criteria and processes discussed in Chapter 14 worthy of close consideration in staffrooms, they also noted the need to involve stakeholders in extra-peer accountability processes in order to raise the level of construct validity, reliability and policy legitimation.

A third key concern was the narrow range of leadership accountability processes in use. They were limited to self-initiated peer appraisal and networking processes, the annual planning cycle and the school review process. The potential leadership development practices discussed in Chapter 14 were regarded as worthy of immediate attention by principals' associations and other stakeholders who valued leadership development in school communities.

In sum, the obligations involved in leadership development clarified by principals traversed four dimensions:

- there are many appraisal and reporting mechanisms available and areas worthy of improvement concerning the quality of leadership services in school communities,
- raising leadership capacities in school communities is essential to improving the quality of teaching, parenting and learning,
- executive teams must accept initial responsibility for developing leadership services in their school community, even in advance of stakeholders' expectations until such time that governance capacities develop, and
- school community government needs to value, plan for, develop and celebrate proactive leadership.

Educative Accountability for Administrators

The interpretation of findings provided by DEA central and district personnel were affected by structural and political turbulence, periods of policy paralysis and tightening constraints on public expenditure. On the other hand, they found it helpful to have an auxiliary research process that added ideas to the ongoing process of policy review.

The core duties of central DEA officials had been modified by a new contractual accountability relationship between government and executive. They were engaged more than ever in 'the business' of government, 'repositioning' the government's service profile, and rationalizing services. Sustaining the professional commitment of colleagues was a major challenge in such circumstances. Another was building the capacities that would enable policy communities to become 'learning organizations'. A third challenge was mediating across the cultural boundaries between public administration and education. A fourth was advocating equity and excellence as equally important educational policy principles.

Whatever the priorities of the wider public service, the DEA valued leadership that celebrated the plurality of Tasmanian society. While it answered regularly according to productivity and political criteria, it also promoted means of developing the philosophical and strategic capacities of school policy communities. Hence, when the DEA had adopted the touchstone as *de facto* policy, and accepted supported items as 'best practices,' four forms of support services were subjected to critical review.

School support services were evaluated in terms of their capacity to (a) promote professional development, (b) facilitate school community development, (c) aggregate needs and (d) raise the effectiveness of delivery systems. Second, testing policies, programmes and practices were reformed. Third, impetus was added to the review of equity policies, programmes and practices. Fourth, myths related to leadership services were identified as impeding school community development:

- responsibility for providing leadership is vested in positions,
- teachers' leadership responsibilities are limited to the classroom,
- parents and students have no leadership responsibilities in schools, and
- 'the school' is a professional community requiring professional leadership.

The DEA took the view that current leadership in schools could build the forms of governance, strategic leadership and planning and change management capacities required to challenge these myths and the minority view valuing 'professional autonomy.' They believed that the concept of mutual accountability would continue to 'build on the high ground.' They repeatedly emphasized a social and communal rationale for accountability, coherence between philosophy and action, and the value of reflective action or Aristotelian *praxis*. Given the limits of collegialism summarized in Table 16.3, they concluded that opportunities for positive co-governance and co-management experiences and access to external perspectives required a much higher priority.

A new three-year school review cycle was designed, therefore, to gain greater coherence between internal and external accountability criteria and processes, to provide supportive privileges to 'under-performing' schools and to collect systemic accountability data that would also provide comparative benchmarks. The model is summarized in Table 16.5.

The research reported above suggests that there are substantial limits to this model while acknowledging that it will constitute a major advance on previous practices. It is not as compelling as it might have been due to the way it has been 'pre-loved' and perhaps 'over-owned' by district superintendents. While it was designed by these 'in-house' experts who did consult extensively, the process traded off the benefits that would have accrued to real stakeholder participation in (say) a Tasmanian Policy Advisory Board of Education. Action research allocates equal political trust in stakeholders, deliberately suspends authority and builds a co-operative policy community as it creates new policy knowledge and commitment to changed structures and practices.

The focus of the model is biased towards the quality of learning and leadership with progressively less emphasis on teaching, parenting and governance. The focus is not on the growth of school community policy knowledge or theories. It favours collecting standardized empirical data and objectified perceptions over building each school community's theory building capacities. It is more positivist that post-positivist in its methodology.

There are other problems that will surface in time. The limited access that district superintendents have to resources suggests that the model will be shown to have a restricted capacity to be remedial. This will tend to compound rather than reconstruct cynicism and individualistic professionalism in some schools. The reliance on hierarchical line management runs counter to the principles of subsidiarity, pluriformity and complementarity valued by Tasmania stakeholders. It can be expected that some communities will find aspects of the model intolerant of local values, diversity and local co-operative policy making. The model could also encounter political controversy whenever it marginalizes key stakeholders. It has already predefined evaluation categories and criteria and will probably partition involvement and knowledge in an arbitrary manner during operation. It has yet to declare its methodology for programme evaluation.

On the other hand, there will be many opportunities for stakeholders to raise their voices as the ERU and district superintendents seek their co-operation in foreshadowed initiatives:

Table 16.5: Tasmania's three-year school review cycle to be introduced in 1997

Year	Activity	Processes, data and criteria
1	Assisted school review	Empirical 'best practice indicators' data on teaching and learning, leadership and management, and professional and staff development.
		Perceptions; evaluative feedback from parents, students and staff.
		Empirical data; examination results, DART scores, key intended numeracy and literacy outcomes data, retention rates and student outcomes.
		Recommendations for action embedded in the school improvement plan which is to be approved by the district superintendent.
		'Under-performing' schools may be subjected to 'next-year audit'.
2	Implement school improvement plan	Provide School Charter, a statement of vision and intent, long-term plan of priorities, activities and schedules, short-term plan of programs and objectives, and an annual report to the District Office and the School Council, Parents and Friends or other representative body.
	Action research evaluation	Define evaluation together, define evaluation in context, review wider knowledge, evaluate consequences of options, select policies and strategies, action, evaluation of outcomes, redefine evaluation.
3	Repeat Year 2	
4	Repeat Year 1	

- designing data collection instruments and methods and negotiate licenses to use and modify survey materials and software,
- creating 'clearing house' infrastructure to share school self-review resources,
- creating a networked accountability information system that will offer site and systemic comparative performance analysis,
- enabling school communities and DEA personnel to contribute to and draw data from a systemic accountability information system,
- enabling school communities to help design, participate in and respond to internal-external evaluations of learning, teaching, leadership, parenting and governance,
- enabling school communities to use action research methods and cycles and planning processes that develop and attend to site-specific priorities, and
- selecting school community development personnel expert in the areas of co-governance with stakeholders, strategic leadership, evaluation, co-planning and change management, in a context of contracting budgets.

Future Research

Two directions are suggested. One project would aim to help Australasian systems and schools to develop more educative accountability policies and practices. Another would aim to better understand the international relationships between power and accountability in education. These agendas are discussed briefly in turn.

Most of the nine devolved systems in Australasia appear to be gradually replacing public, moral, professional, responsive and communitarian forms of accountability with market, political, productivity and contractual mechanisms. Bureaucratic, professional and client views continue to compete unevenly in policy discourse. The importance of creating alternative discourses has never been more urgent. The stakes are high. As argued above, accountability helps determine the 'moral economy' of systems and schools, the only organizations in society expected to both pass on yet improve societies' values. A lot, therefore, hangs on the definitions and assumptions used in accountability policy research.

The study reported above has provided some preliminary reasons for defining an 'educative' accountability policy as an expression of theories about what constitutes valuable knowledge, how learning can be demonstrated and improved and how obligations should be discharged between stakeholders. It has highlighted why the consequences entailed should be handled as community projects.

It has also been shown that substantial practical advances may be made when accountability policies are developed in terms of practical processes (procedures, actions or methods) and criteria (standards, benchmarks or indicators). It has also demonstrated that the processes should be those used to collect data, report on and improve performances and services. The criteria are to be those used to evaluate the quality of learning, teaching, parenting, leadership and governance. In this way, 'educative accountability' can be seen as a school community and system policy project intended to both 'prove' and 'improve' education through the active and critical engagement of stakeholders.

The study above has also demonstrated the potency of policy derived from pragmatic theories; mutual accountability, communitarian professionalism, educative leadership and democratic governance. Habermasian arguments that relate accountability to legitimacy, structure and commitment in complex modern democracies also proved critical. Future research in Australasia might therefore consider methodology that employs the Lakatosian idea that policy knowledge may be built on the touchstone between seemingly incompatible perspectives. It might also use Quine and Ullian's view that policy knowledge exists an improvable web of belief.

This is to suggest that Australasian systems would be well advised to use post-positivist and post-paradigmatic methodology. It might also do better than the study reported above by maintaining a better balance between five conjoint research activities; mapping the theories currently being used to justify policies and practices, building them through the articulation of the touchstone between perspectives, providing constant comparison with other theories, applying empirical tests, and subjecting them to vigorous conjecture and refutation.

Such research need not begin again at the beginning. It might extend and refine the policy options developed in Tasmania for school and state policy communities. Instrumentation, for example, may be moderated by the evolution of touchstone elsewhere. Policy actors may be interviewed and policy documents reviewed in other systems. School stakeholders may be surveyed to provide them each with voice. Reports to policy communities could provide relevant community, state and Australasian profiles to ensure that each policy community has the information needed to maintain its own policy research process.

Such an Australasian study is also advised to adopt a more systematic sample design that used in the study reported above. A three-stage stratified sampling strategy is suggested. First, all nine public school systems in Australasia might be invited to participate. Second, each system might be asked to select randomly five urban and five rural schools in each of four categories of schools; small primary, medium-sized primary, large primary, district high or area, and secondary. With attrition, this should gain a minimum of three schools in each of the ten cells. Third, each school principal might be asked to invite up to ten teachers and ten parents most active in school policy making, and all executive team members, to respond to the instrument. This means that up to 25 people in up to 50 school communities in each of nine systems may respond. Machine-read forms would be necessary since this sampling would yield up to 11,250 responses.

More appropriate data analysis and interpretation methods would be required in such a multi-state study. Item and scale reliability coefficients could be used to verify validity and reliability. Confirmatory factor analysis could validate the structure of the instrument. Analysis of variance could identify significant differences on the ideal scale within and between schools and systems to measure degrees of support for policy proposals. Semantic differentials could be calculated to measure the extent to which practices differ from ideals and the potential need for intervention. Discriminant analysis could identify the groups of items that account for significantly different policy preferences between groups of stakeholders, schools and systems.

Interpretation, as noted above, could be conducted bilaterally with each policy community involved to help build local theory and theory building capacities. Such research could refine iteratively the ninety processes and criteria that Tasmanian stakeholders supported. Their touchstone of 'educative accountability' stressed communitarian values of mutual accountability, co-operative politics and the importance of building governance capacities. The research proposed here could provide feedback to each participating school community and state policy forum. The feedback could be differentiated by current theories in use, with other options and theories attached for consideration.

This means that theory building concerning the nature of 'educative accountability' could advance simultaneously at school, state and project level, moderated by coherence and statistical tests. It might be predicted that policy making would become more epistemically critical in each setting, accountability policies would cohere more closely with evaluation and strategic planning, and that zones of multiple legitimacies would be developed to accommodate the pluralism of each setting and system. Such predictions also indicate the need for a second research project.

Given the effect that governance structures, political processes and political ideologies were found to have on accountability policy development in the United States and in England and Wales, a separate follow-up study is warranted in international settings. The purpose of this project would be to examine how the politics of education are contributing to the reconstruction of accountability policies at site and systemic levels. The international policy context is characterized by changing powers and restructures, conceptual disarray, methodological confusion, multiple 'reform' strategies, blunt instruments, interest group politics, and plural political ideologies. And yet, despite this, accountability strategies and measures are being used increasingly to reform, control and administer educational institutions and systems around the world.

Hence, the objective of the second research project would be to reconnect the policy problem of accountability to learning, teaching, parenting, leading and governing by considering political activity. The primary international audiences for such a project would be policy communities in schools, educational networks and systems, and those engaged in policy studies and research.

Given these ends and audiences, case and policy studies and theoretical reflections would be required to examine the politics and purposes of accountability in a context of pluralism, differentiated powers and contested structures. It would also be important to identify how strategic options and priorities concerning accountability processes and criteria are being generated as the collection and use of performance and service data evolve. It would also be important to understand better how accountability politics help reconstruct norms, beliefs and educational, management and governance practices. This assumes that accountability policy making is intimately linked to the politics of legitimation, access and equity, productivity, choice, excellence and community. The project would encourage research concerning the politics of accountability in classrooms, school communities, districts, regions, states or nations, using group, institutional, systemic and comparative perspectives.

To conclude this section, chapter and text, funding is being sought to pursue the improvement of Australasian educative accountability practices. The latter agenda is already being attended to. Two premier journals, the *International Journal of Educational Research* and *Educational Policy*, are issuing edited collections of accountability policy research in 1996 and 1997. The Politics of Education Association of the American Educational Research Association has commissioned the 1997 Yearbook and it is to be entitled *The Politics of Accountability: Educative and International Perspectives.* Hopefully, these and other projects may further advance the growth of knowledge about educative accountability.

Bibliography

Aitchison, J. (1995). School management in the market place: A secondary school perspective. In A. Macbeth, D. McCreath and J. Aitchison (Eds.) *Collaborate or compete? Educational partnerships in a market economy*, pp. 77–87. (London: Falmer Press).
Allen, G. and Martin, I. (Eds.) (1992). *Education and community: The politics of practice*. (London: Cassell).
Anderson, D. (1993). The interaction of public and private school systems, *Australian Journal of Education*, 36, 2, 213–236.
Angoff, W. H. (1971). *The College Board admissions testing program: A technical report on research and development activities relating to the Scholastic Aptitude Test and Achievement tests.* (New York: College Entrance Examination Board).
Angus, L. (1992). 'Quality' schooling, conservative educational policy, and educational change in Australia. *Journal of Education Policy*, 7, 4, 379–397.
Apple, M. W. (1982). *Education and power*. (Melbourne: Routledge and Kegan Paul).
Archbald, D. (1996). Measuring school choice using indicators, *Educational Policy*, 10, 1, 88–108.
Argyris, C. and Schön, D. A. (1978). *Organizational learning: A theory of action perspective*. (Reading, MA: Addison-Wesley).
Arrow, J. J. (1950). A difficulty in the concept of social welfare, *Journal of Political Economy*, 58, 328–346.
Aspin, D., Chapman J. D. and Wilkinson, V. (1994). *Quality schooling*. (London: Cassell).
Australian Council of State School Organisations and Australian Parents Council (1996). *Assessing and reporting student achievement: A report of the national parent consensus.* (Canberra: DEET).
Barber, C. (1992). Accountability in Systems of Government Schools. Unpublished BEd(Hons) thesis, University of Tasmania.
Barber, M. and Fuller, P. (1995). *Inspection quality 1994/1995*. (London: OFSTED).
Bates, R. J. (1980). Perspective: Educational administration, the sociology of science and the management of knowledge, *Educational Administration Quarterly*, 16, 2, 1–20.
Bates, R. J. (1982). Toward a critical practice of educational administration, *Studies in Educational Administration*, 27, 1–15.
Bates, R. J. (1983). *Educational administration and the management of knowledge*. (Geelong, Victoria: Deakin University Press).
Bates, R. J. (1988). Is there a new paradigm in educational administration? Paper given to the Special Interest Group on Organisation Theory, American Educational Research Association, 5–10 April, New Orleans.

Bates, R. J. (1990). Leadership and the rationalization of society. Paper given to the Special Interest Group on Organization Theory, American Educational Research Association, 16–20 April, Boston.

Bates, R. J. (1994). The bird that sets itself on fire: Thomas Greenfield and the renewal of educational administration. Paper presented to the 8th International Intervisitation Program, Ontario Institute for Studies in Education, May, Toronto.

Beare, H. (1995). New patterns for managing schools and school systems. In C. W. Evers and J. D. Chapman (Eds.) *Educational administration: An Australian perspective*, pp. 132–152. (Melbourne: Allen and Unwin).

Beare, H. and Boyd, W. L. (Eds.) (1993). *Restructuring schools: An international perspective on the movement to transform the control and performance of schools.* (London: Edward Arnold).

Beare, H. Caldwell, B. J. and Millikan, R. H. (1989). *Creating an excellent school.* (London: Routledge).

Becher, T. (1979). Self-accounting, evaluation and accountability. *Educational Analysis*, 1, 1, 63–65.

Becher, T. and Eraut, M. R. (1977). *Accountability in the middle years of schooling.* (Swindon: SSRC).

Becher, T. and Maclure, S. (Eds.) (1978). *Accountability in education.* (Windsor, Berkshire: NFER).

Belsey, A. (1986). The New Right, social order and civil liberties. In R. Levitas (Ed.) *The ideology of the New Right*, pp. 26–38. (Cambridge: Polity Press).

Bender, F. (Ed.) (1970). *Marx. The essential writings.* (New York: Harper and Row).

Berrell, M. M. and Macpherson, R. J. S. (1995). Educational sociology and educational administration: Parallel problems with paradigms, epistemology, research and theory, *Educational Administration and Foundations*, 10, 1, 9–32.

Berry, B. and Ginsberg, R. (1996). Accountability, school reform, and equity: The troubling case of Sylvan school district one. Paper given to the Annual Meeting of the American Educational Research Association, 8–12 April, New York.

Blackmore, J. (1992). *Making educational history: A feminist perspective.* (Geelong: Deakin University Press).

Blank, R. K. and Schilder, D. (1991). State policies and state role in curriculum. In S. H. Fuhrman and B. Malen, *The politics of curriculum and testing*, pp. 37–62. (London: Falmer).

Bolton, E. (1993). Imaginary gardens with real toads. In C. Chitty and B. Simon (Eds.) *Education answers back: Critical responses to government policy*, pp. 3–16. (London: Lawrence and Wishart).

Boston, K. (1995). The structure and organization of the Department of School Education. Memorandum to all staff. (Sydney: New South Wales Department of Education, 10 August).

Bowles, S. and Gintis, H. (1976). *Schooling in capitalist America.* (New York: Basic Books).

Boyd, R. N. (1991). On the current status of scientific realism. In R. Boyd, P. Gasper and J. D. Trout (Eds.) *The philosophy of science*, pp. 195–222. (Boston, MA: MIT Press).

Boyd, W. L. (1979). The changing politics of curriculum policy-making for American schools. In J. Schaffarisk and G. Sykes (Eds.) *Value conflicts and curriculum issues: Lessons from research and experience*, pp. 73–138. (Berkerley: McCutcheon).

Boyd, W. L. and Halpin, D. (1994). National 'systemic' school reform: Lessons from the British experience. American Educational Research Association's Politics of Education Association, School District of Philadelphia and Temple University's School of Education Conference, Temple University, 27–28 October.

Boyd, W. L., Lugg, C. and Zahorchak, G. I. (1996). Social traditionalists, religious conservatives and the politics of outcomes-based education: Pennsylvania and beyond, *Education and Urban Society*, 28, 3, May, 347–365.

Brighouse, T. (1995). Competition, Devolution, Choice and Accountability: An Education Authority view of the need for diversity and equality. In A. Macbeth, D. McCreath and J. Aitchison (Eds.) *Collaborate or compete? Educational partnerships in a market economy,* pp. 36–47. (London: Falmer Press).
Browne, S. (1979). The accountability of HM Inspectorate. In J. Lello (Ed.) *Accountability,* pp. 35–44. (London: Ward Lock).
Burgess, T. (1992). Accountability with confidence. In T. Burgess (Ed.) *Accountability in schools,* pp. 3–14. (Essex: Longman).
Burgher, W. and Duckett, W. (1992). *Polling attitudes of community on education (PACE) manual.* (Bloomington, IN: Phi Delta Kappan).
Burns, R. (1994). *Introduction to research methods.* (Melbourne: Longman).
Burrell, G. and Morgan, G. (1981). *Sociological paradigm and organisational analysis: Elements of the sociology of corporate life,* Second edition. (London: Heinemann).
Bush, T., Kogan, M. and Lenney, T. (1989). *Directors of education—Facing reform.* (London: Jessica Kingsley).
Caldwell, B. J. and Spinks, J. M. (1988). *The self-managing school.* (Lewes: Falmer).
Caldwell, B. J. and Spinks, J. M. (1992). *Leading the self-managing school.* (Lewes: Falmer).
Callaghan, J. (1987). *Time and chance.* (London: Collins).
Callahan, R. E. (1962). *Education and the cult of efficiency.* (Chicago: University of Chicago Press).
Campbell, R. F., Fleming, T., Newell, L. J. and Bennion, J. W. (1987). *A history of thought and practice in Educational Administration.* (New York: Teachers College Press).
Carnegie Forum on Education and the Economy (1986). *A nation prepared: Teachers for the 21st century,* (Washington, DC: Carnegie).
Casey, P. M. (1993). Subsidiarity, collegiality and consensus: Systems within the Australian Catholic school sector. In A. Walker, R. J. S. Macpherson and C. Dimmock (Eds.) *Framing Research in Educational Administration,* pp. 166–179. (Melbourne: ACEA).
Chapman, J. (Ed.) (1990). *School-based decision making and management.* (London: Falmer).
Chapman, J. and Dunstan, J. F. (Eds.) (1991). *Democracy and bureaucracy: Tensions in public schooling.* (London: Falmer).
Chapman, J. D. (1995). Set the signals at green! The William Walker oration, *Journal of Educational Administration,* **1,** 1, 4–21.
Charlton, K. (1965). *Education in renaissance England.* (London: Routledge and Kegan Paul).
Chin, R. and Benne, K. D. (1976). General strategies for effecting changes in human systems. In W. Bennis (Ed.) *The planning of change,* pp. 67–86. (New York: Holt, Rinehart and Winston).
Chitty, C. (1989). *Towards a new education system: The victory of the New Right?* (Lewes: Falmer Press).
Chitty, C. (1994). Consensus to conflict: the structure of educational decision-making transformed. In D. Scott (Ed.) *Accountability and control in educational settings,* pp. 8–31. (London: Cassell).
Churchland, P. S. (1985). The ontological status of observables: In praise of the super-empirical virtues. In P. Churchland and C. Hooker (Eds.) *Images of Science.* (Chicago: University of Chicago Press).
Cibulka, J. G. (1991). Educational accountability reforms: Performance information and political power. In S. Fuhrman and B. Malen (Eds.) *The politics of curriculum and testing,* pp. 181–201. (London: Falmer).
Cibulka, J. G. and Derlin, R. L. (1995). State educational performance reporting policies in the US: Accountabilities many faces, *International Journal of Educational Research,* **23,** 6, 479–492.

Cibulka, J. G. and Derlin, R. L. (1996). Accountability reforms and systemic initiatives in two states: From policy adoption to policy sustainability. Paper given to the Annual Meeting of the American Educational Research Association, 8–12 April, New York.

Cibulka, J. G., Reed, R. J. and Wong, K. K (Eds.) (1992). *The politics of urban education in the United States.* (Washington, DC: Falmer).

Ciccarelli, A. M. (1996). Towards a new discourse of educational organisations: Theorising educational organisations from a critical-pragmatist perspective. Unpublished EdD dissertation, University of Sydney.

Clark, B. (1996). Personal correspondence and interview. Department of School Education, Sydney, NSW, April 4.

Clark, D. and Astuto, T. (1994). Redirecting reform: Challenges to popular assumptions about teachers and schools. *Phi Delta Kappan*, March, 513–520.

Clune, W. (1993). The best path to systemic educational policy: Standard/centralized of differentiated/decentralized, *Educational Evaluation and Policy Analysis*, 15, 3, 233–254.

Clune, W. (1994). The shift from equity to adequacy in school finance, *Educational Policy*, 8, 4, 376–395.

Codd, J. and Gordon, L. (1991). School charters: The contractualist state and education policy, *New Zealand Journal of Educational Studies*, 26, 1, 21–34.

Cohen, A. (1987). Instructional alignment: Searching for the magic bullet, *Educational Researcher*, 16, 8, 16–20.

Cohen, D. K. (1995). What is the system in systemic reform? *Educational Researcher*, 24, 9, 11–19.

Coleman, J. S., Campbell, E. Q., Hobson, C. J., McPartland, J., Mood, A. M. and Weinfield, F. D. (1966). *Equality of educational opportunity.* (Washington, DC: US Government Printing Office).

Cooley, W. W. and Bernauer, J. A. (1991). School comparisons in state-wide testing programs. In R.E. Stake (Ed.) *Advances in program evaluation.* (Greenwich, CT: JAI Press).

Coopers & Lybrand Ltd (1988). *Local management of schools: A report to the Department of Education and Science.* (London: HMSO).

Coordingley, P. and Wilby, P. (1987). *Opting out of Mr. Baker's proposals.* Ginger Paper 1. (London: Education Reform Group).

Corbett, H. D. and Wilson, B. L. (1993). *Testing, reform, and rebellion.* (Norwood, NJ: Ablex).

Cresap Ltd. (1990). *Cresap's final report: Review of the Department of Education and the Arts.* (Hobart, Tasmania: Cresap).

Cronbach, L. G., Ambron, S. R., Dornbush, S. M., Hess, R. D., Hornik, R. C. and Phillips, D. C. (1980). *Towards reform of program evaluation: Aims, methods, and institutional arrangements.* (San Francisco: Jossey Bass).

Crowson, R. and Porter-Gehrie, C. (1980). The discretionary behaviour of principals in large-city schools, *Educational Administration Quarterly*, 16, 1, 45–68.

Crowson, R. L., Boyd, W. L. and Mawhinney, H. B. (Eds.) (1996). *The politics of education and the new institutionalism: Reinventing the American school.* (London: Falmer Press).

Crowther, F. and Gibson, I. (1990). Research as problem discovery: Discovering self-as-researcher through naturalistic inquiry. In R. J. S. Macpherson and J. Weeks (Eds.) *Pathways to knowledge in educational administration: Methodologies and research in progress in Australia*, pp. 39–48. (Armidale, NSW: ACEA).

Crowther, F. and Ogilvie, D. (Eds.) (1992). *The new political world of educational administration.* Melbourne: Australia Council of Educational Administration.

Crump, S. J. (1992). Pragmatic policy development: Problems and solutions in educational policy making, *Journal of Educational policy*, 5, 4, 1–15.

Crump, S. J. (1995). Towards action and power: Post-Enlightenment pragmatism? *Discourse: Studies in the Cultural Politics of Education*, 16, 2, 203–217.
Cuban, L. (1986). Principaling: Images and roles, *Peabody Journal of Education*, 63, 1, 107–119.
Cuban, L. (1988). *The managerial imperative and the practice of leadership in schools.* (Albany: State University Press of New York).
Culbertson, J. A. (1981). The antecedents of the theory movement, *Educational Administration Quarterly*, 17, 1, 25–47.
Culbertson, J. A. (1988). A century's quest for knowledge base. In N. J. Boyan (Ed.) *Handbook of research in educational administration*, pp. 3–24. (New York: Longman).
Cusack, B. O. (1993). Political engagement in the restructured school: The New Zealand experience, *Educational Management and Administration*, 21, 2, 107–114.
Cuttance, P. (1995). Quality assurance and quality management in education systems. In C. W. Evers and J. D. Chapman (Eds.) *Educational administration: An Australian perspective*, pp. 296–316. (Melbourne: Allen and Unwin).
Darling-Hammond, L. and Ascher, C. (1992). *Creating accountability in big city schools.* (Columbia, NY: ERIC and NCREST).
Darling-Hammond, L. and Snyder, J. (1992). Curriculum inquiry. In P. Jackson (Ed.) *Handbook of research on curriculum*, (New York: MacMillan).
Darling-Hammond, L. and Snyder, J. (1993). Accountability and the changing context of teaching. In L. Darling-Hammond, J. Snyder, J. Ancess, L. Einbender, A. L. Goodwin and M. B. Macdonald (Eds.) *Creating learner-centred accountability*, pp. 1–20. (Teachers College, NY: NCREST).
Darling-Hammond, L. (1989). Accountability for professional practice, *Teachers College Record*, 91, Fall, 59–80.
Darling-Hammond, L. (1993). Reframing the school reform agenda: Developing capacity for school transformation, *Phi Delta Kappan*, June, 753–761.
Darling-Hammond, L., Snyder, J., Ancess, J., Einbender, L., Goodwin, A. L. and Macdonald, M. B. (Eds.) (1993). *Creating learner-centred accountability.* (Teachers College, NY: NCREST).
David, M. (1989). Education. In M. McCarthy (Ed.) *The new politics of welfare.* (London: Macmillan).
Davies, H. (1993). Educative accountability: The perspective from the districts. Unpublished BEd Project, University of Tasmania.
Davis, B. (1993). Establishment of a Review Unit. Circular memorandum to principals and teachers-in-charge of all schools and colleges, Hobart: Department of Education and the Arts, 29 March.
Dearing, R. (1993). *The national curriculum and its assessment.* Interim Report (London: SEAC and NCC).
Deem, R. (1994). School governing bodies: public concerns and private interests. In D. Scott (Ed.) *Accountability and control in educational settings*, pp. 58–72. (London: Cassell).
Department of Education and Science (1981). *The school curriculum.* (London: HMSO).
Dewey, J. (1916). *Democracy and education.* (New York: Macmillan).
Dewey, J. (1938). *Logic: The theory of inquiry.* (New York: Henry Holt).
Dewey, J. (1963). *Experience and education.* (London: Collier-Macmillan).
Donlan, T. (Ed.) (1984). *The College Board technical handbook for the Scholastic Aptitude Test and Assessment Tests.* (New York: College Entrance Examination Board).
Donoughue, B. (1987). *Prime Minister: The conduct of policy under Harold Wilson and James Callaghan.* (London: Jonathan Cape).
Dornun, J., Jenkins, K. and Berry, B. (1995). *Barriers to reform.* (Greensboro, NC: Southeastern Regional Vision for Education).

Duignan, P. A. and Bhindi, N. (1995). A quest for authentic leadership. Paper given to the British Educational Management and Administration Society Annual Conference, Balliol College, Oxford, September.

Duignan, P. A. and Macpherson, R. J. S. (1991). *Educative leadership for the corporate managerialist world of educational administrators and managers.* (Melbourne: Australian Council of Educational Administration).

Duignan, P. A. and Macpherson, R. J. S. (1993). Educative leadership: A practical theory, *Educational Administration Quarterly*, 29, 1, 8–33.

Duignan, P. A. and Macpherson, R. J. S. (Eds.) (1992). *Educative leadership: A practical theory for new administrators and managers.* (Basingstoke: Falmer Press).

Dunleavy, P. and O'Leary, B. (1987). *Theories of state: The politics of liberal democracy.* (Basingstoke: Macmillan).

Earley, P. (1994). *School governing bodies: Making progress?* (Berkshire: NFER).

East Sussex Accountability Project (1980). Accountability in the middle years of schooling. Final report to the Social Science Research Council, Part 1. Brighton: University of Sussex.

Elam, S. M. (1990). The 22nd annual Gallup poll of the public's attitudes towards the public schools. *Phi Delta Kappan*, 72, 1, 41–55.

Elam, S. M., Rose. L. C. and Gallup, A. M. (1993). The 25th annual Phi Delta Kappan/Gallup polls of the public's attitudes towards the public schools, *Phi Delta Kappan*, 57, 2, 137–152.

Elliott, J., Bridges, D., Ebbutt, D., Gibson, R. and Nias, J. (1981). *School accountability: The SSRC Cambridge accountability project.* (London: Grant McIntyre).

Elmore, R. F. (1994). Thoughts on program equity: Productivity and incentives for performance in education, *Educational Policy*, 8, 4, 453–459.

Elmore, R. F. (1995). Structural reform in educational practice, *Educational Researcher*, 24, 9, 23–26.

Elmore, R. F. and Associates (1990). *Restructuring schools: The next generation of educational reform.* (San Francisco: Jossey Bass).

Eraut, M. (1978). Accountability at school level. In T. Becher and S. Maclure (Eds.) *Accountability in education*, pp. 152–199. (Windsor, UK: NFER).

Evers, C. W. and Lakomski, G. (1991). *Knowing educational administration: Contemporary methodological controversies in educational administration research.* (Oxford: Pergamon Press).

Evers, C. W. (1979). Analytic philosophy of education: From a logical point of view, *Educational Philosophy and Theory*, 11, 2, 1–16.

Evers, C. W. (1984). Epistemology and justification: From classical foundationalism to Quinean coherentism and materialist pragmatism. In C. W. Evers and J. C. Walker (Eds.) *Epistemology, semantics, and educational theory.* (Sydney: Department of Education, University of Sydney).

Evers, C. W. (1987a). Philosophical research in educational administration. In R. J. S. Macpherson (Ed.) *Ways and meanings of research in educational administration*, pp. 53–78. (Armidale: University of New England Press).

Evers, C. W. (1987b). Ethics and ethical theory in educative leadership—a pragmatic and holistic approach. In C. W. Evers (Ed.) *Moral theory for educative leadership*, pp. 3–30. (Melbourne: Ministry of Education).

Evers, C. W. (1988). Educational administration and the new philosophy of science, *Journal of Educational Administration*, 26, 1, 3–22.

Evers, C. W. in association with Dillon, D., Duignan, P. A., Long, M., Macpherson, R. J. S., Norman, M. and Williams, J. (1987). Ethics and ethical theory in educative leadership—A pragmatic and holistic approach. In C. W. Evers (Ed.) *Moral theory for educative leadership*, pp. 3–30. (Melbourne: Ministry of Education).

Ewington, J. (1996). Parents' perceptions of school effectiveness: An investigation into parents' perceptions of the effectiveness of Tasmanian public schools. Unpublished PhD thesis, University of Tasmania.
Falk, I. (1995) A reconceptualisation of learning: Why put critical literacy into practice? *Critical Forum*, 3, 2 and 3, 64–81.
Feyerabend, P. (1975). *Against method: Outlines of an anarchistic theory of knowledge*. (London: New Left).
Fidler, B., Earley, P. and Ouston, J. (1995). OFSTED inspections and their impact on school development. Paper presented to the European Conference on Educational Research, University of Bath, 14–17 September.
Finn, C. E. Jr. (1991). *We must take charge: Our schools and our future*. (New York: The Free Press).
Foster, W. (1986a). *The reconstruction of leadership*. (Geelong: Deakin University Press).
Foster, W. (1986b). *Paradigms and promises: New approaches to educational administration*. (Buffalo: Prometheus).
Fowler, G. (1978). Participation in educational government. In *Participation and accountability, E361 Education and the urban environment (Block IV)*. (Milton Keynes: The Open University).
Frazer, M., Dunstan, J. and Creed, P. (1988). *Perspectives on organizational change: Lessons from education*. (Melbourne: Longman Cheshire).
French, R. L. and Bobbett, G. (1995). A detailed analysis of report cards on schools produced in eight eastern states and a synthesis of report card studies in nineteen states. Paper given to the Annual Meeting of the American Educational Research Association, 18–22 April, San Francisco, CA.
Fuhrman, S. H. (Ed.) (1993). *Designing coherent education policy: Improving the system*. (San Francisco: Jossey Bass).
Fuhrman, S. H. and Malen, B. (Eds.) (1991). *The politics of curriculum and testing*. (London: Falmer).
Fullan, M. G. (1991). *The new meaning of educational change*. (New York: Teachers College Press).
Fullan, M. G. (1993). Coordinated school and district development in restructuring. In J. Murphy and P. Hallinger, *Restructuring schooling: Learning from ongoing efforts*, pp. 143–164. (Newbury Park, CA: Corwin).
Fullan, M. G. (1996). Turning systemic thinking on its head. *Phi Delta Kappan*, 77, 6, February, 420–423.
Gallacher, N. (1995). Partnerships in education. In A. Macbeth, D. McCreath and J. Aitchison (Eds.) *Collaborate or compete? Educational partnerships in a market economy*, pp. 16–24. (London: Falmer Press).
Galton, M. (1995). *Crisis in the classroom*, (London: David Fulton).
Galtung, J. (1967). *Theory and methods of social research*. (New York: Columbia University Press).
Gee, J. (1990). *Social linguistics and literacies: Ideology in discourses*. (London: Falmer).
Giddens, A. (1979). *Central problems in social theory* (Chapter 2). (London: Macmillan).
Giddens, A. (1984). *The constitution of society: Outline of the theory of structuration*. (Berkeley and Los Angeles, CA: University of California Press).
Glascock, C. H., Franklin, B. J., Hutsell, W. W. and Byram, G. H. (1995). A national survey of all fifty state departments of education comparing educational indicators reports and data verification techniques. Paper given to the Annual Meeting of the American Educational Research Association, 18–22 April, San Francisco, CA.
Glass, G. V. (1972) The many faces of educational accountability, *Phi Delta Kappan*, 53, June, 636–639.

Glatter, R. (1995). Partnership in the market model: Is it dying? In A. Macbeth, D. McCreath and J. Aitchison (Eds.) *Collaborate or compete? Educational partnerships in a market economy*, pp. 25-35. (London: Falmer Press).
Goddard, D. (1992). Evaluation for improvement. In T. Burgess (Ed.) *Accountability in schools*, pp. 75-88. (Harlow: Longman).
Goertz, M. (1986). *State educational standards.* (Princeton: Educational Testing Service).
Gold, K. (1996). Opted out, run out of options. *The Observer*, 24 March, p. 9.
Goldstein, H. (1979). Consequences of using the Rasch model for educational assessment. *British Educational Research Journal*, 5, 2, 211-220.
Goldstein, H. (1980). Dimensionality, bias, independence and measurement scale problems in latent trait score models. *British Journal of Mathematical and Statistical Psychology*, 33, 2, 234-246.
Golensky, M. (1993). The board-executive relationship in non-profit organizations: Partnership or power struggle. *Non-profit Management and Leadership*, 4, 2, 177-191.
Goodlad, J. I. (1984). *A place called school.* (New York: McGraw-Hill).
Gordon, L. (1995). Controlling education: Agency theory and the reformation of New Zealand schools, *Educational Policy*, 9, 1, 54-74.
Government of Australia, Department of Prime Minister and Cabinet (1994). *The working nation.* (Canberra: Commonwealth Department of Prime Minister and Cabinet).
Government of Great Britain, Department for Education (1994). *The parent's charter: Publication of information about secondary schools.* Circular 14/94. (London: Department for Education).
Government of Great Britain, Department of Education and Science (1975). *A language for life.* The Bullock Report. (London: HMSO).
Government of Great Britain, Department of Education and Science (1977). *Education in schools: A consultative document.* Green Paper. (London: HMSO).
Government of Great Britain, Department of Education and Science (1988a). *Task group on assessment and testing: A report.* (London: HMSO).
Government of Great Britain, Department of Education and Science (1988b). *Three supplementary reports to the TGAT.* (London: HMSO).
Government of Great Britain, Department of Education and Science/Welsh Office (1977). *A new partnership for our schools.* Chairman W. Taylor. (London: HMSO).
Government of Great Britain, Office for Standards in Education (1992). *Framework for the inspection of schools.* (London: HMSO).
Government of Great Britain, Office for Standards in Education (1993). *Handbook for the inspection of schools.* (London: HMSO).
Government of Great Britain, Office for Standards in Education (1994). *A focus on quality.* (London: HMSO).
Government of Great Britain, Parliament, House of Commons (1981). *The secondary school curriculum and examinations: with special reference to the 14 to 16 year old age group.* Second report from the Education, Science and Arts Committee. (London: HMSO).
Government of Great Britain, Parliament, House of Commons. (1991). *Education (Schools) Bill.* (London: HMSO).
Government of Great Britain, Statutes. (1988). *Education Reform Act 1988.* Chapter 40. (London: HMSO).
Government of Great Britain, Statutes. (1992). *Education (Schools) Act 1992.* Chapter 38. (London: HMSO).
Government of New Zealand, Taskforce to Review Education Administration (1988). *Administering for excellence: Effective administration in education.* Chairman Brian Picot. Wellington, (New Zealand: Government Printer).

Government of Tasmania (1981) *Review of Tasmanian government administration. Report of phase II of the review.* Chairman Sir George Cartland. (Hobart: Government Printer).
Government of Tasmania (1982). *Review of efficiency and effectiveness of the Education Department.* Report by Phillip Hughes. (Hobart: Government Printer).
Government of Tasmania, Department of Education and the Arts (1984). *Secondary education and the future.* (Hobart: Department of Education and the Arts).
Government of Tasmania, Department of Education and the Arts (1991a). *Our children and the future.* (Hobart: Department of Education and the Arts).
Government of Tasmania, Department of Education and the Arts (1991b). *Policy statement: Staff appraisal.* (Hobart: Department of Education and the Arts).
Government of Tasmania, Department of Education and the Arts (1991c). *School and college councils: Interim guidelines.* (Hobart: Department of Education and the Arts).
Government of Tasmania, Department of Education and the Arts (1993a). *Accountability policy for the Division of Education.* (Hobart: Department of Education and the Arts).
Government of Tasmania, Department of Education and the Arts (1993b). *Local school leadership and management.* (Hobart: Department of Education and the Arts).
Government of Tasmania, Department of Education and the Arts (1993c). *School planning 1994.* (Hobart: Department of Education and the Arts).
Government of Tasmania, Department of Education and the Arts (1993d). *School resource package.* (Hobart: Department of Education and the Arts).
Government of Tasmania, Department of Education and the Arts (1993e). *Framework for curriculum provision K-12.* (Hobart: Department of Education and the Arts).
Government of Tasmania, Department of Education and the Arts (1995). *Interim guidelines for school and college councils.* (Hobart: Department of Education and the Arts).
Government of Tasmania, Department of Education and the Arts (1996). *School plan implementation review 1996.* (Hobart: Department of Education and the Arts).
Government of Tasmania, Department of Education and the Arts, Bowen District (1996). *Bowen District Office: 1995 Report.* (Rosney, Tasmania: Department of Education and the Arts, April).
Government of Tasmania, Statutes (1994). *Education Act 1994.* (Hobart: Government Printers).
Government of Tasmania, Tasmanian Education Council (1993). *Report to the Minister for Education and the Arts on Reporting to Parents.* Chairperson Roy Swain. (Hobart: Tasmanian Education Council, September).
Government of the United States of America, Department of Education, Special Study Panel on Education Indicators (1991). *Education Counts: An indicator system to monitor the nation's educational health.* (Washington, DC: National Center for Educational Statistics).
Government of the United States of America, Department of Education, National Center for Education Statistics (1989). *The condition of education.* (Washington, DC: US Department of Education).
Government of the United States of America, Department of Education, National Commission on Excellence in Education (1983). *A nation at risk: Imperative for reform.* (Washington, DC: US Department of Education).
Government of the United States of America, Department of Education (1994). *Goals 2000: Educate America Act.* (Washington, DC: US Department of Education).
Grace, G. (1995). *School leadership: Beyond education management. An essay in policy scholarship.* (London: Falmer).
Grady, N., Macpherson, R. J. S., Mulford, W. R. and Williamson, J. (1994). *Australian school principals: Profile 1994.* Report to the Australian Principals Associations Professional Development Council, May. (University of Tasmania: School of Education).

Gray, J. (1995). *After social democracy.* (London: Demos).

Gray, J. and Wilcox, B. (1994). Performance indicators: Flourish or perish? In K.A. Riley and D. L. Nuttall (Eds.) *Measuring quality: Education indicators—United Kingdom and international perspectives,* pp. 69–86. (London: Falmer Press).

Gray, J. and Wilcox, B. (1995). *Good school, bad school: Evaluating performance and encouraging improvement.* (Buckingham: Open University Press).

Greenfield, T. and Ribbins, P. (1993). *Greenfield on Educational Administration: Towards a humane science.* (London: Routledge).

Greenfield, T. (1975). Theory about organisations: A new perspective and its implications for schools. In M.G. Hughes (Ed.) *Administering education.* (London: Althone Press).

Greenfield, T. (1978). Reflections on organization theory and the truths of irreconcilable realities, *Educational Administration Quarterly,* **14,** 2, 1–23.

Greenfield, T. (1979a). Ideas versus data: How can the data speak for themselves? In G. Immegart and W. Boyd (Eds.) *Problem-finding in Educational Administration: Trends in research and theory,* pp. 167–190. (Lexington: Lexington Books).

Greenfield, T. (1979b). Organisation theory as ideology. *Curriculum Inquiry,* **9,** 2, 97–112.

Greenfield, T. (1980). The man who comes back through the door in the wall: Discovering Truth, discovering self, discovering organisations, *Educational Administration Quarterly,* **16,** 3, 26–59.

Greenfield, T. (1988). The decline and fall of science in educational administration. In D. E. Griffiths, R. T. Stout and P. B. Forsyth (Eds.) *Leaders for America's schools: The report and papers of the National Commission on Excellence in Educational Administration.* (Berkeley: McCutchan).

Griffith, J. E. (1990). Indicators and accountability in the USA. In T. J. Wyatt and A. Ruby (Eds.) *Education indicators for quality, accountability and better practice: Papers from the [second] national conference on indicators and quality in education,* (Sydney: Australian Conference of Directors-General of Education, 1–13).

Griffiths, D. E. (1957). Towards a theory of administrative behaviour. In R. F. Campbell and R. Gregg (Eds.) *Administrative behavior in education,* pp. 354–390. (New York: Harper).

Griffiths, D. E. (1983). Evolution in research and theory: A study of prominent researchers, *Educational Administration Quarterly,* **19,** 3, 201–221.

Griffiths, D. E. (1986). Can there be a science of organization? In G. S. Johnston and C. C. Yeakey (Eds.) *Research and thought in administrative theory: Developments in the field of Educational Administration,* pp. 133–143. (Lanham: University Press of America).

Griffiths, D. E. (1988). Administrative theory. In N. J. Boyan (Ed.) *Handbook on research in Educational Administration,* pp. 27–51. (New York: Longman).

Griffiths, D. E. (1993). School administrators and the educational reform movement in the United States. In Y. M. Martin and R. J. S. Macpherson (Eds.) *Restructuring administrative policy in public schooling: Canadian and international case studies,* pp. 35–55. (Calgary, Alberta: Detselig).

Griffiths, D. E. (1995). Book review—Greenfield on Educational Administration: Towards a humane science. *Educational Administration Quarterly,* **31,** 1, 151–158.

Griffiths, D. E., Stout, R. T. and Forsyth, P. B. (Eds.) (1988). *Leaders for America's schools: The report and papers of the National Commission on Excellence in Educational Administration.* (Berkeley: McCutchan).

Griggs, C. (1989). The New Right and English secondary education. In Lowe, R. (Ed.) *The changing secondary school,* pp. 99–128. (Lewes: Falmer Press).

Grimshaw, W. A. (1990). Legislating for education quality and the New South Wales strategy. Paper given to The Education Summit, Sydney, 13–14 November.

Guba, E. (Ed.) (1990). *The paradigm dialogue.* (Newbury Park: Sage).
Guthrie, J. W. (1989). The evolving political economy of education and the implications of educational evaluation. Paper given to the British Educational Management and Administration Society, Leicester University, 15–17 September.
Guthrie, J. W., Hayward, G. C., Kirst, M. W. and Odden, A. R. (1988). *Conditions of education in California 1988.* (Berkeley, CA: Policy Analysis for California Education, University of California).
Habermas, J. (1970). *Towards a rational society.* (Boston: Beacon Hill).
Habermas, J. (1971). *Knowledge and human interests.* (Boston: Beacon Hill).
Habermas, J. (1978). Problems of legitimation in late capitalism. In P. Connorton (Ed.) *Critical sociology.* (Harmondsworth: Penguin).
Habermas, J. (1979). *Communication and the evolution of society.* (Boston: Beacon Hill).
Haertel, G. D., Katzenmeyer, C. G. and Haertel, E. H. (1989). Capturing the quality of schools: Approaches to evaluation. Paper given to the Annual Meeting of the American Educational Research Association, March, San Francisco.
Haladnya, T. M., Nolen, S. B. and Haas, N. S. (1991) Raising standardized achievement test scores and the origins of test score pollution. *Educational Researcher,* 20, 5, 2–7.
Halpin, A. W. (1958). The development of theory in educational administration. In A. W. Halpin (Ed.) *Administrative theory in education,* pp. 1–19. (Chicago: University of Chicago Press).
Halstead, M. (1994). Accountability and values. In D. Scott (Ed.) *Accountability and control in educational settings,* pp. 146–165. (London: Cassell).
Haney, W. and Raczek, A. (1994). Surmounting outcomes accountability in education. Paper prepared for the US Congress Office of Technological Assessment, Chestnut Hill, MA: Centre for the Study of Testing, Evaluation and Educational policy, Boston College.
Haney, W. M. (1993). Cheating and escheating on standardized tests. Paper presented at the Annual Meeting of the American Educational Research Association, Atlanta, GA, April.
Haney, W. M., Madaus, G. F. and Lyons, R. (1993). *The fractured marketplace for standardized testing,* (Boston, MA: Kluwer Academic Publishing).
Hannaway, J. and Crowson, R. (Eds.) (1989). *The politics of reforming school administration.* (New York: Falmer).
Hargreaves, D. H. and Hopkins, D. (1991). *The empowered school: The management and practice of development planning.* (London: Cassell).
Harrington, G. (1992). Educational accountability policies in locally managed schools. Circular memorandum to principals of all high, district high and primary schools, senior officials, schools councils and parent's associations, 3 September.
Herrington, C. D. (1993). Accountability, invisibility and the politics of numbers: School report cards and race. In C. Marshall (Ed.) *The new politics of race and gender,* pp. 36–47. (London: Falmer).
Hesse, M. (1974). *The structure of scientific inference.* (London: Macmillan).
Hesse, M. (1980). *Revolutions and reconstructions in the philosophy of science.* (London: Harvester).
Hicks, J. (1981). *Wealth and welfare: Collected essays on economic theory. Volume 1.* (Cambridge, MA: Harvard University Press).
Hill, P. T., Harvey, J. and Praksac, A. (1993). Pandora's box : Accountability and performance standards in vocational education. RAND Report R-4271. (Santa Monica, CA: Rand).
Hocking, H. and Langford, J. (1990). Accountability and quality: A school-based perspective. In T. J. Wyatt and A. Ruby (Eds.) *Education indicators for quality, accountability and better practice: Papers from the [second] national conference on indicators and quality in education,* pp. 131–138. (Sydney: Australian Conference of Directors-General of Education).

Hodgkinson, C. (1978). *Towards a philosophy of administration*. (Oxford: Oxford University Press).
Hodgkinson, C. (1981). A new taxonomy of administrative process, *Journal of Educational Administration*, **19**, 141–152.
Hodgkinson, C. (1983). *The philosophy of leadership*. (London: Basil Blackwell).
Hodgkinson, C. (1991). *Educational leadership: the moral art*. (New York: SUNY Press).
Holmes Group. (1986). *Tomorrow's teachers: A report of the Holmes Group*. (Lansing: Holmes).
Honig, B. (1994). How can Horace best be helped? *Phi Delta Kappan*, **75**, 10, June, 790–796.
House, E. R. (1972). The domain of economic accountability, *Educational Forum*, **37**, 1, 13–23.
House, E. R. (1975). The price of productivity: Who pays? In W. J. Gephardt (Ed.) *Accountability: A state, a process, or a product?* pp. 49–57. (Bloomington, IN.: Phi Delta Kappan).
Hoy, W. K. and Miskel, C. G. (1978). *Educational administration: Theory, research, and practice*, Second edition. (New York: Random House).
Hoyle, E. (1980). Evaluation of the effectiveness of educational institutions. *Educational Administration*, **8**, 2, 159–178.
Ibrahim, A. B. (1995). Assessment of accountability systems in Malaysian education. *International Journal of Educational Research*, **23**, 531–544.
Jacobson, S. L. and Berne, R. (1993). *Reforming education: The emerging systemic approach*. (Thousand Oaks, CA: Corwin).
Jeffrey, B. and Woods, P. (1995). Reconstructions of reality: Schools under inspection. Paper presented at the European Conference on Educational Research, University of Bath, 14–17 September.
Jones, K. (1989). *Right turn: The Conservative revolution in education*. (London: Hutchinson Radius).
Kaagan, S. S. and Coley, R. (1989). *State education indicators: Measured strides and missing steps*. (New Brunswick, NJ: Centre for Policy Research in Education, Rutgers University).
Kay, B. (1976). The Assessment Performance Unit: Its task and rationale, *Education 3–13*, 4, 2.
Kemmis, S. and McTaggart, R. (1988). *The action research planner*, Third edition. (Geelong, Victoria: Deakin University Press).
Kenway, J. (1990). *Gender and education policy: A call for new directions*. (Geelong, Victoria: Deakin University Press).
Keys, S. and Fernandes, C. (1990). *A survey of school governing bodies: Report for the DES*. (Slough: NFER).
Kirst, M. W. (1988). *Who should control our schools? Reassessing current policies*. (Stanford, CA: Centre for Educational Research).
Kirst, M. W. (1990). *Accountability: Implications for state and local policymakers*. (Washington, DC: US Department of Education).
Kirst, M. W. (1994). The politics of nationalizing curricular content. *American Journal of Education*, **102**, 4, 383–393.
Kirst, M. W. (1995). Recent research on intergovernmental relations in education policy, *Educational Researcher*, **24**, 9, 18–22.
Kogan, M. (1978). The impact and policy implications of monitoring procedures. In T. Becher and S. Maclure (Eds.) *Accountability in education*, pp. 113–126. (Windsor, UK: NFER).
Kogan, M. (1986). *Education accountability: An analytic overview*. London: Hutchinson.
Kogan, M., Johnson, D., Packwood, T. and Whitaker, T. (1984). *School governing bodies*. (London: Heinemann).
Kuhn, T. S. (1957). *The Copernican revolution*. (Cambridge, MA: Harvard University Press).
Kuhn, T. S. (1970). *The structure of scientific revolutions*, Second edition. (Chicago: University of Chicago Press).
Lakatos, I. (1970). Falsification and the methodology of scientific research programmes. In

I. Lakatos and A. Musgrave (Eds.) *Criticism and the growth of knowledge.* (Cambridge: Cambridge University Press).

Lakomski, G. (1987). The cultural perspective in educational administration. In R. J. S. Macpherson (Ed.) *Ways and meanings of research in educational administration,* pp. 115–138. (Armidale: University of New England Press).

Lange, D. (1988). *Tomorrow's schools: The reform of education administration in New Zealand.* (Wellington: Government Printer).

Lather, P. (1991). *Feminist research in education: Within/against.* (Geelong, Victoria: Deakin University Press).

Leithwood, K. and Montgomery, D. (1982). The role of the elementary school principal in program improvement, *Review of Educational Research,* **52,** 3, 309–339.

Leithwood, K. A. (1992). The move towards transformational leadership, *Educational Leadership,* **49,** 5, 8–12.

Lello, J. (1979). *Accountability in education.* (London: Ward Lock).

Lello, J. (1993). *Accountability in practice.* (London: Cassell).

Lessinger, L. (1970). *Every kid a winner: Accountability in education.* (Palo Alto, CA: Science Research Associates).

Levacic, R. (1992). Local management of schools: Aims, scope and impact, *Educational Management and Administration,* **20,** 1, 16–29.

Levacic, R. (1995). *Local management of schools: Analysis and practice.* (Buckingham: Open University Press).

Levacic, R. and Glover, D. (1995). The relationship between efficient resource management and school effectiveness. Paper presented to the European Conference on Educational Research, University of Bath, 14–17 September.

Levin, B. (1994). Improving educational productivity through a focus on learners. *Studies in Educational Administration,* **60,** Winter, 15–22.

Levit, M. (1973). The ideology of accountability in schooling. In R. Leight (Ed.) *Philosophers speak out on accountability in education,* pp. 37–50. (Danville, IL: Interstate Publishers and Printers).

Lincoln, Y. S. (1984). Of wakes, monarchies and positivists: The emergent paradigm debate. *Organisation Theory Dialogue,* **4,** 2, 13–16.

Linton, M. (1995). Britons lose faith in Westminster. *The Guardian Weekly,* **152,** 23, June 4, p. 1.

Lister, G. (1992). Managing change. In T. Burgess (Ed.) *Accountability in schools,* pp. 109–116. (Harlow: Longman).

Lortie, D. (1986). Teacher status in Dade County: A case of structural strain? *Phi Delta Kappan,* **67,** 8, 568–575.

Lyotard, J. F. (1986). *The post-modern condition, a report on knowledge.* (Manchester: Manchester University Press).

Macbeth, A. (1995). Partnership between parents and teachers in education. In A. Macbeth, D. McCreath and J. Aitchison (Eds.) *Collaborate or compete? Educational partnerships in a market economy,* pp. 48–63. (London: Falmer Press).

Macbeth, A., McCreath, D. and Aitchison J. (Eds.) (1995). *Collaborate or compete? Educational partnerships in a market economy.* (London: Falmer Press).

MacDonald, B. (1978). Accountability, standards and the process of schooling. In T. Becher and S. Maclure (Eds.) *Accountability in education,* pp. 127–151. (Windsor, UK: NFER).

MacGregor, J. (1990). *Speeches on education: National curriculum and assessment.* (London: DES).

Maclure, S. (1978). Introduction—background to the accountability debate. In T. Becher and S. Maclure (Eds.) *Accountability in education* (pp. 9–25). (Windsor, UK: NFER).

MacPherson, C. B. (1962). *The political theory of possessive individualism*. (London: Oxford University Press).
Macpherson, R. J. S. and Caldwell, B. J. (1992). The skills and attributes required of aspiring and current Australian primary and secondary school principals over the next decade: A preliminary analysis. A stimulus paper commissioned by the Project on Leadership and Management Training for School Principals, a Commonwealth Project of National Significance managed by the Australian Secondary Principals' Association.
Macpherson, R. J. S. and Taplin, M. (1995). Principals' policy preferences concerning accountability: Implications for key competencies, performance indicators and professional development, *Journal of School Leadership*, 5, 5, 448–481.
Macpherson, R. J. S. (1984). On being and becoming an educational administrator: Methodological Issues, *Educational Administration Quarterly*, 20, 4, 58–75.
Macpherson, R. J. S. (1988). Talking up organization: The creation and control of knowledge about being organized. In D. E. Griffiths, R. T. Stout and P. B. Forsyth (Eds.) *Leaders for America's schools: The report and papers of the National Commission on Excellence in Educational Administration*, pp. 160–182. (CA: McCutchan).
Macpherson, R. J. S. (1991a). The restructuring of administrative policies in Australia and New Zealand state school education systems: Implications for practice, theory and research, *Journal of Educational Administration*, 29, 4, 51–64.
Macpherson, R. J. S. (1991b). The politics of Australian curriculum: The third coming of a national curriculum agency in a neo-pluralist state. In Fuhrman, S. and Malen B. (Eds.) *The politics of curriculum and testing*, pp. 203–218. (Basingstoke: Falmer Press).
Macpherson, R. J. S. (1992a). Educative leadership and the co-option of corporate managerialism and oligarchic politics. In T. Simpkins, L. Ellison and V. Garrett (Eds.). *Implementing educational reform: The early lessons*. (Harlow: Longman).
Macpherson, R. J. S. (1992b). Educative accountability policies for systems of 'self-managing' schools: Towards domesticating near-feral schools. Paper given to the Australian Council for Educational Administration Conference, 5–8 July, Darwin.
Macpherson, R. J. S. (1992c). History, organisation and power: A preliminary analysis of educational management in Tasmania, *School Organisation*, 12, 3, 269–288.
Macpherson, R. J. S. (1993a). Administrative reforms in the Antipodes: Self-managing schools and the need for educative leaders. *Educational Management and Administration*, 21, 1, 40–52.
Macpherson, R. J. S. (1993b). Challenging provider capture with radical changes to educational administration in New Zealand. In Y. M. Martin and R. J. S. Macpherson (Eds.) *Restructuring administrative policy in public schooling: Canadian and international case studies*, pp. 243–262. (Calgary, Alberta: Detselig).
Macpherson, R. J. S. (1994a). *Educative accountability policies for locally managed schools: Preliminary research findings*. Report to the Secretary, Department of Education and the Arts, 21 October.
Macpherson, R. J. S. (1995). *Educative accountability policies for Tasmania: A record of the April policy seminar for stakeholders*. Report to the Secretary, Department of Education and the Arts, 30 April.
Macpherson, R. J. S. (1996a). Accountability: Towards reconstructing a 'politically incorrect' policy issue, *Educational Management and Administration*, 24, 2, 139–150.
Macpherson, R. J. S. (1996b). Educative Accountability Policy Research: Methodology and Epistemology, *Educational Administration Quarterly*, 32, 1, 80–106.
Macpherson, R. J. S. (1996c). Educative accountability policies for Tasmania's locally managed schools: Interim policy research findings, *International Journal of Educational Reform*, 5, 1, 35–55.

Macpherson, R. J. S. (1996d). Accountability: More than my job's worth, *Educational Leadership Perspectives*, **2**, May, 1–3.
Malen, B. and Fuhrman, S. H. (1991). The politics of curriculum and testing: Introduction and overview. In S. Fuhrman and B. Malen (Eds.) *The politics of curriculum and testing*, pp. 1–9. (London: Falmer).
Mandelson. P. and Liddle, R. (1996). *The Blair revolution: Can New Labour deliver?* (London: Faber and Faber).
Margonis, F. and Parker, L. (1995). Choice, privatisation, and unspoken strategies of containment, *Educational Policy*, **9**, 4, 375–403.
Marland, S. (1972) Accountability in education, *Teachers College Record*, **73**, February, 339–345.
Marquand, D. (1988). *The unprincipled society: New demands and old politics*. (London: Jonathan Cape).
Marshall, C. (Ed.) (1993). *The new politics of race and gender*. (Washington, DC: Falmer).
Martin, D. T., Overholt, G. O. and Urban, W. J. (1976). *Accountability in American education: A critique*. (Princeton, NJ: Princeton).
Martin, I. S. (1992). Community education: LEAs and the dilemmas of possessive individualism. In G. Allen and I. Martin (Eds.) *Education and community: The politics of practice*, pp. 28–33. (London: Cassell).
Martin, W. and Willower, D. (1981). The managerial behaviour of high school principals, *Educational Administration Quarterly*, **17**, 1, 69–90.
Martin, Y. M. and Macpherson, R. J. S. (Eds.) (1993). *Restructuring administrative policy in public schooling: Canadian and international case studies*. (Calgary, Alberta: Detselig).
Masterman, M. (1970). The nature of a paradigm. In I. Lakatos and A. Musgrave (eds.) *Criticism and the growth of knowledge*. (Cambridge: Cambridge University Press).
Mawhinney, H. B. (1995). Systemic reform or new wine in an old bottle: Generating accountability through Ontario's *New Foundations in Education* reforms, *International Journal of Educational Research*, **23**, 505–518.
Maychell, K. and Keys, W. (1994). *Under inspection: LEA evaluation and monitoring*. (Windsor, UK: NFER).
McBrien, R. P. (1980). *Catholicism*. (Melbourne: Dove Communications).
McCarthey, M. C. (1988). Research in Educational Administration: Promising signs for the future. *Educational Administration Quarterly*, **22**, 3, 3–20.
McDonnell, L. and Fuhrman, S. (1986). The political context of education reform. In V. D. Mueller and M. McKeown (Eds.) *The fiscal, legal, and political aspects of state reform of elementary and secondary education*, pp. 43–64. (Cambridge, MA: Ballinger).
McDonnell, L. M. (1994). *Policy-makers' views of student assessment*. (Santa Monica, CA: Institute of Education and Training, Rand).
McDonnell, L. M. (1989). The policy context. In R. J. Shavelson, L. M. McDonnell and J. Oakes, *Indicators for monitoring mathematics and science education*. (Santa Monica, CA: The RAND Corporation).
McPherson, A. and Raab, C. D. (1988). *Governing education: A sociology of policy since 1945*. (Edinburgh: Edinburgh University Press).
Mitchell, D. (1988). Educational politics and policy: The state level. In N. J. Boyan (Ed.) *Handbook of research on educational administration*, pp. 453–466. (New York: Longman).
Mitchell, D. E. and Goertz, M. E. (Eds.) (1990). *Education politics for the new century*. (London: Falmer).
Moffat, J. (1994). On to the past: Wrong-headed school reform. *Phi Delta Kappan*, April, 584–595.
Monk, D. H. and Roellke, C. F. (1995). Accountability, resource allocation and the production of educational outcomes, *International Journal of Educational Research*, **23**, 493–503.

Monk, D. H., Roellke, C. F. and Nusser, J. L. (1996). Resource-based indicators for educational accountability and school improvement. Paper given to the Annual Meeting of the American Educational Research Association, 8–12 April, New York.

Morgan, G. (1980). Paradigms, metaphors and puzzle-solving in organizational theory, *Administrative Science Quarterly,* **25,** 4, 605–622.

Morris, B. (1991). Schoolteacher appraisal: Reflections on recent history, *Educational Management and Administration,* **19,** 3, 166–171.

Mortimore, P. (1995). *Effective schools: Current impact and future potential.* (London: Institute of Education, University of London).

Mortimore, P. and Mortimore, J. (1991). Teacher appraisal: Back to the future, *School Organization,* **11,** 2, 125–143.

Murphy, J. (1993). Restructuring: In search of a movement. In J. Murphy and P. Hallinger (Eds.) *Restructuring schooling: Learning from ongoing efforts,* pp. 1–31. (Newbury Park, CA: Corwin).

Murphy, J. and Hallinger, P. (Eds.) (1993). *Restructuring schooling: Learning from ongoing efforts.* (Newbury Park, CA: Corwin).

Murphy, J. T. and Cohen, D. K. (1974). Accountability in education—The Michigan experience, *Public Interest,* **36,** Summer, 53–81.

Nisbet, J. (1978). Procedures for assessment. In T. Becher and S. Maclure (Eds.) *Accountability in education,* pp. 95–112. (Windsor, Berkshire: NFER).

Nolan, Lord (1995). *The report of the committee on standards in public life.* Chairman Lord Nolan. (London: HMSO).

Northfield, J. R., Duignan, P. A. and Macpherson, R. J. S. (Eds.) (1987). *Educative leadership and the quality of teaching.* (Sydney: NSW Department of Education).

Nuttall, D. L. (1981). *School self-evaluation: Accountability with a human face?* (London: Schools Council).

Nuttall, D. L. (1982). *Accountability. Part 1, accountability and evaluation. E364, Block 1.* (Milton Keynes: Open University).

Oakes, J. (1986). *Educational indicators: A guide for policy makers.* (New Brunswick, NJ: Centre for Policy Research in Education, Rutgers University).

Oakes, J. (1989). What educational indicators? The case for assessing school context. *Educational Evaluation and Policy Analysis,* **11,** 2, 181–199.

Odden, A. and Clune, W. (1995). Improving educational productivity and school finance, *Educational Researcher,* **24,** 9, 6–10.

Odden, A. (1990a). Educational indicators in the United States: The need for analysis, *Educational Researcher,* **19,** 1, 24–29.

Odden, A. (1990b). Making Sense of Education Indicators: The Missing Ingredient. In T. J. Wyatt and A. Ruby (Eds.) *Education indicators for quality, accountability and better practice: Papers from the [second] national conference on indicators and quality in education,* pp. 33–57. (Sydney: Australian Conference of Directors-General of Education).

OERI State Accountability Study Group (1988). *Creating responsible and responsive accountability systems. Report of the OERI State Accountability Group.* Washington, DC: US Department of Education, Office of Educational Research and Information, ED 299 706.

Openshaw, R. (1995). *Unresolved struggle: Consensus and conflict in state post-primary education.* (Palmerston North: Dunmore).

Organisation for Economic Co-operation and Development (1994). *School: A matter of choice,* (Paris: OECD).

Organisation for Economic Co-operation and Development (1995). *Schools under scrutiny,* (Paris: Centre for Educational Research and Innovation, OECD).

Osborne, D. and Gaebler, T. (1992). *Reinventing government: How the entrepreneurial spirit is transforming the public sector*. (New York: Addison-Wesley).

Ouston, J. and Klenowski, V. (1995). Parents' responses to school inspections. Paper presented to the European Conference on Educational Research, University of Bath, 14–17 September.

Ouston, J., Fidler, B. and Earley, P. (1995). School improvement through school inspection? Paper given to the Annual Meeting of the American Educational Research Association, 18–22 April, San Francisco, CA.

Oxford University Press (1990). *The Oxford dictionary of quotations*, Third edition. (Oxford: Oxford University Press).

Papadakis, E. and Taylor-Goodby, P. (1987). *The private provision of public welfare*. (Hemel Hempstead: Wheatsheaf Books).

Paris, S. G., Lawton, T. A., Turner, J. C. and Roth, J. L. (1991). A developmental perspective on standardized achievement testing, *Educational Researcher*, **20**, 5, 12–20.

Park, S. H. (1995a). Towards epistemological unity of educational administration, *International Studies in Educational Administration*, **23**, 1, 46–57.

Park, S. H. (1995b). Understanding Australian naturalism in educational administration, *Educational Administration and Foundations*, **10**, 2, 39–59.

Park, S. H. (1995c). Naturalised epistemology and connectionism. In R. Cotter and S. J. Marshall (Eds.) *Research and practice in educational administration*, pp. 11–19. (Hawthorn, Melbourne: ACEA).

Park, S. H. (1997a). Australian naturalism and its critics, *Educational Management and Administration*.

Park, S. H. (1997b, in press). An evaluation of T. B. Greenfield's Subjectivism in educational administration, *Leading and Managing*.

Parsons, T. (1951). *The social system*. (Glencoe: The Free Press)

Parsons, T. (1966). *Societies: Evolutionary and comparative perspectives*. (Englewood Cliffs, NJ: Prentice-Hall).

Passow, A. H. (1988). Whither school reform? *Educational Administration Quarterly*, **24**, 3, 246–256.

Pateman, T. (1978). Accountability, values and schooling. In T. Becher and S. Maclure (Eds.) *Accountability in education*, pp. 61–94. (Windsor, UK: NFER).

Petch, P. (1992). Theory and practice. In T. Burgess (Ed.) *Accountability in schools*, pp. 89–98. (Harlow: Longman).

Phillips, D. C. (1983). After the wake: Postpositivistic educational thought, *Educational Researcher*, **12**, 5, 4–12.

Pipho, C. (1989). Stateline, *Phi Delta Kappan*, **70**, 9, 662–663.

Pitner, N. (1986). Substitutes for principal leader behaviour: A exploratory study, *Educational Administration Quarterly*, **22**, 2, 23–42.

Pitner, N. (1988). The study of administrative effects and effectiveness. In N. J. Boyan (Ed.) *Handbook of research in educational administration*. (New York: Longman).

Plano, J. C. and Greenberg, M. (1972). *The American political dictionary*, Third edition. (Hinsdale, IL: Dryden).

Popper, K. (1972). *The logic of scientific discovery*, Sixth edition, originally published as *Logik der Forschungi* in 1935. (London: Hutchinson).

Powell, B. and Steelman, L. C. (1984). Variation in state SAT performance: Meaningful or misleading? *Harvard Educational Review*, **54**, 389–412.

Pringle, M. (1992). Embodying the school. In T. Burgess (Ed.) *Accountability in schools*, pp. 63–74. (Harlow: Longman).

Pritchard, A. (1996). Research on quality assurance in Western Australian state secondary schools using a post-positivist/empiricist methodology. Research-in-progress seminar given to the School of Education, University of Tasmania, 3 June.

Pusey, M. (1991). *Economic rationalism in Canberra: A nation building state changes its mind.* (Oakleigh: Cambridge University Press).

Quine, W. V. O. and Ullian, J. S. (1978). *The web of belief,* Second edition. (New York: Random House).

Quine, W. V. O. (1951). Two dogmas of empiricism. *Philosophical Review,* **60,** 20–43.

Quine, W. V. O. (1960). *Word and object.* (MA: Technology Press MIT, and New York: John Wiley).

Quine, W. V. O. (1961). *From a logical point of view,* Second edition. (New York: Harper).

Radnor, H. A., Ball, S. J. and Vincent, C. (1996). Local educational government, accountability and democracy in the UK: Paper given to the Annual Meeting of the American Educational Research Association, 8–12 April, New York.

Radnor, H. A., Ball, S. J., Henshaw, L. and Vincent, C. (1995). Whither democratic accountability in education? An investigation into head teachers' perspectives on accountability in the 1990s in England and Wales. Paper presented to the European Conference on Educational Research, 14–17 September, University of Bath.

Rae, K. (1996). Devising New Zealand measures to report on Tomorrow's Schools, *International Studies in Educational Administration,* **24,** 1, 29–37.

Ribbins, P. (1985). Organisation theory and the study of educational institutions. In M. Hughes, P. Ribbins and H. Thomas (Eds.) *Managing education: The system and the institution,* pp. 223–261. (London: Holt, Reinhart and Winston).

Riley, K. A. and Nuttall, D. L. (1994). Measuring performance—National contexts and local realities. In K. A. Riley and D. L. Nuttall (Eds.) *Measuring quality: Education indicators —United Kingdom and international perspectives,* pp. 122–131. (London: Falmer Press).

Ritzer, G. (1975). *Sociology, a multiple paradigm science.* (Boston: Allyn and Bacon).

Ritzer, G. (1981). *Toward an integrated sociological paradigm: The search for an exemplar and an image of the subject matter.* (Boston: Allyn and Bacon).

Ritzer, G. (1991). *Metatheorizing in sociology.* (Lexington: D. C. Heath).

Rizvi, F. (1986). *Administrative leadership and the democratic community as a social ideal.* (Geelong: Deakin University Press).

Rizvi, F. (1990). *Approaches to the study of Educational Administration.* (Geelong: Deakin University Press).

Rizvi, F. (Ed.) (1985). *Ethics and Educational Administration.* (Geelong: Faculty of Education, Deakin University).

Rizvi, F. A., Duignan, P. A. and Macpherson, R. J. S. (Eds.) (1990). *Educative leadership in a multicultural society.* (Sydney: Department of Education).

Robinson, V. M. (1989). The nature and conduct of critical dialogue, *New Zealand Journal of Educational Studies,* **24,** 2, 175–187.

Robinson, V. M. (1993). *Problem-based methodology: Research for the improvement of practice.* (Oxford: Pergamon Press).

Robinson, V. M. (1994). The promise of critical research in educational administration, *Educational Administration Quarterly,* **30,** 1, 56–76.

Robinson, W. S. (1950). Ecological correlations and the behaviour of individuals, *American Sociological Review,* **15,** 351–357.

Rockmore, T., Colbert, J. G., Gavin, W. J. and Blackeley, T. J. (1981). *Marxism and alternatives: Towards a conceptual interaction among Soviet Philosophy, neo-Thomism, Pragmatism and Phenomenology.* (Dortrecht: Reidel).

Rorty, R. (1979). *Philosophy and the mirror of nature*. (Princeton: Princeton University Press).
Rorty, R. (1989). *Contingency, irony and solidarity*. (Cambridge: Cambridge University Press).
Rosenhead, J. (1992). Into the swamp: the analysis of social issues. *Journal of the Operational Research Society*, **43**, 4, 293–305.
Rothman, R. (1995). *Measuring up: Standards, assessment and school reform*. (San Francisco, CA: Jossey-Bass).
Sackney, L. and Dibski, D. (1994). School-based management: A critical perspective, *Educational Management and Administration*, **22**, 2, 104–112.
Salganik, L. H. (1994). Apples and apples: Comparing performance indicators for places with similar demographic characteristics, *Educational Evaluation and Policy Analysis*, **16**, 2, 125–141.
Sallis, J. (1977). *School managers and governors: Taylor and after*. (London: Ward Lock).
Sallis, J. (1979a). The parent: Schools must earn parents' trust, *Education*, 150.
Sallis, J. (1979b). Powers and duties of school governors: Past, present and future, *Where?* 150.
Sallis, J. (1991). *School governors: Your questions answered*. (London: Hodder and Stoughton).
Sashkin, M. (1974). Models and roles of change agents. In J. W. Pfeiffer and J. E. Jones (Eds.) *The 1974 annual handbook for group facilitators*. (La Jolla, CA: University Associates).
Schumpeter, J. A. (1954). *History of economic analysis*. (Oxford: Oxford University Press).
Scott, B. (1990). *School-centered education: Building a more responsive state school system*. (Sydney: Management Review, NSW Education Portfolio).
Scott, D. (1994). Making schools accountable: Assessment policy and the Education Reform Act. In D. Scott (Ed.) *Accountability and control in educational settings*, pp. 41–57. (London: Cassell).
Seldon, R. (1994). How indicators have been used in the USA. In K. A. Riley and D. I. Nuttal (Eds.) *Measuring quality*. (London: Falmer).
Shavelson, R. J., McDonnell, L. M. and Oakes, J. (1989). *Indicators for monitoring mathematics and science education*. (Santa Monica, CA: The RAND Corporation).
Shaw, M., Brindlecombe, N. and Ormston, M. (1995). Teachers' perceptions of inspection: the potential for improvement in professional practice. Paper presented to the European Conference on Educational Research, University of Bath, 14–17 September.
Shepard, L. (1991). Will national tests improve student learning? *Phi Delta Kappan*, **73**, 3, 232–238.
Simey, M. (1995). *Government by consent: The principles and practice of accountability in local government*. (London: Bedford Square Press).
Simkins, T. (1992). Policy, accountability and management: Perspectives on the implementation of reform. In T. Simkins, L. Ellison and V. Garrett (Eds.) *Implementing educational reform: The early lessons*, pp. 3–13. (London: Longman and BEMAS).
Simkins, T. (1995). The equity consequences of educational reform, *Educational Management and Administration*, **23**, 4, 221–232.
Simon H. A. (1976). *Administrative behavior*, 1976 Edition. (New York: Free Press).
Sizer, T. R. (1984). *Horace's compromise: The dilemma of the American high school*, (Boston, MA: Houghton Mifflin).
Smith, M. S. (1988). Educational indicators. *Phi Delta Kappan*, **69**, 7, 487–491.
Smith, M. S. and O'Day, J. (1991). Systemic school reform. In S. Fuhrman and B. Malen (Eds.) *The politics of curriculum and testing*, pp. 233–268. (New York: Falmer Press).
Sockett, H. (Ed.) (1980). *Accountability in the English educational system*. (London: Hodder and Stoughton).

South Australian Commission for Catholic Schools (1987). *Annual Report 1986/87.* (Adelaide: Catholic Education Office).

Stake, R. E. (1976). *Evaluating educational programs: The need and the response.* (Paris: OECD/CERI).

Stenhouse, L. (1977). A proposed experiment in accountability, *Times Educational Supplement,* 13 May.

Stoten, M. (1992). A director's assessment. In T. Burgess (Ed.) *Accountability in schools,* pp. 99–108. (Essex: Longman).

Stroud, B. (1992). Logical positivism. In J. Dancy and E. Sosa (Eds.) *A companion to epistemology,* pp. 262–265. (Oxford: Basil Blackwell).

Sullivan, M. (1995). The perspective from an urban primary school. In A. Macbeth, D. McCreath and J. Aitchison (Eds.) *Collaborate or compete? Educational partnerships in a market economy,* pp. 88–97. (London: Falmer Press).

Sweetman, J. (1994). A week in May to make or break. *Guardian Education,* 15 November, p. 16.

Sykes, G. and Elmore, R. F. (1989). Making schools manageable: policy and administration for tomorrow's schools. In J. Hannaway and R. Crowson (Eds.) *The politics of reforming school administration.* (London: Falmer).

Tabberer, R. (1995). *Parents' perceptions of OFSTED: Research summary.* (Slough: NFER).

Tarnas, R. (1991). *The passion of the Western mind: Understanding the ideas that have shaped our world view.* (New York: Ballantyne).

Taylor, F. W. (1911). *Shop management.* (New York: Harper and Bros.).

Taylor, F. W. (1947). *Scientific management.* (New York: Harper and Bros.).

Taylor, W. (1978). Values and accountability. In T. Becher and S. Maclure (Eds.) *Accountability in education,* pp. 26–60. (Windsor, UK: NFER).

Theobald, P. and Mills, E. (1995). Accountability and the struggle over what counts, *Phi Delta Kappan,* 76, 6, 462–465.

Thody, A. (1995). The governor-citizen: Agent of the state, the community of the school? In A. Macbeth, D. McCreath and J. Aitchison (Eds.) *Collaborate or compete? Educational partnerships in a market economy,* pp. 123–136. (London: Falmer Press).

Thorndike, R. L. and Hagen, E. P. (1969). *Measurement and evaluation in psychology and education.* (New York: John Wiley).

Timar, T. B. (1990). The politics of school restructuring. In D. E Mitchell and M. E. Goertz (Eds.) *Education politics for the new century,* pp. 55–74. (London: Falmer).

Times Education Supplement (1976). James Callaghan's speech at Ruskin College. 22 October.

Times Education Supplement (1994). Back with the teachers. Editorial. 11 November, p. 14.

Tyler, R. W. (1983). Educational assessment, standards, and quality: Can we have one without the others? *Educational measurement: Issues and Practice,* 2, 2, 21–23.

United States of America, National Governors' Association (1986) *A time for results.* (Washington, DC: Author).

Vann, B. J. (1993). A personal reflection on school improvement in an era of increasing public accountability. Unpublished MA thesis, Leicester University.

Vann, B. J. (1995). The accountability and assessment of schools in England and Wales: Current issues for practitioners. *Leading and Managing,* 1, 3, 180–192.

Walker, J. C. and Evers, C. W. (1982). Epistemology and justifying the curriculum of educational studies, *British Journal of Educational Studies,* 30, 2, 312–319.

Walker, J. C. and Evers, C. W. (1984). Towards a materialist pragmatist philosophy of education, *Education Research and Perspectives,* 11, 1, 23–33.

Walker, J. C. in association with Duignan, P. A., Flynn, P., Francis, D., Ikin, R., Macpherson,

R. J. S., Maxwell, B. and Wade, B. (1987). A philosophy of leadership in curriculum development: A pragmatic and holistic approach. In J. C. Walker (Ed.) *Educative leadership and curriculum development*, pp. 1–41. (Canberra, ACT: ACT Schools Authority).
Walker, R. J. (1985). Cultural domination of Taha Maori: The potential for radical transformation. In J. Codd, R. Harker and J. Nash (Eds.) *Political issues in New Zealand education*, pp. 73–82. (Palmerston North: Dunmore).
Ward, K. and Taylor, R. (Eds.) (1986). *Adult education and the working class*. (London: Croom Helm).
Warwick, D. (1995). Schools and businesses. In A. Macbeth, D. McCreath and J. Aitchison (Eds.) *Collaborate or compete? Educational partnerships in a market economy*, pp. 171–182. (London: Falmer Press).
Watkins, P. (1985). *Agency and structure: Dialectics in the administration of education*. (Geelong, Victoria: Deakin University Press).
Watkins, P. (1986). *A critical review of leadership concepts and research: The implications for educational administration*. (Geelong, Victoria: Deakin University Press).
West-Burnham, J. (1992). *Managing quality in schools*. (Harlow: Longman).
West-Burnham, J. and Davies, B. (1994). Quality management as a response to educational change. *Studies in Educational Administration*, 60, Winter, 47–52.
Whitfield, R. C. (1976). *Curriculum planning, teaching and educational accountability*. (Birmingham: University of Aston).
Willms, J. D. (1992). *Monitoring school performance: A guide for evaluators*. (Lewes: Falmer Press).
Willower, D. J. (1979). Ideology and science in organization theory, *Educational Administration Quarterly*, 15, 3, 20–42.
Willower, D. J. (1981). Educational administration: Some philosophical and other considerations, *Journal of Educational Administration*, 19, 2, 115–139.
Willower, D. J. (1986). Mystifications and mysteries in educational administration. In G. S. Johnston and C. C. Yeakley (Eds.) *Research and thought in administrative theory: Developments in the field of Educational Administration*, pp. 37–56. (Lanham, MD: University Press of America).
Willower, D. J. (1988). Synthesis and projection. In N. J. Boyan (Ed.) *Handbook of research in educational administration*, pp. 729–747. (New York: Longman).
Willower, D. J. (1993). Whither Educational Administration: The post-positivist era, *Journal of Educational Administration and Foundations*, 8, 2, 13–31.
Willower, D. J. (1994). Values, valuation and explanation in school organizations, *Journal of School Leadership*, 4, 5, 466–483.
Wilson, S. M. (1994). Appropriate forms of accountability for Tasmanian state schools: Teachers' policy preferences. Unpublished MEd thesis, University of Tasmania.
Wintour, P. and Clouston, E. (1995). Tories crash to by-election defeat. *The Guardian Weekly*, 152, 23, 4 June, p. 11.
Wintour, P., Hencke, D. and Bates, S. (1995). Tories attack PM over Nolan 'mess'. *The Guardian Weekly*, 152, 22, 28 May, p. 12.
Wise, A. E. (1979). *Legislated learning: The bureaucratization of the American classroom*. (Berkeley, CA: University of California Press).
Wise, A. E. and Gendler, T. (1989). Rich schools, poor schools: The persistence of unequal education, *The College Board Review*, 151, 12–27.
Wohlstetter, P. and Odden, A. (1992). Rethinking school-based management policy and research, *Educational Administration Quarterly*, 28, 4, 529–549.
Wohlstetter, P., Wenning, R. and Briggs, K. L. (1995). Charter schools in the United States: The question of autonomy, *Educational Policy*, 9, 4, December, 331–358.

Wong, K.-C. (1995). Education accountability in Hong Kong: Lessons from the school management initiative, *International Journal of Educational Research*, 23, 519–529.

Wong, K. K. and Moulton, M. H. (1996a). Developing institutional performance indicators for Chicago schools: Conceptual and methodological issues considered. In Wong, K. K. (Ed.) *Advances in educational policy: Rethinking school reform in Chicago*. (Greenwich, CT: JAI Press).

Wong, K. K. and Moulton, M. H. (1996b). A report card on institutional performance of the Chicago public schools. Paper given to the Annual Meeting of the American Educational Research Association, 8–12 April, New York.

Wong, K. K. and Sunderman, G. L. (1994). Redesigning accountability at the system wide level: The politics of school reform in Chicago. Paper given to the Association for Public Policy Analysis and Management Research Conference, Chicago, 27–29 October.

Wylie, C. (1995). Finessing site-based management with balancing acts, *Educational Leadership*, 53, 4, 54–59.

Yanguas, J. and Rollow, S. G. (1994). The rise and fall of adversarial politics on the context of Chicago school reform: Parent participation in a Latino school community. Paper given to the Annual Meeting of the American Educational Research Association, 18–22 April, San Francisco, CA.

Yates, L. (1990). *Theory/practice dilemmas: Gender, knowledge and education*. (Geelong, Victoria: Deakin University Press).

Yeatman, A. (1990). *Bureaucrats, technocrats, femocrats: Essays on the contemporary Australian state*. (Sydney: Allen and Unwin).

Young, M. F. D. and Whitty, G. (1977). *Society, state and schooling*. (Ringmer: Falmer Press).

Young, M. F. D. (Ed.) (1971). *Knowledge and control*. (London: Collier-Macmillan).

Young, R. E. (1995). Education's telos. *Discourse: Studies in the Cultural Politics of Education*, 16, 3, 429–432.

Young, R. E. (1996). *Intercultural Communication—Pragmatics, genealogy, deconstruction*. (Avon, UK: Multilingual Matters).

Author Index

Aitchison, J. 49
Allen, G. 59
Ambron, S. R. 86
Ancess, J. 140
Anderson, D. 7
Angoff, W. H. 108
Angus, L. 9
Apple, M. W. 183
Archbald, D. 115–116
Argyris, C. 189
Arrow, J. J. 133
Ascher, C. 114, 126, 135, 138, 140, 141
Aspin, D. 188
Astuto, T. 135
Australian Council of State School Organisations and Australian Parents Council (ACCSO/APC) 217–219, 234, 320

Ball, S. J. 72, 76, 78, 79, 80, 175
Barber, C. 165, 169
Barber, M. 72
Bates, R. J. 181, 184
Bates, S. 77
Beare, H. 2, 4, 9, 13, 16, 17, 18, 66, 217
Becher, T. 31, 33, 34, 53
Belsey, A. 48
Bender, F. 183
Benne, K. D. 14
Bennion, J. W. 185
Bernauer, J. A. 107
Berne, R. 4
Berrell, M. M. 19, 183, 186
Berry, B. 135, 136, 137
Bhindi, N. 189–190

Blackeley, T. J. 183
Blackmore, J. 181
Blank, R. K. 121
Bobbett, G. 109
Bolton, E. 57
Boston, K. 9
Bowles, S. 183
Boyd, R. N. 181
Boyd, W. L. 1, 4, 13, 87, 123, 130, 131
Bridges, D. 30, 39, 41, 42
Briggs, K. L. 22, 216
Brighouse, T. 49
Brindlecombe, N. 69
Browne, S. 65
Burgess, T. 42, 53–54
Burgher, W. 108
Burns, R. 48, 146
Burrell, G. 184–185
Bush, T. 6
Byram, G. H. 113, 114

Caldwell, B. J. 9, 16, 17, 66, 152, 277
Callaghan, J. 29
Callahan, R. E. 86
Campbell, E. Q. 86
Campbell, R. F. 178
Carnegie Forum on Education and the Economy 88, 89
Casey, P. M. 216
Chapman, J. D. 9, 15, 188
Charlton, K. 26
Chin, R. 14
Chitty, C. 27, 30, 48, 49
Churchland, P. S. 190

Cibulka, J. G. 15, 121, 122, 123–124, 127, 142, 143, 144, 145
Ciccarelli, A. M. 188
Clark, B. 17
Clark, D. 135
Clouston, E. 77
Clune, W. 22, 124
Codd, J. 9
Cohen, A. 100
Cohen, D. K. 22, 87
Colbert, J. G. 183
Coleman, J. S. 86
Coley, R. 98
Cooley, W. W. 107
Coopers & Lybrand Ltd 49, 52
Coordingley, P. 49
Corbett, H. D. 120, 121
Creed, P. 14
Cresap Ltd. 15, 152, 162
Cronbach L. G. 86
Crowson, R. L. 15, 99, 131
Crowther, F. 9, 166, 167
Crump, S. J. 189
Cuban, L. 99
Culbertson, J. A. 179, 185
Cusack, B. O. 9
Cuttance, P. 16

Darling-Hammond, L. 114, 126, 135, 137, 138, 139, 140, 141
David, M. 59
Davies, B. 7
Davies, H. 165
Davis, B. 158, 162
Dearing, R. 10, 63
Deem, R. 76
Department of Education and Science 42
Derlin, R. L. 122, 123–124, 127, 142, 143, 144, 145
Dewey, J. 19, 21, 181, 183
Dibski, D. 11, 18
Dillon, D. 17
Donlan, T. 108
Donoughue, B. 30
Dornbush, S. M. 86
Dornun, J. 135
Duckett, W. 108
Duignan, P. A. 9, 17, 177, 178, 182, 187, 189–190, 277
Dunleavy, P. 319
Dunstan, J. 14
Dunstan, J. F. 9, 15

Earley, P. 63, 64, 69, 72
East Sussex Accountability Project 31, 40
Ebbutt, D. 30, 39, 41, 42
Einbender, L. 140
Elam, S. M. 108
Elliot, J. 30, 39, 41, 42
Elmore, R. F. 22, 99, 100–101
Elmore, R. F. and Associates 6, 7, 21, 141, 163, 175
Eraut, M. R. 31, 48
Evers, C. W. 9, 17, 19, 177, 185, 188, 190, 191
Ewington, J. 9, 165, 169

Falk, I. 277
Fernandes, C. 64
Feyerabend, P. 179
Fidler, B. 69, 72
Finn, C. E. 90
Fleming, T. 185
Flynn, P. 9, 17, 177, 187
Forsyth, P. B.
Foster, W. 141, 181
Fowler, G. 27
Francis, D. 9, 17, 177, 187
Franklin, B. J. 113, 114
Frazer, M. 14
French, R. L. 109
Fuhrman, S. H. 2, 15, 84, 102, 123
Fullan, M. G. 2, 6, 11, 15, 50, 131, 132
Fuller, P. 72

Gaebler, T. 123
Gallacher, N. 49
Gallup, A. M. 108
Galton, M. 10
Galtung, J. 107
Gavin, W. J. 183
Gee, J. 276
Gendler, T. 138
Gibson, I. 166, 167
Gibson, R. 30, 39, 41, 42
Giddens, A. 182, 191
Ginsberg, R. 135, 136, 137

Gintis, H. 183
Glascock, C. H. 113, 114
Glass, G. V. 86
Glatter, R. 49
Glover, D. 72
Goddard, D. 54–56
Goertz, M. E. 15, 102
Gold, K. 76
Goldstein, H. 29
Golensky, M. 63
Goodlad, J. I. 88, 99
Goodwin, A. L. 140
Gordon, L. 9
Government of Australia, Department of Prime Minister and Cabinet 161
Government of Great Britain, Department for Education 66, 67
Government of Great Britain, Department of Education and Science 28, 31, 60
Government of Great Britain, Department of Education and Science/Welsh Office 31, 32
Government of Great Britain, House of Commons, Education, Science and Arts Committee 32
Government of Great Britain, Office for Standards in Education (OFSTED) 67, 70, 71, 72
Government of Great Britain, Parliament, House of Commons 43
Government of Great Britain, Statutes 49, 50–51, 53, 58, 60, 61, 63, 67, 68
Government of New Zealand, Taskforce to Review Education Administration 7
Government of Tasmania 152
Government of Tasmania, Department of Education and the Arts (DEA) 152, 153, 154, 155, 157, 159, 160, 302
Government of Tasmania, Statutes 157, 158, 160
Government of Tasmania, Tasmanian Education Council (TEC) 156
Government of the United States of America, Department of Education, National Commission on Excellence in Education 88
Government of the United States of America, Department of Education, National Center for Education Statistics (NCES) 93, 95, 112
Grace, G. 182
Grady, N. 277
Gray, J. 66, 78, 81
Greenberg, M. 3, 4
Greenfield, T. 141, 179–180
Griffith, J. E. 93, 95–96, 97, 98
Griffiths, D. E. 88, 90, 179, 181, 184
Griggs, C. 48
Grimshaw, W. A. 16
Guba, E. 185
Guthrie, J. W. 97, 98

Haas, N. S. 108
Habermas, J. 182, 183
Haertel, E. H. 92
Haertel, G. D. 92
Hagen, E. P. 19
Haladyna, T. M. 108
Hallinger, P. 1, 5, 6
Halpin, A. W. 179
Halpin, D. 1, 130
Halstead, M. 72, 73–75, 175, 216
Haney, W. M. 85, 87, 106, 107, 108, 133
Hannaway, J. 15
Hargreaves, D. H. 66
Harrington, G. 151, 171
Harvey, J. 108
Hayward, G. C. 97
Hencke, D. 77
Henshaw, L. 72, 76, 175
Herrington, C. D. 142
Hess, R. D. 86
Hesse, M. 187
Hicks, J. 133
Hill, P. T. 108
Hobson, C. J. 86
Hocking, H. 152
Hodgkinson, C. 50, 178, 181
Holmes Group 89
Honig, B. 131–132
Hopkins, D. 66
Hornik, R. C. 86
House, E. R. 87, 106
Hoy, W. K. 181
Hoyle, E. 36
Hutsell, W. W. 113, 114

Ikin, R. 9, 17, 177, 187

Jacobson, S. L. 4
Jeffrey, B. 69
Jenkins, K. 135
Johnson, D. 63, 64
Jones, K. 48

Kaagan, S. S. 98
Katzenmeyer, C. G. 92
Kay, B. 28
Kemmis, S. 7, 182
Kenway, J. 181
Keys, S. 64
Keys, W. 43
Kirst, M. W. 13, 22, 93, 97, 130
Klenowski, V. 69
Kogan, M. 6, 10, 44–45, 46, 58, 63, 64, 177
Kuhn, T. S. 179, 185

Lakatos, I. 177, 330
Lakomski, G. 19, 177, 181, 185, 188, 190, 191
Lange, D. 3
Langford, J. 152
Lather, P. 185
Lawton, T. A. 108
Leithwood, K. 99
Leithwood, K. A. 11
Lello, J. 56–57
Lenney, T. 6
Lessinger, L. 86
Levacic, R. 10, 52, 53, 65, 72
Levin, B. 7
Levit, M. 86
Liddle, R. 81
Lincoln, Y. S. 185
Linton, M. 76
Lister, G. 56
Long, M. 17
Lortie, D. 99
Lugg, C. 123
Lyons, R. 87, 107, 108
Lyotard, J. F. 186

Macbeth, A. 49
MacDonald, B. 46, 48
Macdonald, M. B. 140
MacGregor, J. 61

Maclure, S. 28, 29, 33, 34
MacPherson, C. B. 60
Macpherson, R. J. S. 2, 4, 6, 9, 11, 16, 17, 19, 21, 152, 169, 170, 171, 177, 178, 182, 183, 186, 187, 189, 193, 211, 225, 277
Madaus, G. F. 87, 107, 108
Malen, B. 15, 102
Mandelson, P. 81
Margonis, F. 21
Marland, S. 86
Marquand, D. 30
Marshall, C. 15
Martin, D. T. 88
Martin, I. 59
Martin, I. S. 60
Martin, W. 99
Martin, Y. M. 4, 11
Masterman, M. 185
Mawhinney, H. B. 131
Maxwell, B. 9, 17, 177, 187
Maychell, K. 43
McBrien, R. P. 216–217
McCarthey, M. C. 185
McCreath, D. 49
McDonnell, L. M. 84, 91–92, 98, 148
McPartland, J. 86
McPherson, A. 6
McTaggart, R. 7, 182
Millikan, R. H. 9, 16, 17, 66
Mills, E. 19
Miskel, C. G. 181
Mitchell, D. E. 15
Moffat, J. 21
Monk, D. H. 134–135
Montgomery, D. 99
Mood, A. M.
Morgan, G. 16, 184–185
Morris, B. 10
Mortimore, J. 10
Mortimore, P. 10, 69
Moulton, M. H. 110, 111, 125
Mulford, W. R. 277
Murphy, J. T. 1, 5, 6, 87

Newell, L. J. 185
Nias, J. 30, 39, 41, 42
Nisbet, J. 33, 34, 46, 47
Nolan, Lord 77

Nolen, S. B. 108
Norman, M. 17
Northfield, J. R. 9
Nusser, J. L. 134
Nuttall, D. L. 29, 31, 34, 78

O'Day, J. 123, 124, 143
O'Leary, B. 319
Oakes, J. 91–92, 135, 136
Odden, A. R. 6, 18, 22, 92, 93, 95, 96, 97, 98, 99, 130, 216
OERI State Accountability Study Group (SASG) 114, 126
Ogilvie, D. 9
Openshaw, R. 5
Organisation for Economic Co-operation and Development (OECD) 67, 80, 84, 106, 118, 119–120, 124, 129, 147
Ormston, M. 69
Osborne, D. 123
Ouston, J. 69, 72
Overholt, G. O. 88
Oxford University Press 4

Packwood, T. 63, 64
Papadakis, E. 59
Paris, S. G. 108
Park, S. H. 178, 189
Parker, L. 21
Parsons, T. 179
Passow, A. H. 88
Pateman, T. 45–46
Petch, P. 49–50
Phillips, D. C. 86
Pipho, C. 86, 89
Pitner, N. 99
Plano, J. C. 3, 4
Popper, K. 179
Porter-Gehrie, C. 99
Powell, B. 107
Pratsac, A. 108
Pringle, M. 54
Pritchard, A. 189
Pusey, M. 161

Quine, W. V. O. 179, 187, 188, 330

Raab, C. D. 6

Raczek, A. 85, 106, 107, 108, 133
Radnor, H. A. 72, 76, 78, 79, 80, 175
Rae, K. 5
Reed, R. J. 15
Ribbins, P. 180, 185
Riley, K. A. 78
Ritzer, G. 184
Rizvi, F. 9, 180, 181, 182
Rizvi, F. A. 182
Robinson, V. M. 23, 50, 184, 188
Robinson, W. S. 107
Rockmore, T. 183
Roellke, C. F. 134–135
Rollow, S. G. 110
Rorty, R. 185
Rose, L. C. 108
Rosenhead, J. 39
Roth, J. L. 108
Rothman, R. 123

Sackney, L. 11, 18
Salganik, L. H. 137, 138
Sallis, J. 42, 63
Sashkin, M. 145, 146
Schilder, D. 121
Schön, D. A. 189
Schumpeter, J. A. 133
Scott, B. 16
Scott, D. 60, 69
Seldon, R. 91, 106
Shavelson, R. J. 91–92
Shaw, M. 69
Shepard, L. 135
Simey, M. 3
Simkins, T. 57, 80
Simon, H. A. 179
Sizer, T. R. 94
Smith, M .S. 91, 123, 124, 143
Snyder, J. 135, 140
Sockett, H. 10
South Australian Commission for Catholic Schools 217
Spinks, J. M. 9, 16, 152
Stake, R. E. 34
Steelman, L. C. 107
Stenhouse, L. 26
Stoten, M. 56
Stout, R. T. 90

Stroud, B. 179
Sullivan, M. 49
Sunderman, G. L. 110
Sweetman, J. 10
Sykes, G. 99, 100–101

Tabberer, R. 69
Taplin, M. 169, 170
Tarnas, R. 186
Taylor, F. W. 19
Taylor, R. 59
Taylor, W. 27, 45
Taylor-Goodby, P. 59
Theobald, P. 19
Thody, A. 49
Thorndike, R. L. 19
Timar, T. B. 15
Times Education Supplement 2, 29
Turner, J. C.
Tyler, R. W. 92

Ullian, J. S. 187, 188, 330
United States of America, National Governors' Association 89
Urban, W. J. 88

Vann, B. J. 10, 78
Vincent, C. 72, 76, 78, 79, 80, 175

Wade, B. 9, 17, 177, 187
Walker, J. C. 9, 17, 177, 186, 187
Walker, R. J. 182
Ward, K. 59

Warwick, D. 49
Watkins, P. 181
Weinfield, F. D. 86
Wenning, R. 22, 216
West-Burnham, J. 7
Whitaker, T. 63, 64
Whitfield, R. C. 74
Whitty, G. 183
Wilby, P. 49
Wilcox, B. 66, 78
Wilkinson, V. 188
Williams, J. 17
Williamson, J. 277
Willms, J. D. 61, 116, 117
Willower, D. 99
Willower, D. J. 181, 185, 188
Wilson, B. L. 120, 121
Wilson, S. M. 165, 167
Wintour, P. 77
Wise, A. E. 86, 87, 138
Wohlstetter, P. 6, 18, 22, 130, 216
Wong, K. K. 15, 110, 111, 113, 125
Woods, P. 69
Wylie, C. 9

Yanguas, J. 110
Yates, L. 181
Yeatman, A. 161
Young, M. F. D. 183
Young, R. E. 189

Zahorchak, G. I. 123

Subject Index

A Nation at Risk 88
A Time for Results 89
absolute responsibility 26
absolutism 4, 57
access 124
  to knowledge 148
Accountability Policy Questionnaire (APQ) 167
accountability
  client 314
  collective 305
  communitarian 312
  constitutional 4
  contractual 16, 36, 74, 234, 312, 320, 326
  criteria 11
  cycle 154, 172, 308
  democracy 3–4, 11
  democratic 4, 11
  horizontal 17
  law 4
  line 160
    management 154, 172
  local public 68
  mandated 65
  models 44, 57
  moral 36
  movement 87
  mutual accountability 234, 320, 323, 327
  natural 114
  outcomes 87, 90
  in partnership 33
  intra-professional 40
  mechanism 314
  multiple forms of 46
  outcomes 52, 86, 101
  performance 142
  perspectives on 7, 8
  policy
    Division of Education 154
    making 96, 148
    research 33–36
  political 16, 32, 312
  politics of 105
  processes 5, 11
  productivity 87
  professional 7, 25–26, 36, 37, 42, 140
  public 36
    interest 32–33
  relationships 17, 36
  remedial 95
  resource 132, 133
  responsive 40, 74, 312
  role-holders 73
  schemes 35
  school 46
  structures 16
  summative 10
  surrogate 80
  symbolic 28
  test-based outcomes 106
  theories of competing 175
  token 49
  tripartite relationships 73, 74
accreditation 119
achieve priorities 207
achievement 115
  data collection 95
action plan 72

361

action research 7, 39, 48, 140, 145, 146, 219, 318, 323
  evaluation 328
active engagement 192
add value 67, 97
adequacy 22, 149
administration 5, 90
  Bush 90
administrative impulse 147
adoptability 145
adoption 313
adversarialism 49
advisory
  powers 157
  services 216
aesthetic 29
  purposes 50, 311
age 176
aggregate needs 304, 327
aggregated data 125
Aird, Michael 153
alienation 85, 184
alignment 148
alleged misconduct 77
alternative
  discourse 329
  stories 185
altruism 190, 192
ambiguous purposes 44
American Educational Research Association 331
analysis
  of non-response 168
  of variance 330
analytical
  methods 176
  research 72
ambivalent support 193
anecdotal
  evidence 17
  theory 179
annual
  cycle
    budget 284
    planning 326
  report 155
  school development processes 43
anomie 13

answerability 34, 72
answering to
  clients 312
  colleagues 312
antecedent 13
anti-statism 30
appraisal 200
  and reporting mechanisms 326
  by an independent expert 200
  by parents 200
  by students 200
  by the P and F/School Council 204
  by DEA 204
  by peers 200
  by school colleagues 204
  by the community 204
  discussion 200
  of support and feedback 204
appropriate
  government 77
  partners 44
arbitrary process 22
arbitrating underlying theories in competition 177
Aristotelian *praxis* 311, 327
Arrow's impossibility theorem 133
articles of faith 186
artless empowerment of professionals 84
assent 25
Assessment Performance Unit (APU) 28
assessment 28, 157
  of achievement 43
assisted school
  review 328
  self-review 303
associational life 313
assurance 101
at risk 113
atomization 22
attendance 142, 143
attitudes 115
  of children 203
audiences 34, 87
audited self-evaluation 43
Australia 6, 7, 8, 15–19
Australian
  Council of State School Organisations (ACSSO) 217

Education Act 163
Education Union (Tasmania) 284
naturalism 189
Parents Council (APC) 217
school principals 277
authentic
  assessment 105, 112, 125, 138
  leadership 189, 190
  participation 237
authoritarian 123, 246, 316
authority 184
automatic
  assent 36
  leadership 157
  resistance 124
autonomy 117
averages 66
axioms 186

baby boomers 84
backwash 28
Baker, Jan 158
Baker, Kenneth 48, 61
balanced information 218
Barber, Christine 165
basic 135
  purposes 5
    of education 74
  skills 41
    movement 86, 101
Bates, Richard 181
behaviour
  management skills 203
  objectives 87
  science of management 315
  theory of motivation 318
beliefs 187
benchmark 156
benefits 85
best
  interests of the students 246
  pedagogy 20
  practice 4, 190, 219, 227, 257, 318
Beswick, John 154
Bhindi, Narottam 189
bias 184
bicameral house of legislature 123
big picture 241

bill of rights 76
Black Papers 27
blame 47
blue-ribbon panel 35
Board of
  directors 63
  of Governors 52
  of Professional Training Standards 89
  of Trustees 9, 112
    Boston, Ken 17
boundary disputes 110
Bowen District 159
Boyle, Sir Edward 27
budget
  cuts 190
  deficits 120
  -setting 65
budgeting 207
budgetary problems 283
Bullock report, *A Language for Life* 28
bureaucracies 5, 155, 172, 318
bureaucratic technique 88
bureaucratization 87, 123
Bush's *America 2000* 129
Bush, President 1
business of government 285
business sector 144
by-elections 77

California 96, 98, 105, 113, 118–120
  Learning Assessment System (CLAS) 96, 118
  School Recognition Program 119
  Assessment Program (CalAP) 96
Callaghan, James 29
Cambridge Accountability Project (CAP) 39–42
Cambridgeshire 56
Canada 11, 18
canteens 157
capacities 225
  as learners and researchers 207
capacity building 137, 160, 173, 249
capacity to hear and care 207
capitalism 41
career structure 152
careerism 225
caring 40

Carnegie Forum Report 88
case studies 140
casual
  relationships 74
  stories 115
categories 170
causal stories 92
Central Control Model 74, 75
centralist control 2
CEO 283
certification 89
chain of responsibility model 75
challenges of practice 204
change
  management 235
  normative 14
Chapman, Judith 188
charter schools 21
cheating 108
Chicago 110
  Teachers Union (CUT) 111
Chief Executive Officer (CEO) 63, 158
child comparisons 194
child-centred approach 30
childcare facilities 157
choice 5, 85, 112, 116, 216
Ciccarelli, Anna 188
circuit-breaker 49
circularity 138, 148
citizenry 30
citizens 21
City Technology Colleges 51
civic purposes 312
civil and criminal courts 77
civilized
  community 313
  living 21, 183
class 176
  analysis 81
  conflict 27
  power 183
classroom environment 203
client
  accountability 314
  choice 7
  perspective 7
  relationship 26
  representation 10

climate
  of self review 119, 126
  of surveillance 69
clinical preparatory programmes 90
Clinton's *Goals 2000: Educate America Act* 129
Clinton, President 1
closed management systems 86
closed peer culture 322
closed systems 91
closedness of schools 265
co-governance 183, 237
co-leadership 237
co-management 183, 301
co-operative 200
  policy research 171
  politics 137, 319
  social systems 60
  teacher development 61
  teaching 7, 176
co-ordination 121
  of information 154
co-plan teaching programs 200
co-teachers 5
coaching 108
Coalition for Essential Schools 94
Cocker, Penny 163
code of
  conduct 26, 72
  ethics 26
  expertise 26
  practice 26
coercive strategies 291
cognitive skills 108
coherence 2, 122, 124, 315
  movement 131
  test 177
  of theory 188
Coleman report 86, 107
collaborative
  consortia 14
  decision making 207
  dialogue 50
  planning 1, 2, 7, 176
collaborative strategic analyses 234
collective
  accountability 305
  concern ideology 76

identities 60
identity 323
interpretation 170
collegial
  norms 322
  professionalism 14
collegiality 176
colonized 277
Colorado 123, 144
command economies 14
commissioned national review 43
commitment 117, 308, 311
Committee on Standards in Public Life 77
commodification 22, 161, 173
common
  ground 21, 191
  sense 318
    policy 276, 282
  standards 187
Commonwealth 18
  (federal) election 283
  Department of Prime Minister and Cabinet 161
  Department of Employment, Education and Training (DEET) 161
communication 41, 57
  skills 203
  vacuums 136
communicative
  competence 183
  prerequisites 97
communitarian 229, 235
  accountability 312
  ethic 81
  liberalism 81, 315, 318
  limited government 319
  parenting 307
  professionalism 323
  theory of effective teaching 237
  values 312
communitarianism 13
community 188
  accountability politics 235
  adult education 59
  builders 5
  investment 85
  partnerships 49
  projects 329

school development 286, 297, 304
therapy 137
values 85
comparability 29, 61, 215, 319
  assessment 156, 235
  benchmarks 327
  indicators 161
  power 41
  reporting 241
  research 147
compensatory education 144
competence 26
competence-based education 87
competencies 169
  outlined by department 203
competency-based graduations 143
competing perspectives 46
competition 51, 56, 61
competitive
  advantage 285
  excellence 123
  self-interest 59
  tendering 51
  workforce 89
complaints procedure, formal 209
complexity 99
compliance 125
  regulations 316
complimentarity 215, 217, 319
comprehensive 27, 311
  criteria 219
  information 218
  pragmatism 47
comprehensivization 85
compulsory education 85
computerized accounting 152
conceptual
  analysis 23
  disarray 331
Condition of Education 93
conditional acceptance 29
Conditions of Education in California 97
conferences 29
confidence in public education 42
confidential relationship 25
confirmability 167
conflicting principles 54
confrontation 50

conjecture 173
Connecticut 98, 113, 114
consensus 15
  building 27
  of cynicism 13
  of disenchantment 59
  of the governed 3
consent of the governed 83, 313, 314
consequences 317
consequencialist moral theory 21, 192
conservatism 46
Conservative Party 27
consistency 65, 190
consolidation of portfolios 284
constant comparison 167, 188
constituencies 15
constitutional
  accountability 4
  authority 84
  scepticism 53
construct relevance 111, 125, 316
construct validity 108, 115, 125, 147, 290
construction 176, 191
constructive empiricism 18
constructivism 181
constructivist 141
  theories 9
consultant 145
consultations 10, 140, 143
consumer
  initiative 121
  interest ideology 76
consumerism 2, 7, 139
consumerist 176
  data 61
  metaphors 49
content
  analysis 166
  and process standards 87
contestability 57, 313
  of advice 39
contestable 94
contested
  knowledge claims 177
  policy settlements 314
  theories of knowledge 176
  values 94
context indicators 136

contexts 2
contextual variables 109
contextualized 219, 313
contextualization 65
continuous improvement 2
contract 32
contracted inspectors 67
contractual 32
  accountability 16, 74, 234, 312, 320, 326
contradiction 123, 132
contradictory theories 114
co-operative learning 1
Coopers and Lybrand 49
core
  competencies 285
  curriculum 30
  duties 304
  functional duties 284
  subjects 51
  technologies 6, 99, 175
corporal punishment 142
corporate
  managerialism 7, 215, 319
  managerialist 16
  managers 234, 235
correction 188
cost
  rising 316
  -benefit 85
  effectiveness 85
Council of
  Australian Governments (COAG) 161
  Chief State School Officers (CCSSO) 102
court rulings 110
creative accounting 133
creativity 190, 192, 207
credibility 57, 167
Cresap Ltd 152
Cresap report 15, 162
crisis-orientated decision making 121
criteria development 198
criterion-referenced 61,
criterion-referenced assessment 61, 62
critical
  culture of formative evaluation 285
  dialogue 50, 189, 326
  friend or mentor 153
  friends 105

subjectivism 178
criticism 109
cross-curricular themes 51
cross-level analysis 106
cross-visitation 132
Crump, Stephen 189
cult of efficiency 86
cultural
　boundaries 327
　capital 183
　interests 246
　transformation 78
culture 119, 179
　of unchanging practice 136
current political touchstone 322
curricular criteria 69, 71
curriculum 51, 157
　compliance 121
　development 31
　framework 1
　formal 20
　goals to parents 194
customized leadership services 301
cuts in public expenditure 284
cynicism 4, 39, 190, 313

Dade County, Florida 142
　Public Schools 142
Daly, Elizabeth 302
data 189
　categories 94
　normative types of 190
　reified objective 190
　subjective 190
Davis
　Bruce 162, 283
　Helen 165
de facto policy 286, 304, 327
de-politicised 80
de-skilling teachers 130
DEA 204
　guidelines 207
　officials 164
Dearing, Sir Ron 2, 63
decentralization 4, 85
deep values conflict 94
defensive politics 30
deficit
　analysis 290
　budgets 78
definitions 25
delegation of the authority to govern 313
delivery systems
　effectiveness of 286, 297, 304, 327
demand-side (consumer) 115
democracy 3–4, 216
democratic
　accountability 4, 11
　governance 4, 307
　ideal 4
　pluralism 313
　principles 3
　rhetoric 172
　theory 15
Democrats 112
demographics 84, 113, 137
demonstrable learning 118
departitioning of policy knowledge 314, 317, 319
Department of
　Education and Science (DES) 28, 31
　Education and the Arts (DEA) 151
　Education, Culture and Community Development (DECCD) 283
dependability 167
dependency 59
designing controls 137
Despotism 83
determinist patterns 176
Development Assessment Resource for Teachers (DART) 288
development plan 204
devolution 1–3, 124
　of governance 154
devolved responsibility 2
Dewey, John 19, 183
dialectic 187
dialectical analysis 189
differential impacts 52
differentiated
　power 54
　relevance 110
　zones of legitimacy 313, 318
diploma by exhibition 94

368 EDUCATIVE ACCOUNTABILITY

disaffected minorities 74
disaggregation 114
  of data 126
discipline 40, 59, 108, 135, 157, 186
discourse of
  ingratiation 310
  reform 18
discriminant analysis 331
discriminate 108
discursive
  colonization 277, 282
  repertoire 78
discussion between colleagues 200
disincentives to innovate 100
display identity 276
distorted communications 184
distribution
  failure 59
  formulae 51
distributional
  data 109, 193
  equity 81, 115, 315, 318
    principles 80
distributive justice 76
District Superintendent 154, 155
district
  offices 172
  superintendent 159, 172
  test results 316
districting
  effects of 107
districts 125
diverse public values 317
division of function 53
doctor–client consultation 243
documentation 200
documented 312
domestication 276
dominant social discourse 186
double duty 161
Downing Street 31
  Policy Unit 30
downscaling 15
drop-out rate 89
dropout 142, 143
dry economic policies 18
dualism 19

duel process 73
Duignan, Patrick 189
duty 26

East Sussex Accountability Project (ESAP) 31
Eastern states 109
ecological fallacy 107
ecology 117
economic
  competitiveness 130
  conditions 176
  elites 123
  privatism 60, 311
  prosperity 9
  rationalism 85, 231, 285, 318
  recession 85
  returns 161
economics of poverty 136
economist 133
economy 123, 316
educated
  person 20
  workforce 130
Education
  Act (1944) 27, 157,
  Reform Act (1988) 49, 50, 57, 58, 60, 63
  (Schools) Act (1992) 58
Education
  Establishment 30
  Index 95
  Secretary 61
  Summit 142
education
  indicators 91
  individualized 94
  marketization 5
  new sociology of 183
  outcome-based 87
  teacher 89
  undergraduate 89
Educational Indicators Panel 95
Educational Policy 331
Educational Review Unit (ERU) 158
educational
  administration 188, 191, 318
  adult 59–60
  authenticity 142

SUBJECT INDEX

choice 21
governance 27
leadership 90
management 86
models 44
politics 15
productivity 7, 312
reform movement 88
sociology 186
stakeholders 5
educative
    theory of accountability
        prerequisites 97
    leadership 189, 307
    management
        effective and efficient 231
effectiveness 65, 145, 152, 161, 169, 285
efficacy 117
efficiency 152, 161, 285
egalitarian ethic 59
ego 133, 190, 192
election
    1983 UK 39
elites 123
elitist theory 54
emancipatory conditions 182
emotion 179
empirical
    base 15
    proof 173
    -rational 48
    realities 176
    strategies 291
    test 188
    verification 180
empiricism 181
empowerment 80
    ethic 59
    of teachers 89
enforcing attendance 199
        England 25–36, 39–58
English 42
enrolment market 51
entitlements 44
entrepreneurial problem solving 318
entrepreneurialism 52, 123
epistemic
    capacities 99

critique 98
privileges 191, 192, 310
trap 185
critical 177
    approach 177, 290, 191
epistemically critical review 183
epistemological watersheds 310
epistemology 20, 175, 186
equal opportunities 59
equality 3
    of opportunity 18
equity 18, 100, 112, 123, 138, 183, 184, 285, 316, 327
ethical problems 109
ethics 21, 179, 192
ethnic data categories 142
ethnicity 113, 136
European social democracies 315
evaluating accountability schemes 34
Evaluation and Research Unit 152
evaluation 34, 35, 132, 158, 207, 317
    coherence 204
    goal-free 35
    adversary 35
    models 34, 35
    of teaching 98
evaluation purposes
    summative and formative 234, 235
Evers, Colin 186
evidence 39, 119
evolution of structures 182
evolutionary
    approach 183
    process 192
Ewington, John 165
examinations 33
    11-plus 10, 27
examinations, end-of-school 195
excellence 112, 285
exchange theory 49
executive
    arm 284
    teams 326
expectations of the community 207
expenditure 5
experiment 46
experimentalism 19

expertise 139
  encouragement of 318
  in local governance 315
experts 39
explanatory unity 190
explicit 312
exploitation 184
expulsion 142
external
  communication 207
  review 63
  validity 97

face validity 99
fads 54
fairness 106, 311
faith 76
false
  positives 134
  negatives 134
falsifiable 179
falsification 192
farms 157
fascism 4
Federal
  education evaluation 85
  Elementary and Secondary Education Act (1965) 86
federal
  authorities 161
  budget 285
  fishing expeditions 161
  governments 84
  Labour Government 283
feedback 2, 200, 224, 246, 251
  from parents and students 204
  from peers 200
  from staff 204
feral criteria 22
feral norms 237, 322
fidelity 313
Field, Michael 152
finance 51
financial
  accounting 133
  emergencies 284
first wave 101, 315

fiscal
  resources 115
  restraint 120
fixed term performance contracts 204
flexible and temporary teams 284
Focus on Learning Program 119
focus groups 119, 167
focused curriculum 94
folios 156
follow up study 331
forced development of schools 125
formative evaluation 2, 10, 192, 194
forms of evidence 154
foundational
  claims 186
  premises 186
  subjects 51
foundationalist subjectivism 188
fragmentation 123, 131
frame of reference 187
Framework for Curriculum Provision K-12: 154
frameworks 44
Frankfurt School of Social Theory 183
fraternity 3, 314
free
  economy 48
  market 48
freedom of
  information act 76
frozen policy production 176
fund raising 51
fundamentalist
  religious rights 124
  Right 123
  socialism 30
funds 44
futility 117

gender 113, 142, 176
genealogy 176
generalizations 40, 190
generic functions 284
George Bush 89
Goals 2000: Educate America Act 1
golden age 43, 57
governance 125, 151, 207, 317
  authentic 190

management and 204
powers 4
report card 111
values 319
governed, the 313
governing bodies
  effective 64
governing bodies 32, 61
government 178, 326
  backbenchers 77
  big 123, 316
  by mandate 318
  strong 48
governors 64, 93, 125
grades 194, 195
grammar 27
Grant Maintained Schools (GMS) 51
Great Debate 25, 29, 30, 37, 42
Green
  chapter 14
  Paper 31
Greenfield, Thom 189
gridlock 176
Griffiths, Dan 179
Gronn, Peter 189
grounded theory 40
group theories 187
growth of knowledge 310
guardians 177
guilt 99
gunman 284

Habermas 182
implementation, effective 6, 175
Hanlon, David 302
happiness 40
Harrington, Graham 151, 162
head teachers 44
Her Majesty's Inspectorate's (HMI)
  inspections 29, 43, 65–66, 68
heroic leadership 99
hierarchical
  governance structures 319
  line management 328
hierarchy 26, 30, 139, 176, 231
Hind, Ian 189
HMI *see* Her Majesty's Inspectorate
hobgoblin 248

Hodgkinson 180
holistic test of justification 188
Holmes Group 89
homework 157
honesty 77
horizontal accountability 17
hostels 157
House of Commons 32
human
  betterment 183
  capital investment 125
  relationships 312
  resource
    development 85
    management 285
humanist
  science 180
  subjectivism 178, 180, 184
humanistic qualities 40
hyper-individualism 13
hyper-individualistic professionalism 322
hypotheses 187
hypothesis testing 173

Idaho 123
idealized view of policy making
  177
ideas 299
ideological
  commitment 186, 314
  conflict 27
  partitioning 314
  rhetoric 14
ideology 78
idiosyncratic professionalism 246
ill-structured problem 23
Illinois 122
  State Board of Education 110
image management 41, 57
imaginative interventions 183
imbalances of power 157
implementation 229
implementing curriculum policies
  203
improvement 66, 204, 209
  planning 198
inadequate funding 108
inadmissable policy issues 160

incentives 118
inclusionary
  approach 169, 231
  philosophy 263
incompatible interests 176
incompetent colleague 255
independence 26
independent judgement 188
indicator systems 102, 315
indicators 93, 115, 207
  developed by parent/teacher/student 198
  from research literature 198
individual
  choice 59
  differences 91
  progress measures 198
individualism 3, 314
individualist liberalism 81
individualistic
  philosophy of teaching 237
  professionalism 246, 252
individualized expectations 194
induction training 67
industrial
  disputation 190
  equity 53
  justice 46
  productivity 85
industrialist's curriculum 42
industry and commerce 41
ineffective teachers 43
inequality 30
inexpert 246
infrastructure, supportive 6
influence 29
information 54
  dissemination 115
  systems 152
infrastructure 85
inhibitors 52
inner-city schools 64
inner-state comparisons 94, 98
innovation overload 297
inputs 52, 92
inspections 31, 33, 63, 67, 105, 125
institution self-study 35
institutional
  evaluation 54
  self-evaluation 43
instructional
  expertise 203, 246
  leadership 99, 117
  reform 18
  research 35
  sensitivity 108
instrumental
  ends 315
  and means 88
instrumentalism 181, 318, 330
instruments 109
insurance 312
integration 124
integrity 77
intelligibility 29, 311
intended outcomes 194
intensification 129, 146
inter-school co-operation 53
interdependent limited governments 313
interest
  group politics 331
    interests groups 27
  of children 72
interim policy findings 169
internal
  communication 207
  validity 97
International Journal of Educational Research 331
international
  achievement tests 89
  case studies 4
  relationships 329
  research 187
interpersonal communications 203
interpretation of data 170
interpretivism 185
interstate comparisons 109
intersubjective knowledge 181
intervention 323
  theory and method 145
interviews 167
intra-professional 235
  accountability 40
intra-professionalism 224
intrinsic value 133
intuitions 187

SUBJECT INDEX

investment 98
iron cage of bureaucracy 88
item banks 28

job description 203
joint interpretation of data 218
journals 156
judgement capacities 61
judgements by teachers 198
jurisdiction 124, 319

K-12: 123
   framework 203
Key Stage 2: 10
Key Stage 3: 61
key
   competencies 170, 229, 278, 282, 323
      of leadership services 318
   constituencies 312
   intended numeracy and literacy outcomes 303
   stages 60
Keynesian social democracy 27
kitchen conference 320
knowledge 5, 19
   acquisition 121
   de-partitioning of 182
   domain of knowledge 186
   growth 176
   historical 182
   hermeneutic 182
   partitioning of 182
   production 192
   repartition 184
known solutions 187

Labor Federal Government 18
Labour-Green Award 152
laissez-faire 21, 133
Lakomski, Gabriele 186
Lamar, Alexander 107
language 28
law 91, 93, 102
lay participation 42, 76
lay persons 310
LEA survey 55
leader recruitment 90

leaders
   leadership 203, 206, 225, 249
   relevant qualifications 207
leadership 77, 125, 165,
   capacities 281, 326
   duties 225
   effective 89
   educational 90
   infrastructure 317
   services 165, 203, 204, 207, 327
   vacuums 136
league tables 51, 60
learnable capacities 320
learned
   capacities 291, 312
   incapacities 91, 184, 315
learner 20
   outcomes 87
learner-centred schools 140
Learning attainment targets 51
learning
   attainment targets 60
   authentic 190
   by mastery 94
   by staff and students 207
   capacities 177
   community 76
   conditions 173
   environments 65
   opportunities 112
   organizations 234, 285, 304
   outcomes 125
   student-centred 94
   theory 20
Left, the 144
legacy of cynicism 314
legislated role in accountability 81
legislation 49, 63, 139, 144, 314
legitimacy 3, 122, 183, 318
   of government 313
legitimate stakeholders 72
legitimization crisis 45, 59
Liberal (conservative) government 154
liberal 235
   democracy 319
   values 74
Liberal-National 18
   (conservative) coalition government 283

liberation 176
liberty 3, 116, 314
licensure systems 90
life
  chances 50, 195
  skills 198
  liberty and happiness 83
light sampling 28
Likert terms 167
limits of sovereignty 319
line
  accountability 160
  management 308
    accountability 154, 172
    managers 16, 157, 172
linguistic tactics 276
linkage 145
literacy 120, 161, 196
  functional 89
litmus
  issue 158
  test 43
Local
  Education Authority (LEA)-maintained schools 2, 42–44
  Management of Schools (LMS) initiative 2, 10, 50–53
  School Councils (LSC) 110
  School Leadership and Management 154
local
  consultation 32
  control 123, 144, 316
  public accountability 68
localist
  ethic 60
  self-sufficiency 123
localized selection of leaders 204
locally-managed schools 162
logic 19
logical
  empirical epistemology 141
  empiricism 148, 176, 178, 192
  positivism 178
longitudinal data 107
Lord Nolan 77
Louisiana 113, 114
lower unit-costs 85

MacGregor, John 61
magazines 194
Major, John 57
majority politics 138
management
  analysis 35
  financial contraction 216
  skills 207
managerial
  efficiency 84
  functionalism 85
managerialism 49
mandate 54, 78, 88, 39, 240, 257, 279
mandated
  accountability 65
  agencies 29
  outcomes 138
  systemic priorities 175
mandating 88
manipulating testing procedures 97
manipulative power 231
Manual of Strategies for Release to Press and Public 109
manufactured myth 286
Maori 182
marked work 203
market 4
  forces 9
  incentives 88
  liberalism 30
marketization 5
marks 194, 195
Marxist
  analysis 183
  economic determinism 81
  'working class' 183
    Maryland 142
    Commission on School Performance 143
master discourse 186
mastery learning 123
mathematics 28, 42, 85, 89
matrix sampling 28
Mayor 111
MBO (management by objective) 86
measurement of actual achievement 175
media 123
mentoring process 204
metavalues 192

methodological confusion 331
methodology 34, 65, 143, 169
Miami 142
Michigan 86, 113, 114
milieu 117
mindsets 185
minimal dependencies 108
minimum
   competence testing 101, 123
   targets 143
Minister of Education 27
Ministerial reserve powers 22, 157
minorities 90
minority
   enrolment 138
   rights 4
   students 135
MIS (management information systems) 86
mobility
   teacher and student 136
model
   accountability 44
   community government 79
   consumerist 75
   local management 79
   partnership 75
   professional 75
   regional service 79
   three-year 303
moderation 43, 199, 291
modern languages 42
modernist technology of management 86
modesty 190
monism 319
monist and centralist tendencies 313
monitored work 203
monitoring 16, 28, 105, 125, 172
   compliance 112
   student progress 194
moral 32
   accountability 101, 234, 312
   choices 133
   code 191
   community 50
   conditions 191
   conscience 56
   duty 4
   economy 45, 73, 141, 149, 182, 231, 311

   imperatives 191
   knowledge 191, 310
   obligations 73
   order 50, 276, 312
   philosophy 73
   steerage 73
   stewardship 323
morally accountable educator 31
morale 117
morally
   responsible schooling 131
   right 189
motivation 183
moved on 255
multi-dimensional 187
multi-level democratic structures 234
multi-perspectival 187
multi-skilled staff 94
multiple
   choice 84
   roles 73
   zones of
      accountability policy legitimacy 177
      legitimacy 317
multiple-choice tests 98
mutual accountability 234, 320, 323, 327
mutually exclusive
   epistemologies 141
   perspectives 310
mutually respectful partnership 218
myth 163, 195, 219, 305, 327
mythology 261
   rich 196
myths of bureaucracy 13

Napier, Sue 283
National
   Assessment of Educational Progress
      (NAEP) 94
   Centre for Education Statistics (NCES) 93
   Commission on Excellence in Education
      88
   Commission on Excellence in Educational
      Administration 90
   Foundation for Educational Research
      (NFER) 29
   Governors' Association 89
   Union of Teachers (NUT) 32

national
  assets 60
  competency standards for managers 207
  curriculum 2, 48, 51
    profiles 198
  diplomacy documents 198
  goals 1, 90, 130
  liabilities 60
  performance standards 84
  policy community 18
  policy discourse 276, 309
  political interest 316
  prosperity 85
  standards 66
Nations' Governors 89
natural
  accountability 114
  justice 34
naturalism 181
naturalistic
  coherentism 189, 190, 191
  research 166, 167
naturalize 190
nature of knowledge 175
need for an audience 72
needs aggregation 286, 297
needs-based resourcing 17
negotiated
  performance contracts 204
  roles 50
negotiating
  industrial agreements 112
  new goals 200
negotiations 310
neo-centralism 2, 319
neo-centralist 215, 235
neo-conservatism 48, 50
neo-liberalism 48, 50, 81
neo-Marxism 30
neophyte educational administrators 91
networking 131, 200, 326
networks 132
New Compact for Learning 106
New Labour rhetoric 81
New Movement 179
New Right 25
New Right myth 216

New South Wales 16
  Department of School Education 284
New York 98, 105
  School Quality Review 105
  State Department of Education 134
New Zealand 6, 9
new goals for professional development 199
newsletters 194
Nolan Report 77
non-conformist teacher 46
non-damaging 311
non-empirical data 184
non-foundational
  epistemology 187
  naturalistic coherentism 178
non-foundationalism 191
non-linearity 131
non-remedial 123, 316
norm-referenced
  assessment 61, 62
  tests 61, 143
normative
  model 54
  order 183
  strategies 291
  theory 191
norms 122, 239
Northfield, Jeff 189
numbers-based strategy 142
numeracy 155, 156, 196

objective 311
  assessment results 198
  performance data 176
objectivity 77, 167
obligations 2, 109, 192, 307, 311, 317
observation 181
  reports 19
obsolescence of tests 125
obsolete political touchstone 322
occupational ethics 13
Office for Standards in Education (OFSTED) 43, 67
  inspectors 67
Office of Educational Research in Information (OERI) 114
official memorandum 163

SUBJECT INDEX 377

Oklahoma 113, 114
OPEC 27
open 312
openness 50, 54, 77
   and climate/tone of the school 207
opportunistic samples 167
opportunity to learn (OPL) 116–120, 126
opting out 216
Organization for Economic Cooperation and
    Development (OECD) 95
Organizational Change Study 134
organizational
   criteria 69, 71
   culture 18, 285
   dynamics 99
   learning 285, 312
   micropolitics 231
   norms 50
   rationality 311
   skills 203, 207
   studies 91
orthodoxy 69
OTL 316
Our Children and the Future 152
outcomes
   accountability 52, 86, 101
   -based education (OBE) 123, 316
   production function 315
outputs 92
overdue 318
overlapping memberships 34
overload 131
overseas exchanges 204

paradigm trap 184, 191
paradox 11, 187
parent
   activists 42
   consultation 200
   education 156
   minority 87
   opportunities 200
   perceptions 69
   power 59
   rights of the 72
   support of school leader 207
   /teacher
      interviews 156, 194

   partnerships 59
   /student discussion 194
   views 194
   parental
   choice 138
   criteria 40
   expectations 198
   involvement 31
   needs 217
   participation 10
   preferences 45
parental, professional and DEA involvement
    204
Parents
   as Teachers Program 321
   Charter 66
parents 5, 40
   reporting to 156
Parents' and Friends' Associations (P and F)
    155, 193, 194, 204
Pareto's immeasureability of utility thesis
    133
Park Sun Hyung 189
participation 115
   rates 290
participative policy making 1
participatory culture 80
parties 25
partisan politics 120
partition knowledge 186
partitioned
   curriculum 235
   of knowledge 2
partitions 314
partnership 314
   responsibilities 49
paternalism 59
Patmore, Peter 152
pedagogical
   criteria 69, 71
   leadership 294
pedagogy 9, 216
Peer Mediation Program 321
peer
   appraisal 194, 204
   -based legitimization 42
   networks 204, 299
   review 105, 125

pencil and paper norm-referenced tests 61
Pennsylvania 109, 123
People 83
people, the 314
Performance Review Summary 118
performance
  accountability 142
  accreditation systems 86
  criteria 16
  data 7, 63, 109
  incentives 89
  indicators 66, 91, 112, 169, 170, 192, 198, 207, 227, 278, 318
  management 16
  reporting 123
periodic whole-school 55
permanent reflective self-appraisal 106
persistent dilemmas 169
personal and social development 29
personnel management 86
Pettitt, David 189
philosopher kings 98
philosophers of science 178, 179
physical development 29
Picot Report 7, 9
plan
  long-term 155
  outcomes 207
  short-term 155
planned
  change 145
  development
    of classrooms 200, 204
    of teachers 200
planning 132, 172, 200
plural
  constituencies 27
  political ideologies 331
  society 177
  values 188
  views of reality 179
pluraliformity 319
pluralism 28, 229, 315
pluralist philosophy of administration 235
plurality of irreconcilable views 74
pluriformity 215, 217
polarized political climates 53
Policy Analysis for Californian Education 97

Policy Statement: Staff Appraisal 153
policy 93, 102
  action research 126
  actors 96, 111, 125, 310, 330
  adaption 149
  advice 156
  adequacy 144
  adoption 149
  advisory board 98
  analysis 91
  building 163
  change 110
  choice 190
  community 21, 319
    of stakeholders 323
  contradictions 9
  design 122
  discourse 123, 183
  elites 180, 191
  explanation 194
  gridlock 180, 191
  impact 122
  impasse 175
  implementation 110
  knowledge 163, 176, 191
  legitimacy 22, 308, 311
  myth 216
  options 48, 176
  orchestration 131
  paralysis 12, 326
  partners 5
  preference 309
  research 163, 178
    methodology 146, 175
    phases 173
  responsiveness 7
  seminars 309
  settlement 23, 122
  steerage 19
  vacuums 11
policy-making
  strategies 204
  capacities 141
political
  accountability 16, 32, 312
  action 176
  acumen 57
  alignments 30

authority 3
blameworthiness 90
controversy 10, 93
credibility 313, 315
criteria 327
cultures 121, 127
cycles 127, 284
demand 121
high ground 78
history 124
ideology 98
imagination 290
impulse 88, 121
incorrectness 5, 7, 13–23, 169
interest 146
intervention 27
mechanisms 4
noise 32
peace 158
pressure 122
profiles 127
realities 98
relativism 33
relief 286
significance 193
stakes 93
system 3
turbulence 190, 326
turf 94
turmoil 120
twilight world 78
politically
  critical
    conditions 6
    strategies 123
    Politics of Education Association 331
Politics of Accountability: Educative and International Perspectives, The 331
politics 91, 179, 192
  of accountability 125
  of public education 24
  of race 136
politics of education 9, 15
poor teaching 135
Pope Pius XI 217
popularity 122
populist impediments 123
Port Arthur 284

portfolio-wide goals 284
positional authority 235
positive public relations 194
positivism 19, 191, 310
positivist
  decision science 178
  foundations 181
  methodologies 317
possessive individualism 60
post-modern 184–186
post-modernism 191
post-paradigmatic methodology 330
post-positivist
  ideas 310
  policy research 307
post-structuralism 184–186, 191
post-structuralist 184
post-war
  consensus 25, 27
  reconstruction period 25
potency 190
poverty 120, 136, 137
power 14
  relationships 45
  strategies 291
power-coercive 48
power-coercive change strategies 14
powers 44
PPBS (Planning, programming, budgeting system) 86
practical
  meaning 56
  solutions 187
  theory 189, 192
pragmatic 235
  moral theory 17
  theories 330
pragmatism 46, 181
praxis 300, 302, 305
preferred attitudes 41
premature pressure 98
President 93
presidential
  coalition of governors 84
  legitimacy 130
  politics 30, 129, 130, 142, 161
  of rhetoric 84
presidents 178

press for achievement 148
pressure 120
primary educational structure 235
primary heads 64
principles 218
  intuition 39
  of line management 164
  of professionalism 164
priorities 79, 132
priority sets 46
Pritchard, Alan 189
private morality 29
privatization 7, 85
privatized inspections 43
privileged status 42
pro-active leadership 271
probity 316
problem-context-solution 187
problem-solving 229
problem-solving capacities 49
problems 177
procedural forms of equity 80, 115
process
  model of accountability 46
  standards 95
processes 92
production function 14, 88, 92
productions functions 2
productivity 16, 39, 127, 207, 327
  research 22
professional 32
  advice 73
  authority 42
  autonomy 42
  accountability 37, 42
  commitment 285
  competencies 318
  development 29, 169, 170, 173, 200, 207, 286, 297, 327
    program 200, 204
  discretion 63
  ethics 26
  expertise 42
  freedom 45
  impediments 123
  judgement 32
  knowledge 246
  learning community 43
  perspective 7
  practices 18
  responsibility 2
professionalism 6, 26, 44, 237, 318,
  isolationist 41
  rational 41
  participatory 41
professionally affirming 257
professionally enhancing leadership 252
professionals
  competent 140
Program Quality Review 119
programme
  activity 284
  evaluation 329
progressive education 30
  projects 235
promote professional development 304
promotion 17, 200
properties 191
proscribed outcomes 125
protocols 17
provider capture 2, 59, 310
provision and generation of school vision, appraisal of 204
proxy 52, 116, 199, 267
prudence 190
psychic prisons 182
psychology 20
Public Accounts Inquiry 161
public 32
  accountability 1, 22, 32, 312
  administration 285, 327
  attitudes 108
  choice 18
  confidence 85
  consent 94
  disapproval 42
  discourse 99
  dissatisfaction 87
  domain 171, 311
  expectations 17, 22
  interest 32, 312
  legitimacy 157, 172
  management 285
  myth 135
  obligations 52

organizations 311
policy discourse 312
project 81
property 4, 44
scrutiny 99
service 285
services 26
values 108
pulse of the portfolio 297
pupil records 33
purposes 318
　of education 87

qualified teachers 89
qualitative data schedule 166
Quality Assurance (QA) 16–17
　policy 189
Quality Assurance Review and Educational Audit Unit 284
Quality Indicator Reports 96
quality 27, 216
　evidence 67
　of external liaison 204
　of pedagogy 326
　of physical environment 207
　of society 313
　of teaching 200, 207, 209
　schooling 188
quantification 161, 173
quasi-autonomous non-government organizations (QUANGOs) 76

race 136, 142, 176
radical
　constructivism 179
　humanism 185
　humanist perspective 87
Rae, Hon Peter 152
Rand
　report 91, 101
　Rasch statisiical procedures 29
Rash analysis 111
rates of participation 198
rational
　conduct 72
　goal setting 52
　management 123
　strategies 291

rational-empirical 14
rationality 183, 308, 318
ratonal trap 131
raw data 61
raw performance data 70
re-educative 14
re-educative strategies 291
re-politicise 80
readiness 89
reading 155, 156
recentralization 4
reciprocal obligations 81
recommendations for school reviews 207
reconciliation dynamics 32
reculturing 131, 132
recurrent mediation 176
refutation 173, 190
regional debates 30
regression 138
regularatory approach to teaching 100
regulations 112
relationships
　authentic 190
　mutual 188
　voice and 141
relativism 50, 186
relativistic ethic 33
relativity 177
reliability 46, 67, 80, 106, 167, 235, 330
　of inspectors 313
remedial 311
　accountability sub-system 95
　help 73
repetitious advice 126
Report Card 96
report learning 194
reporting 156, 157, 172
　to community 204
　to parents 204
reporting outcomes 154
reposition 285
representative democracy 27, 36, 72
repression 184
Republican 111
research 198
　development and diffusion 145
　findings 193
　future 307, 329

research (*continued*)
  literature 207
  methodologies 186
  paradigms 184
  questions 173, 309
    research-based theories 185
researcher 145
reserve powers 25
residual practices 80
resistance 57
  by professionals 39
resource
  accountability 132, 133
  agreements 152
  allocations 153
  management 172
responsibilities 44, 317
responsible
  autonomy 81
  groups 173
responsive accountability 40, 74, 312
responsiveness 39, 52
restricted capacity 328
restructures 190
restructuring 5, 100, 131
retention rates 97, 303
return on investment 161
reverse take-over 276
reviewing 172
reviews 194
revolutionary change 183
rhetoric 87, 124
rhetoric, rich 316
rhetorical profile 142
Right, the 144
rights 44
  unalienable 83
Rizvi, Fazal 189
Robinson, Vivian 184
role
  clarity 63
  confusion 153
role-holding 80
routine compliance 111
running records 194
rural 168
Ruskin College 29

safety 83
saliency 37
sampling regime 168
SBM *see* school-based management
sceptical 261
Scholisic Aptitude Test (SAT) 94
School
  Accountability Report Card 96, 119
  Charter 9, 155
  Council 194, 204
    guidelines 321
  of Education 171
  Plan Implementation Review 302
  Planning 154
  Resource Package 154
School and College Councils: Interim Guidelines 153, 157
School Curriculum, The 42
School Plan Implementation Review 1996 160
school 43, 140
  accountability 46
  achievement reports 17
  autonomy 162
  board of trustees 320
  budgets 52
  charter 9, 153, 216, 320
  choice 115, 116
  climate, survey of 204
  community 9, 159, 198, 225, 234, 255, 234
    activities 203
    accountability policy research 146
  curriculum 42
  development 16
  districts 84, 140
  effectiveness 115, 126
    governing bodies 216
  effective 89
  effectiveness 52
  effects of 107
  evaluation 42–44, 65–72
  governance 59, 63–64
  governing bodies 32
  governors 52
  improvement 29, 109, 111, 124, 173, 176
    plan 303, 328
    team 143

SUBJECT INDEX

inspections 121, 147
  panel 159
organization 141
performance 316
personnel 119
policy 257
profiles 109
processes 52
productivity 22
renewal 116
report cards 106, 109
reports 109
responsively accountable 87
restructuring movement 1, 5, 15
review 16, 118, 160
  cycle
    one-year 293, 289, 305, 306
    three-year 327, 328
  process 200, 224, 326
  self- 316
  self-evaluation 55
  stakeholders 330
  vision 204
school-based
  budgeting 2
  evaluation 67
  management 5, 6, 18, 138
    programme budgeting 152
Schools
  Board of Tasmania 156
  Curriculum and Assessment Authority (SCAA) 2
Schools Bill (1991) 43
science 28, 42, 85, 89, 130
  of education 87, 88
science, good 19, 185
scientific
  management 19, 20
  realism 181
Scottish 6
  policy settings 49
second
  chances 84
  wave 88, 101, 315
Secondary Colleges Staff Association 153
Secondary Education and the Future 152
secondary modern 27
secrecy 76

Secretary of Education 158, 283
Secretary of State 51, 60
secular support 77
security 58, 311
segregation 117
selection and transfer 17
selection of
  principals 204
  teachers 200
self assessment 33
self-accounting 40, 41
  model 74, 75
self-appraisal 204
self-criticism 46
self-evaluation 4, 31, 200
self-examination 116
self-initiated peer appraisal 326
self-management 308
self-managing 215
  school 152
self-referentially coherent 311
selflessness 77
semantic differentials 331
seminar 171, 211
Senate Bill 813: 96
Senate Select Committee 161
Senior Superintendent (Equity) 289
senior
  executives 193
  management service 152
  school colleagues 200
sense of community 59
sensed data 187
service
  data 15
  delivery 171, 285
servitude 189
SES *see* socio-economic status
set work 203
shares of responsibility 140
significance 58, 311
  political 170
  statistical 170
simplicity 190
simplistic
  dualism 40
  policy dualism's 39
sites of reform 99

situational
  analysis 48
  ethics 50
size of classes 108
skewed distributions of power 183
skill
  development programs 204
  mastery 147
slack personal standards 76
sleaze factor 76
social
  action 9
  attitude 108
  authoritarianism 48
  capital 99
  consensus 87
  construction 184
  democracy 81
  disruption 85
  distance 231
  equity 9, 216
  group 26
  indicators 113
  interaction and diffusion 145
  justice 21, 53, 182, 184
  policy analysis 35
  reality 98
  services 110
  system 14
  welfare 59, 85
socialism 56
socialization 26
socially critical
  action research 182
  subjectivism 182, 184
  purposes 176

societal
  capacities 149
  views 125
society 33
socio-economic
  disadvantages 61
  status (SES) 66, 113
    effects 80, 313
    -adjusted data 69
sociological imagination 290

sociology
  educational 186
solidarity 26, 58, 311
South Australian Commission for Catholic Schools 217
South Carolina 98, 122
Southern states 109
sovereignty 319
Spalding, Ralph 303
Special Study Panel on Education Indicators: *Education counts* 112
special nature of district high schools 266, 268
specific review 55
speculations 187
sponsoring jurisdiction 320
Special Purpose Funds 161
staff
  appraisal 153
  development 132
  support of leader 207
  -supportive budgeting 94
stakeholder 2, 52
  advisory council 113
  economies 315
  economy 81
  participation 43
  positions 177
  sensitive 143
standard assessment tasks (SATs) 61
standardized
  resource management 235
  testing programmes 87
  tests 17, 86, 113, 156, 196
    of numeracy and literacy 194
    results of 200
Standards
  and Assessment Development and Implementation Council (SADI) 144
  for Educational and Psychological Tests, 1985 107
standards 59, 66
  epistemic 191
  public life 80
  service 33
  valued 191
starting points 42

State
  Accountability Study Group (SASG) 114
  Board of Education, Maryland 143
  Department of Education, Maryland 143
  Education Assessment Centre 95, 106
  Education Indicators 95
state 140
  control 30
  education policies 173
  governors 130, 178
  political culture 316
  steerage 51
state curriculum 130
state, strong 48
state, the 30
state-wide testing programmes 316
statistical
  analysis 109
  norms 66, 118
status 26
statutory
  responsibilities 78
  roles 50
steady state funding 158
Stenhouse 26
step-wise trailing 167
stewards 3, 313
stewardship 11, 189
strategic
  evaluation 21
  planning 16, 152
    stratified
    sample 166
    sampling 330
structural
  ambiguities 16
  functionalism 86, 185
  impediments 121
structuralism 16
structuration 191, 192
structure 93, 102
student
  achievement 28, 89, 96
    levels 203
  assessment 59, 141
  at risk 200
  attitude 198
  background 115
  development 69, 71
  diversity 100
  growth 142
  learning 6, 193, 197, 219, 241
    reports of 194
  mobility 120
  moderation meetings 198
  morale 207
  motivation 207
  outcome 96, 143, 200, 303
  profiles 2
  progress 203
  self-assessment 194
  self-esteem 198
  work sampling 194
subjectivism 179, 181, 186, 191
subjects 41
subsidiarity 215, 216, 319
success 100
  criteria 66
summative accountability 10
summative reports 200
superintendents 143
supply-side (producer) 115
support 200
support staff 194
supported 193
Supportive School Environments Program 321
supportive
  classroom and staffroom environments 231
  infrastructure 6
  interdependence 229
surrogate accountability 80
suspension 142
sustainability 145
symbolic
  accountability 28
  democratic rhetoric 157
  policy 153
  politics 124, 142, 318
  reform 129, 142, 316
  revisionism 30
  structures 36
system level 200
system restructuring 11

386 EDUCATIVE ACCOUNTABILITY

systematic
  action research 131
  evaluating 194
  planning 194
  self-monitoring 45
systemic
  action research 162, 172
  policies 219
  policy making 151
  reflection 309
  reform movement 123
  structure 22
  systems
    model 155
    theory 2, 10, 52, 91, 125, 192
tacit moral theories 276
Taplin, Margaret 165
Task Group on Assessment and Testing (TGAT) 60–61
Tasmania 15
Tasmanian
  Council of State School Parents' and Friends' Association 163
  Council of State Schools Parents and Friends Association (TCSSPand F) 321
  Education Council (TAC) 155
  educational culture 273
  Industrial Commission 153
  Liberal Government 152
  Teachers Federation 153
taxpayer resistance 123
Taylor Report 31, 32
teacher 140, 194
  appraisal 17, 200, 293
  attitude 203
  autonomy 246
  classroom development strategy and 200
  contribution 200
  -designed tests 194
  development 109
  empowerment 1, 5, 11
  encouragement 200
  evaluation 194
  identification of intended outcomes 194
  incompetence 111
  knowledge 203
  morale 207, 216
  motivation 207
  observation 194
  participation 203
  planning 194
  qualifications 112
  quality 115
  teaching 199, 202, 244
  understanding 148
  unions 30, 111, 200
  written records
    effectiveness of 203
teacher/student conferencing 194
teachers
  newly appointed 200
teaching 5
  practice 157
  for understanding 5
technical
  capacity 49
  core 65
  incapacity 27, 36
  merit 160
  perspective 6
  rationality 123
  regression 28
  -rational language 316
techniques 125
technological myopia 108
technology 42, 89
  evaluation 78
telecommunication 132
telephone
  follow-up 166
  polling 113
test-based outcomes accountability 106
testing 87
tests 33, 136
  basic competency 88
Thatcher, Margaret 37
The Paideia Network 90
theorem
  Pareto's immeasureability 176
  Arrow's impossibility 176
Theory Movement 179
  theory
    of building 331
    of change 60
    of choice 187
    of competition 188

SUBJECT INDEX

of knowledge 185
of ladenness 180
of organization 91
of state 9, 44
theory overlap 187
think-tanks 30
third wave 89, 101, 315
threshold 134
   attributes 96
token accountability 49
Tomorrow's Teachers 89
Total Quality Management (TQM) 7, 16
totalitarianism 4
touchstone 21, 173, 177, 186, 187, 190, 192, 199, 219, 319
tractable variables 116
traditional
   curricular 123
   roles of LEAs 52
trainer 145
training 200
transaction-observation 35
transferability 167
transition program 200
transparency 85
transparent 219
Treasury 171, 173
trend analysis 66
Trial State Assessment 107
tripartite partnership 27
trust 26
trustees board 320
   councils 153, 320
   decision making 153, 157
   delivery standards 130
   development 200
      plan 153
      program 204
trustful interaction 74
trustworthiness of policy options 176
trustworthy databases 218
truth 28
truths 186

uncritical
   expansionism 84
   faith 215
   mass of human resources 136

under-achievement 28
under-performing schools 306, 327
unemployment, high 120
unfairness 116, 123, 124, 316
unfreezing 99
uniform
   accounting systems 86
   concept 229
uniformity 215, 319
unintended consequences 22
union policy 257
unitary concept 229, 235
United States 5–6, 19, 83–103
universal
   proxy 12, 84
   values 94
University
   Council of Educational Administration 91
   of New England 189
   of Tasmania 171
unlimited progress 85
unnatural liabilities 311
unsupported 193
urban 168
US
   Congress Office for Technology Assessment 106
   Department for Education 114
   public education 83–103
useless reform strategy 285
user-payers 51
utility 133
   of education 87
valid 311
   information 218
validity 46, 67, 106
   external 167
   internal 167
valuable knowledge 307
value 14
   adding 14, 69, 126, 147, 161, 313
   framework 33
value-for-money 85
values 63, 192
   advocacy 121
   base 78, 79
   vacuum 46
venality 76

verification 114, 179, 311
veto 29, 243
vicious regress 186
victimization 209
Vienna Circle 179
Vietnam War 123
virus-like effects 184
vision 119
visitations 43
vocational
 skills 108
 training 85
vocationalism 161, 173
voice 5, 81
vulnerability 27

Wales 25–36, 39–58
Walker, Jim 189
Wall Chart 93
Wardlaw, Rosemary 302
warranted assertability 181, 188
webs of belief 50, 192, 276, 310, 330
welfare capitalism 27
Western Association of Schools and Colleges 119
Western Australia 189

Westminster 14
White chapter 14
whole-school evaluation 121
willingness to engage in development 203
Willower, Don 181, 188
Wilson
 Harold 30
 Pete 97
 Sue 165
women 90
Working Nation, The 161
working consensus 59
workshops 132, 167, 189
written
 checklist 194
 constitution 76
 records 203

Young, Bob 188
youth
 poverty 138
 unemployment 85

zones
 of assurance 16
 of legitimacy 46, 81, 313, 319